Balancing Human Rights, Social Responsibility, and Digital Ethics

Maja Pucelj
Faculty of Organisation Studies, University of Novo Mesto, Slovenia

Rado Bohinc
University of Ljubljana, Slovenia

A volume in the Advances in Human and Social Aspects of Technology (AHSAT) Book Series

Published in the United States of America by
    IGI Global
    Engineering Science Reference (an imprint of IGI Global)
    701 E. Chocolate Avenue
    Hershey PA, USA 17033
    Tel: 717-533-8845
    Fax: 717-533-8661
    E-mail: cust@igi-global.com
    Web site: http://www.igi-global.com

Copyright © 2024 by IGI Global. All rights reserved. No part of this publication may be reproduced, stored or distributed in any form or by any means, electronic or mechanical, including photocopying, without written permission from the publisher.
Product or company names used in this set are for identification purposes only. Inclusion of the names of the products or companies does not indicate a claim of ownership by IGI Global of the trademark or registered trademark.

                Library of Congress Cataloging-in-Publication Data

CIP Pending
ISBN: 979-8-3693-3334-1
EISBN: 979-8-3693-3335-8

British Cataloguing in Publication Data
A Cataloguing in Publication record for this book is available from the British Library.

All work contributed to this book is new, previously-unpublished material.
The views expressed in this book are those of the authors, but not necessarily of the publisher.

For electronic access to this publication, please contact: eresources@igi-global.com.

Advances in Human and Social Aspects of Technology (AHSAT) Book Series

Mehdi Khosrow-Pour, D.B.A.
Information Resources Management Association, USA

ISSN:2328-1316
EISSN:2328-1324

MISSION

In recent years, the societal impact of technology has been noted as we become increasingly more connected and are presented with more digital tools and devices. With the popularity of digital devices such as cell phones and tablets, it is crucial to consider the implications of our digital dependence and the presence of technology in our everyday lives.

The **Advances in Human and Social Aspects of Technology (AHSAT) Book Series** seeks to explore the ways in which society and human beings have been affected by technology and how the technological revolution has changed the way we conduct our lives as well as our behavior. The AHSAT book series aims to publish the most cutting-edge research on human behavior and interaction with technology and the ways in which the digital age is changing society.

Coverage

- Cyber Behavior
- Digital Identity
- Technoself
- End-User Computing
- ICTs and human empowerment
- Cultural Influence of ICTs
- Philosophy of technology
- Technology and Social Change
- Technology and Freedom of Speech
- Gender and Technology

IGI Global is currently accepting manuscripts for publication within this series. To submit a proposal for a volume in this series, please contact our Acquisition Editors at Acquisitions@igi-global.com or visit: http://www.igi-global.com/publish/.

The (ISSN) is published by IGI Global, 701 E. Chocolate Avenue, Hershey, PA 17033-1240, USA, www.igi-global.com. This series is composed of titles available for purchase individually; each title is edited to be contextually exclusive from any other title within the series. For pricing and ordering information please visit http://www.igi-global.com/book-series/advances-human-social-aspects-technology/37145. Postmaster: Send all address changes to above address. Copyright © IGI Global. All rights, including translation in other languages reserved by the publisher. No part of this series may be reproduced or used in any form or by any means – graphics, electronic, or mechanical, including photocopying, recording, taping, or information and retrieval systems – without written permission from the publisher, except for non commercial, educational use, including classroom teaching purposes. The views expressed in this series are those of the authors, but not necessarily of IGI Global.

Titles in this Series
For a list of additional titles in this series, please visit: www.igi-global.com/book-series

Exploring Youth Studies in the Age of AI
Zeinab Zaremohzzabieh (Women and Family Studies Research Center, University of Religions and Denominations, Qom, Iran) Rusli Abdullah (Faculty of Computer Science and Information Technology, Universiti Putra Malaysia) and Seyedali Ahrari (Women and Family Studies Research Center, University of Religions and Denominations, Iran)
Information Science Reference • copyright 2024 • 523pp • H/C (ISBN: 9798369333501)
• US $295.00 (our price)

Social Innovations in Education, Environment, and Healthcare
Harish Chandra Chandan (Independent Researcher, USA)
Information Science Reference • copyright 2024 • 483pp • H/C (ISBN: 9798369325698)
• US $275.00 (our price)

Intersections Between Rights and Technology
Amit Anand (REVA University, India) Akanksha Madaan (REVA University, India) and Alicia Danielsson (University of Bolton, UK & Hume Institute for Postgraduate Studies, Lausanne, Switzerland)
Information Science Reference • copyright 2024 • 475pp • H/C (ISBN: 9798369311271)
• US $385.00 (our price)

Bridging Human Rights and Corporate Social Responsibility Pathways to a Sustainable Global Society
Maja Pucelj (Faculty of Organisation Studies, University of Novo Mesto, Slovenia) and Rado Bohinc (Scientific Research Centre Koper, Slovenia)
Information Science Reference • copyright 2024 • 324pp • H/C (ISBN: 9798369323250)
• US $245.00 (our price)

701 East Chocolate Avenue, Hershey, PA 17033, USA
Tel: 717-533-8845 x100 • Fax: 717-533-8661
E-Mail: cust@igi-global.com • www.igi-global.com

Table of Contents

Preface xiv

Chapter 1
A Bibliometric Analysis of Digital Ethics and Human Rights 1
    Rafiq Idris, Universiti Malaysia Sabah, Malaysia
    Rizal Zamani Idris, Universiti Malaysia Sabah, Malaysia
    Noor Syakirah Zakaria, Universiti Malaysia Sabah, Malaysia
    Mohammad Ikhram Mohammad Ridzuan, Universiti Malaysia Sabah, Malaysia
    Azizan Morshidi, Univerisiti Malaysia Sabah, Malaysia
    Azueryn Annatassia Dania Aqeela Azizan, University of New South Wales, Australia
    Razlina Jamaludin, MRSM Kota Kinabalu, Malaysia

Chapter 2
Artificial Intelligence: A New Tool to Protect Human Rights or an Instrument to Subdue Society by the State 35
    Abhishek Benedict Kumar, Symbiosis International University (Deemed), Pune, India
    Aparajita Mohanty, Symbiosis Law School, Symbiosis International University (Deemed), Pune, India

Chapter 3
Interplay of Artificial Intelligence and Recruitment: The Gender Bias Effect 63
    Shikha Saloni, University School of Business, Chandigarh University, India
    Neema Gupta, University School of Business, Chandigarh University, India
    Ambuj Kumar Agarwal, Sharda University, India
    Raj Gaurang Tiwari, Chitkara University, India
    Vishal Jain, Sharda University, India

Chapter 4
Socially Responsible Application of Artificial Intelligence in Human
Resources Management 82
    Ana Marija Gričnik, Faculty of Economics and Business, University of
        Maribor, Slovenia
    Matjaž Mulej, Faculty of Economics and Business, University of
        Maribor, Slovenia
    Simona Šarotar Žižek, Faculty of Economics and Business, University of
        Maribor, Slovenia

Chapter 5
Legal Aspects of Digital Ethics in the Age of Artificial Intelligence 144
    Hemendra Singh, Jindal Global Law School, O.P. Jindal Global
        University, Sonipat, India

Chapter 6
Sovereignty Over Personal Data: Legal Frameworks and the Quest for
Privacy in the Digital Age... 168
    Andreja Primec, University of Maribor, Slovenia
    Gal Pastirk, University of Maribor, Slovenia
    Igor Perko, University of Maribor, Slovenia

Chapter 7
Cherish Data Privacy and Human Rights in the Digital Age: Harmonizing
Innovation and Individual Autonomy 199
    Bhupinder Singh, Sharda University, India

Chapter 8
Digitalization in Corporations: Integrating Utility of Digital Technology
With Accessibility and Privacy of Data.. 227
    Siddharth Kanojia, O.P. Jindal Global University, India

Chapter 9
Corporate Data Responsibility in East Africa: Bridging the Gap Between
Theory and Practice 246
    Ripon Bhattacharjee, National Law University, Tripura, India
    Aditya Agrawal, O.P. Jindal Global University, India
    Madhulika Mishra, GLA University, Mathura, India
    Akash Bag, Adamas University, India

Compilation of References 283

About the Contributors 329

Index 332

Detailed Table of Contents

Preface.. xiv

Chapter 1
A Bibliometric Analysis of Digital Ethics and Human Rights 1
    Rafiq Idris, Universiti Malaysia Sabah, Malaysia
    Rizal Zamani Idris, Universiti Malaysia Sabah, Malaysia
    Noor Syakirah Zakaria, Universiti Malaysia Sabah, Malaysia
    Mohammad Ikhram Mohammad Ridzuan, Universiti Malaysia Sabah, Malaysia
    Azizan Morshidi, Univerisiti Malaysia Sabah, Malaysia
    Azueryn Annatassia Dania Aqeela Azizan, University of New South Wales, Australia
    Razlina Jamaludin, MRSM Kota Kinabalu, Malaysia

In order to expand the current literature assessing the quantity and quality of the worldwide research production in the subject of digital ethics and human rights, a bibliometric analysis was conducted for the period 1966–2024. The Scopus.com database was consulted with the "digital ethics and human rights" keywords searched. The database identified 537 documents. In terms of the publication production in the subject over the studied period, the United States of America demonstrated the highest number of publications across the world, with the United Kingdom in the second place. The bibliometric data provided is comprehensive but calls for understanding how the research outcomes in digital ethics and human rights field has impacted society. Therefore, the future research in the field of digital ethics and human rights is expected to be based on the transition from theorising to practical development aspects, per se.

Chapter 2
Artificial Intelligence: A New Tool to Protect Human Rights or an Instrument to Subdue Society by the State.. 35

    Abhishek Benedict Kumar, Symbiosis International University (Deemed), Pune, India
    Aparajita Mohanty, Symbiosis Law School, Symbiosis International University (Deemed), Pune, India

With each technological revolution, mankind experiences a shift in the pace of history, leading to the emergence of a new age that fundamentally redefines human beliefs. Artificial intelligence (AI) is a technological subject that is causing a profound shift in human society towards a state characterised by robots and machines. AI encompasses several technologies such as machine learning, natural language processing, big data analytics, algorithms, and other related fields. Nevertheless, as human intelligence is marked by inherent bias in decision-making, AI products that rely on human-generated intelligence also display similar characteristics. The existence of bias and discrimination poses a threat to universal human rights, since these phenomena are deeply embedded in social institutions and have many technological origins. Indeed, because of its facilitation of prejudice, AI disproportionately affects the human rights of marginalized individuals and communities, hence giving birth to a new type of oppression rooted in technology.

Chapter 3
Interplay of Artificial Intelligence and Recruitment: The Gender Bias Effect 63
Shikha Saloni, University School of Business, Chandigarh University, India
Neema Gupta, University School of Business, Chandigarh University, India
Ambuj Kumar Agarwal, Sharda University, India
Raj Gaurang Tiwari, Chitkara University, India
Vishal Jain, Sharda University, India

In this chapter, the authors focus on different ways in which AI is incorporated in the process of recruitment. Along with the above stated objective, they also explore the forms of AI based recruitment, the benefits of AI based recruitment, and the challenges that might be encountered during the process with an emphasis on gender bias. In the findings, they aim to describe the gender bias in professional functions in businesses. On the other hand, they hope to gain insight into potential gender discrepancies between operational and leadership positions, as well as between departments. The findings of this chapter will benefit researchers, academics, and managers in analyzing gender-related practices and policies. Organizations can become more aware of their gendered practices, which affect the recruitment procedure and the varied roles and responsibilities assigned to men and women, by giving voice to the prejudices that generate gender biases. Along with this, they provide the implications, limitations, and future scope of the study.

Chapter 4
Socially Responsible Application of Artificial Intelligence in Human
Resources Management 82
    Ana Marija Gričnik, Faculty of Economics and Business, University of
        Maribor, Slovenia
    Matjaž Mulej, Faculty of Economics and Business, University of
        Maribor, Slovenia
    Simona Šarotar Žižek, Faculty of Economics and Business, University of
        Maribor, Slovenia

Humankind faces growing artificial intelligence (AI) and AI-based applications, influencing almost every activity, including human resource management (HRM), revolutionizing humans' work nature and content, workers, workplaces, HRM processes, etc. AI can support various HRM practices, such as candidate selection, employee training, data analysis, evaluation, etc. If organizations appropriately utilize AI, they can enhance productivity and their general/individual work performance, streamline processes, and increase efficiency, ultimately improving employee engagement and well-being. Hence, organizations can use AI to stay ahead of their competitors and help develop an innovative sustainable socially responsible society (ISSRS) to overcome crises. AI should only be used as a tool and not to replace humans, which is essential for a creative, efficient, satisfying, and successful work environment.

Chapter 5
Legal Aspects of Digital Ethics in the Age of Artificial Intelligence 144
    Hemendra Singh, Jindal Global Law School, O.P. Jindal Global
        University, Sonipat, India

This chapter explores the legal dimensions of digital ethics amidst the proliferation of artificial intelligence technologies. It explores the evolution of legal frameworks governing digital technologies and their adaptation to address emerging ethical dilemmas in artificial intelligence. Focusing on the European Union's Artificial Intelligence Act and OECD's Principles on Artificial Intelligence, it assesses their implications for global AI governance and effectiveness in addressing ethical concerns. The intersection of data protection laws and artificial intelligence ethics is analyzed, emphasizing their role in safeguarding human rights. The chapter also examines legal challenges and solutions in ensuring the ethical use of AI, particularly regarding liability and accountability issues. By identifying emerging challenges and advocating for collaborative governance approaches, it outlines pathways to enhance legal frameworks in addressing evolving ethical concerns.

Chapter 6
Sovereignty Over Personal Data: Legal Frameworks and the Quest for
Privacy in the Digital Age 168
    Andreja Primec, University of Maribor, Slovenia
    Gal Pastirk, University of Maribor, Slovenia
    Igor Perko, University of Maribor, Slovenia

The contribution focuses on an individual's right to control their personal data and the right to privacy, emphasizing the legal aspects and developments in the context of modern technological and digital challenges. These rights become especially critical in the context of swift advancements in technology and digitalization, enabling the accumulation, examination, distribution, and sharing of extensive data volumes in previously unattainable manners. There has been a surge in sensors embedded within smart devices, gathering data from various aspects of our professional and personal lives. Furthermore, artificial intelligence (AI) has emerged, capable of analysing and interpreting data in unthinkable ways. These shifting dynamics present a challenge to reconsider how data is regulated. It's not merely about updating the rules themselves; it involves reimagining the process of creating regulatory frameworks.

Chapter 7
Cherish Data Privacy and Human Rights in the Digital Age: Harmonizing
Innovation and Individual Autonomy 199
    Bhupinder Singh, Sharda University, India

Data privacy encompasses the safeguarding and control individuals exercise over their personal information and data. It revolves around ensuring the confidentiality and security of sensitive data, including financial records, health information, and unique identifiers. In the era of extensive data collection, storage, and sharing in the digital landscape, preserving data privacy has become imperative to uphold individuals' rights and shield them from potential harm. From artificial intelligence and machine learning to internet of things (IoT) devices, innovative solutions have transformed our way of life and work. However, every innovation brings the responsibility to safeguard the privacy and security of individuals whose data is collected and processed. So, balancing innovation and personal security is a nuanced task. While innovation offers substantial benefits, it also poses risks to personal privacy without adequate regulation. This chapter dives into the diverse exploration of human rights protection concerning data and privacy of individuals in the digital arena.

Chapter 8
Digitalization in Corporations: Integrating Utility of Digital Technology
With Accessibility and Privacy of Data 227
    Siddharth Kanojia, O.P. Jindal Global University, India

Corporations are quickly learning how to benefit from digitalization in various facets of their operations whilst placing a significant focus on the requirements and experiences of their clients. Furthermore, by utilizing sophisticated analytical models and effective risk-management techniques, digitalization is also being used as a potent tool to improve internal governance and decision-making processes. Digitalization has made it easier for companies to fetch and access the data of various stakeholders through surveys or customer interactions, purchasing it from third-party data brokers, or using public data sources such as social media or government records. Hence, it also raises important questions about data privacy and security, and the appropriate balance between technological innovation and the protection of legal rights and obligations. Accordingly, this chapter intends to analyze the efficacy and utility of the latest amendments in the provisions of corporate law which intends to take due cognizance of these technological advancements.

Chapter 9
Corporate Data Responsibility in East Africa: Bridging the Gap Between
Theory and Practice 246
    Ripon Bhattacharjee, National Law University, Tripura, India
    Aditya Agrawal, O.P. Jindal Global University, India
    Madhulika Mishra, GLA University, Mathura, India
    Akash Bag, Adamas University, India

East Africa's rapid digital connectivity growth without proper regulatory frameworks, notably in data protection, places regulatory obligations on technology actors. The chapter uses corporate social responsibility (CSR) and corporate digital responsibility (CDR) to examine how firms handle data privacy and ethical issues in the digital world. The research uses various case studies to illuminate the practical obstacles of data responsibility solutions. The lack of specific CSR and CDR techniques remains a major implementation gap despite actors' awareness and efforts to create complete policies. This emphasizes the need for context-specific guidelines to connect policy development and real-world application to ensure that the growing digital economy aligns with human rights and state goals.

Compilation of References 283

About the Contributors 329

Index 332

Preface

The edited monograph *Balancing Human Rights, Social Responsibility, and Digital Ethics* represents comprehensive research of the complicated relationship between digital technologies and ethical concerns in contemporary society. As editors, we are pleased to present this volume, which delves into the moral and legal implications of digital advancement, especially in perspective of the new EU law on artificial intelligence and the OECD's AI recommendations.

In this edited monograph, we address a variety of challenging issues, including the ethical considerations of artificial intelligence, data protection, the digital divide, protection against online threats and the impact of digital transformation on various sectors such as social media, e-commerce and digital healthcare. By integrating ethical principles into the comprehensive guidelines and regulatory standards of the EU AI Act and the OECD AI Principles, this work significantly expands the field of digital ethics.

Our aim is to examine the link between digital ethics and human rights and to emphasize the impact of digital technologies on fundamental rights and social responsibility. We deal with issues related to privacy, freedom of speech and access to information. The European Union, particularly the European Commission, plays a central role in incorporating digital ethics in its legal framework, as demonstrated by the General Data Protection Regulation and the proposed AI law. This harmonization shows that the EU is committed to ensuring that technological progress, particularly in the field of AI, is harmonized with human rights and ethical standards. Furthermore, our monograph tends to be in line with the OECD's AI Principles, which advocate for AI systems that uphold human rights and democratic values.

Balancing Human Rights, Social Responsibility, and Digital Ethics aims to show that digital ethics goes beyond purely moral or technological challenges to encompass broader societal issues. We advocate for regulations and standards that protect human rights in the continuously evolving digital landscape. The European Commission recognizes this need as it strives to create a European Union that supports digital ethics and respects individual rights. These efforts are in line with the global trend to promote responsible and ethical regulation of artificial intelligence as outlined by the OECD. Our monograph strives to become an important guide to understanding these critical changes and their implications for a sustainable and just digital future.

The main aim of this volume is to closely examine the link between digital technology and ethical concerns in the context of evolving regulatory frameworks such as the EU Artificial Intelligence Act and the OECD AI Recommendations. By emphasizing the vital role of ethical principles in guiding technological progress, we aim to improve understanding of the complex relationship between technological developments and fundamental human rights. Research in the present edited monograph goes beyond traditional boundaries and incorporates legal implications and regulatory norms to promote a comprehensive understanding of digital ethics.

Through an in-depth analysis of various topics, including the ethical implications of AI, data privacy and digital healthcare, we strive to offer new insights into the societal and human impact of digitalization. By bridging the gap between theoretical ethical considerations and practical legal applications, we strive to provide a detailed examination of how the European Union and organizations such as the OECD are addressing these challenges.

Furthermore, our study aims to encourage academic discourse, influence policy and the development of corporate strategies by highlighting the importance of harmonizing technological progress with human rights and social responsibility. We hope that this book will become an indispensable resource for promoting a future in which digital ethics is intrinsically linked to global sustainable development, the ethical advancement of artificial intelligence and the preservation of democratic principles in the digital age.

Balancing Human Rights, Social Responsibility, and Digital Ethics is aimed at a diverse audience, including academics and researchers in the fields of digital ethics, law, technology studies, and human rights, as well as policy makers and regulators involved in crafting digital governance frameworks. It will also benefit students in higher education, particularly those studying computer science, law, social sciences and ethics, by providing them with a basic understanding of the ethical implications of digital technology.

This book is particularly relevant for professionals in the technology sector, such as developers, data scientists and corporate executives who have the responsibility to incorporate ethical considerations into the development and implementation of digital technologies. It is also aimed at non-governmental organizations, interest groups and members of civil society who are involved in discussions about digital rights and the ethical use of technology.

Preface

We are confident that this book will be a valuable resource for anyone seeking to navigate through the complex field of digital ethics, human rights and social responsibility in the digital age. By offering both academic insights and practical perspectives, the edited monograph strives to be an extremely useful resource with a broad impact on the society.

ORGANIZATION OF THE BOOK

Chapter 1: A Bibliometric Analysis of Digital Ethics and Human Rights

This chapter, authored by Rafiq Idris, Rizal Zamani Idris, Noor Syakirah Zakaria, Mohammad Ikhram Mohammad Ridzuan, Azizan Morshidi, Azueryn Annatassia Dania Aqeela Azizan, and Razlina Jamaludin, presents a bibliometric analysis of global research production in the field of digital ethics and human rights from 1966 to 2024. Utilizing the Scopus.com database and focusing on the keywords "digital ethics and human rights," the authors identify 537 documents. The analysis reveals that the United States leads in publication output, followed by the United Kingdom. Despite the comprehensive bibliometric data, the chapter calls for further research to transition from theoretical exploration to practical applications, emphasizing the impact of digital ethics and human rights research on society.

Chapter 2: Artificial Intelligence: A New Tool to Protect Human Rights or an Instrument to Subdue Society by the State

Authored by Abhishek Kumar and Aparajita Mohanty, this chapter explores the dual nature of Artificial Intelligence (AI) as both a protector and a potential subjugator of human rights. The authors discuss AI's broad technological scope, including machine learning, natural language processing, and big data analytics, and highlight inherent biases in AI systems derived from human-generated intelligence. These biases pose significant threats to universal human rights, particularly affecting marginalized individuals and communities. The chapter critically examines AI's role in perpetuating new forms of oppression rooted in technology, urging the need for ethical AI deployment to prevent discrimination and protect human rights.

Chapter 3: Interplay of Artificial Intelligence and Recruitment: The Gender Bias Effect

Shikha Saloni, Neema Gupta, Ambuj Agarwal, Raj Tiwari, and Vishal Jain investigate the incorporation of AI in recruitment processes, with a focus on gender bias. The chapter explores various forms of AI-based recruitment, its benefits, and the challenges it presents, particularly regarding gender discrepancies in professional roles. By analyzing gender-related practices and policies, the authors provide insights into the operational and leadership gender gaps within organizations. The findings aim to raise awareness of gender biases in recruitment, offering implications, limitations, and future research directions to promote equitable hiring practices.

Chapter 4: Socially Responsible Application of Artificial Intelligence in Human Resources Management

Authored by Ana Gricnik, Matjaž Mulej, and Simona Šarotar Žižek, this chapter examines the transformative impact of AI on human resource management (HRM). The authors discuss AI's potential to enhance HRM practices, including candidate selection, employee training, and data analysis. Emphasizing the importance of using AI as a tool rather than a replacement for humans, the chapter highlights how AI can improve productivity, streamline processes, and boost employee engagement. The authors advocate for the responsible use of AI to foster an innovative, sustainable, and socially responsible society, ensuring AI's benefits are harnessed without compromising human creativity and satisfaction.

Chapter 5: Legal Aspects of Digital Ethics in the Age of Artificial Intelligence

Hemendra Singh explores the legal dimensions of digital ethics amid the rise of AI technologies. The chapter traces the evolution of legal frameworks governing digital technologies and their adaptation to address ethical dilemmas in AI. Focusing on the EU's Artificial Intelligence Act and OECD's AI Principles, Singh assesses their effectiveness in global AI governance and ethical concerns. The chapter also examines the intersection of data protection laws and AI ethics, highlighting legal challenges and solutions in ensuring ethical AI use, particularly regarding liability and accountability, and proposes collaborative governance approaches to enhance legal frameworks.

Preface

Chapter 6: Sovereignty Over Personal Data: Legal Frameworks and the Quest for Privacy in the Digital Age

Authors Andreja Primec, Gal Pastirk, and Igor Perko discuss the right to control personal data and privacy in the context of modern technological advancements. The chapter addresses the rapid data accumulation and analysis capabilities enabled by AI and other technologies, challenging traditional regulatory frameworks. The authors emphasize the need for reimagining data regulation processes to protect individuals' privacy rights amidst evolving digital landscapes. The chapter advocates for updated legal frameworks that balance technological innovation with robust privacy protections, ensuring individual autonomy in the digital age.

Chapter 7: Cherish Data Privacy and Human Rights in the Digital Age: Harmonizing Innovation and Individual Autonomy

Prof. Bhupinder Singh delves into the importance of data privacy in protecting individuals' rights in the digital era. The chapter discusses the challenges posed by extensive data collection, storage, and sharing, highlighting the responsibility to safeguard personal information amidst technological innovations such as AI and IoT devices. Singh emphasizes the need to balance innovation with adequate regulation to protect privacy and security, exploring diverse approaches to human rights protection concerning data privacy in the digital landscape.

Chapter 8: Digitalization in Corporations: Integrating Utility of Digital Technology With Accessibility and Privacy of Data

Siddharth Kanojia examines how corporations leverage digitalization to enhance operations while prioritizing customer experiences and data privacy. The chapter discusses the use of advanced analytical models and risk-management techniques to improve internal governance and decision-making processes. Kanojia highlights the critical balance between technological innovation and legal obligations to protect data privacy and security. The chapter analyzes recent corporate law amendments addressing technological advancements, emphasizing the importance of aligning digital strategies with regulatory frameworks.

Chapter 9: Corporate Data Responsibility in East Africa: Bridging the Gap Between Theory and Practice

Ripon Bhattacharjee, Aditya Agrawal, Madhulika Mishra, and Akash Bag explore corporate data responsibility in East Africa, focusing on the gap between theoretical frameworks and practical implementation. The chapter uses case studies to examine how firms handle data privacy and ethical issues in the digital economy. Despite awareness and efforts to create comprehensive policies, the authors highlight significant implementation challenges. The chapter calls for context-specific guidelines to bridge the gap between policy development and real-world application, ensuring that the growing digital economy aligns with human rights and state goals.

IN CONCLUSION

In *Balancing Human Rights, Social Responsibility, and Digital Ethics*, we have aimed to offer a comprehensive exploration of the intricate relationship between digital technologies and ethical considerations in modern society. This edited reference book delves into critical themes such as artificial intelligence, data privacy, and the digital divide, all through the lens of human rights and social responsibility. By drawing on the newly established guidelines of the EU Artificial Intelligence Act and the OECD AI recommendations, we have integrated ethical principles with legal implications to present a well-rounded discussion on digital ethics.

As editors, our goal has been to highlight the profound impact of digital technologies on fundamental rights and the importance of embedding ethical considerations into technological advancements. The chapters in this book underscore the necessity of transitioning from theoretical discussions to practical applications, urging for regulatory frameworks that protect human rights in our evolving digital landscape.

The contributions from our esteemed authors provide fresh perspectives and critical insights into the societal and human implications of digitization. From the ethical challenges posed by AI to the legal frameworks governing digital privacy, this book serves as a vital guide for academics, policymakers, professionals, and civil society members. It fosters a deeper understanding of how digital ethics can harmonize with global sustainable development, democratic principles, and the ethical advancement of technology.

We hope that *Balancing Human Rights, Social Responsibility, and Digital Ethics* will inspire further academic discourse, inform policy-making, and encourage the development of business strategies that prioritize ethical considerations. Ultimately, our aspiration is for this book to be a cornerstone resource in promoting a future

Preface

where digital innovation and human rights are intrinsically aligned, ensuring a just and equitable digital era for all.

Maja Pucelj
Faculty of Organisation Studies, University of Novo Mesto, Slovenia

Rado Bohinc
University of Ljubljana, Slovenia

Introduction

As editors, we are pleased to present this comprehensive monograph, titled *Balancing Human Rights, Social Responsibility, and Digital Ethics*, which researches relationship between digital technologies and ethical, moral and legal considerations in the modern world. Presented monograph is being published in the crucial moment, as the mentioned questions gain the importance in the rapidly changing landscape, which is being shaped by the EU Artificial Intelligence Act and the OECD AI recommendations.

In present monograph, we aim to address a wide range of worrying challenges associated with the moral implications of AI, data privacy, the digital divide, protection against online threats, and the transformative effects of digital technologies on social media, e-commerce, and healthcare. By considering ethical principles with legal regulations, we strive to ensure holistic perspective, which integrates these dimensions into a cohesive narrative.

The main focus of the monograph is the intersection of digital ethics and human rights. As digital technologies had a profound impact on fundamental rights and social responsibilities, especially focusing on delicate areas, like privacy, freedom of speech, and access to information, this topic became a vital area for consideration and research. European Union is striving to incorporate digital ethics considerations within its legal framework by ensuring that General Data Protection Regulation (GDPR) and the proposed AI Act ensure that technological advancements, like in AI, align with human rights and ethical standards. Our discussion extends to the OECD's AI Principles, which strive to ensure that AI systems uphold human rights and democratic values, which are crucial for ensuring responsible and ethical regulation of AI on a global level.

Digital ethics surpasses technological challenges, as it addresses also moral, ethical and legal challenges and consequently emerges in broader societal issues. Our monograph emphasizes the importance of an adaptation of regulations and standards that ensure respect of human rights in a constantly evolving digital landscape. The European Commission's efforts to foster a European Union that addresses digital ethics and individual rights reflect the global trend to achieve responsible and ethical AI governance.

Mentioned book want to enhance understanding of the intricate relationship between technological innovation and fundamental human rights, by exposing the crucial role of ethical principles in guiding technological progress. Its goal is to be an essential guide for those, who wish to understand the shift, which is happening in the digital realm and their implications for a sustainable and equitable digital future.

With the help of detail exploration of the ethical, legal, and societal dimensions of digital technologies, we wish to equip readers with the tools and insights necessary to adequately navigate this complex landscape. We wish to contribute to the ongoing dialogue on digital ethics, by offering a multifaced perspective that balances technological advancement with human rights and social responsibility, ensuring that we follow the goal of ensuring a future where technological progress is not only innovative but also ethical, equitable, and aligned with the values that form our society.

Chapter 1
A Bibliometric Analysis of Digital Ethics and Human Rights

Rafiq Idris
Universiti Malaysia Sabah, Malaysia

Rizal Zamani Idris
Universiti Malaysia Sabah, Malaysia

Noor Syakirah Zakaria
Universiti Malaysia Sabah, Malaysia

Mohammad Ikhram Mohammad Ridzuan
Universiti Malaysia Sabah, Malaysia

Azizan Morshidi
https://orcid.org/0000-0001-7786-0322
Univerisiti Malaysia Sabah, Malaysia

Azueryn Annatassia Dania Aqeela Azizan
University of New South Wales, Australia

Razlina Jamaludin
MRSM Kota Kinabalu, Malaysia

ABSTRACT

In order to expand the current literature assessing the quantity and quality of the worldwide research production in the subject of digital ethics and human rights, a bibliometric analysis was conducted for the period 1966–2024. The Scopus.com database was consulted with the "digital ethics and human rights" keywords searched. The database identified 537 documents. In terms of the publication production in the subject over the studied period, the United States of America demonstrated the highest number of publications across the world, with the United Kingdom in the second place. The bibliometric data provided is comprehensive but calls for understanding how the research outcomes in digital ethics and human rights field has impacted society. Therefore, the future research in the field of digital ethics and human rights is expected to be based on the transition from theorising to practical

DOI: 10.4018/979-8-3693-3334-1.ch001

development aspects, per se.

INTRODUCTION

The field of digital ethics and human rights is an evolving field integrated at the point of digital innovation in societal values. The central problem lies in just and equitable integration of emerging technologies within the context of established human rights. The problem is complicated by daily life's deep penetration by digital technologies that give rise to concerns such as digital privacy violation, algorithmic discriminative, the digital divide, erosion of democratic discourse via social media platforms, and so forth. Biological problems like digital privacy, freedom of speech, and the right to access that information that can be simply anchored to human rights. Technology is evolving at a speed that outpaces the current building and aligning human systems. Thus, it is important to ensure digital progress is embedded in the human dignity values and social justice (Linnet Taylor, 2017). Further, the COVID 19 pandemic has created an exceptional dependence on digital technology as virtually every aspect of human life including, work, education, and health such as telehealth and contact tracking have migrated into digital spaces. The rapid shift into digital spaces have exposed the flaws in our technological systems and have necessitated overriding the need to re-evaluate digital ethics and human rights. Prior to the covid pandemic, there was already a pre-existing discussion around the need for digital ethics. However, the post-pandemic era has necessitated a reassessment following most countries' post-pandemic surge in digital surveillance, violation of data privacy, and increased artificial intelligence application in monitoring and containing misinformation and disease (Schwendicke et al., 2020) .

This is why it is critical to explore how digital ethics and human rights have changed from the pre- to post-pandemic times, as the pandemic has become a massive catalyst for radical reform, for the better or worse. The change has brought innovations that shed the light on opportunities digital technologies might bring; however, it has also intensified risks related to privacy, autonomy, and equity – the core of human rights (Sandars, 2009). For instance, as contact-tracing apps have proved their immense benefit to public health, they have also revealed many ethical pitfalls related to surveillance and personal data abuse. To sum up, the digital revolution makes it imperative to make major adjustments to ethical frameworks to secure human rights in the world of digital. Therefore, it is not only academically relevant to discuss the ethics of the pre- and post-pandemic era but also of primary importance to society. The results of this study can be used to formulate responsive measures and strategies to ensure that digital technologies are used for the common good but not against it. Thus, the study is not only valuable for scientific development, but it is also a part of preventive measure to avoid a scenario when the advancement of digital due to pandemic comes at the cost of basic rights and freedoms (Levine et al., 2024). It seems plausible to argue that the reason bibliometric analyses are scarce in this field is that it is new and interdisciplinary. The field of digital rights is a relatively recent school of thought that is constantly developing, and it can be difficult to accurately represent it systematically. In addition, human rights studies seem not to be conducive to quantitative research at all, since, as stated above, this sphere is by definition qualitatively focused on philosophical understanding and legal regulation. Moreover, the large number and variations of publications on the subject further complicate bibliometrics, as it is difficult to aggregate them correctly for quantitative analysis (Wu, 2023).

Therefore, bibliometric analysis is a way to deal with this intricate research problem. When considering clusters of academic and policy-related publications, this approach establishes a quantitative viewpoint on the growth of scholarship around digital ethics and human rights. In terms of citation patterns, collaboration networks, and thematic developments, such an analysis may indicate certain centres of attention, some research gaps, as well as outranked works. Not only does it help us prepare a storyline for the development and present state of research streams – it also provides a factual base for policymakers, technologists, and society to counterweight the ethical facets of digital development (Morshidi et al., 2022). Thus, proactively synthesizing a comprehensive ethical structure becomes a vital endeavour to anchor the relentless development of digital technology in human rights (Montasari, 2022).

Conducting a comprehensive bibliometric analysis of digital ethics and human rights not only enhances academic knowledge in this area but also provides a robust foundation for policymakers, technologists, and society in general to navigate the ethical aspects of digital advancements. Thus, this chapter research questions (RQ) are namely:

RQ1. What is the present trends of publication and patterns of Digital Ethics and Human Rights research publications?

RQ2. Which documents have received the most citations within Digital Ethics and Human Rights studies?

RQ3. Who stands out as the leading entities in Digital Ethics and Human Rights research regarding journals, authors, academic institutions, and nations?

RQ4. What themes have been central to Digital Ethics and Human Rights research before and after the COVID-19 pandemic?

LITERATURE REVIEW

Digital ethics are concerned with navigating the complex and often murky moral terrain of digital technologies and interactions. Theoretically, digital ethics encompasses a wide range of issues, including data privacy, cybersecurity, the digital divide, and the ethical implications of artificial intelligence and algorithmic decision-making, and goes beyond conventional ethical frameworks to tackle the unique challenges posed by the digital environment (Coppola et al., 2019). It is seen as a dynamic and evolving field that requires constant reflection and adaptation as technology advances and societal norms shift (Nemitz, 2018).

In operational terms, digital ethics are translated into guidelines, policies, and standards that govern the behaviour of individuals, organisations, and governments in the digital space. This includes developing ethical codes of conduct for digital professionals, implementing privacy-by-design principles in technology development, and enforcing regulations to protect digital rights and promote fairness (Miller, 2012). From this point of view, digital ethics is an operational discipline that works to ensure digital technologies are created and used to preserve human dignity, achieve social justice, and protect a common good. Essentially, digital ethics operates as a bridge connecting abstract ethical principles with daily choices that contribute to building the digital world (Ahn & Chen, 2022). Human rights in the digital era denote the basic freedoms and securities that all people are entitled to even as digital environments continue to evolve quickly. Human digital rights are an extension of the current established universal human rights specifically to meet the needs for digital privacy, digital service rights, freedom of expression, and right of access to knowledge, as well as protection from internet harm (Coppola et al., 2019). This theory accepts that the Internet and all digital technology rights are essential within the modern world and enhances that people's similar rights in real life should be protected online. The creation of digital rights will require a total reformation of established human rights documents as the new digital realm poses new challenges and opportunities that compel on the inalienability of all rights to match the new age. Operationalizing human rights in the digital era involves turning the theoretical rights into practical guidelines, policies, regulations, and practices that uphold people's rights online (Wu, 2023). This includes passing laws that safeguard digital privacy, formulating measures that secure egalitarian access to the internet, establishing guidelines that ban cyberbullying, digital surveillance, and internet trolling. It also requires the active engagement of governments, international organizations, and non-profits in enforcing and securing these rights, and just like the states, tech companies and other digital players should have a human rights policy infrastructure (Studiawan & Sohel, 2021). Ultimately, with the digital era, operational human rights entails creating a regulatory measure that not only confronts the challenges of the digital era but also capitalizes on the merit of the digital era on human dignity and equity (Suleman et al., 2022).

The discourse between digital ethics and human rights serves as one of the most critical areas of modern scholarship that demands a nuanced discussion of the ethical dimensions of digital technologies and their relevance to human rights (Bankar & Shukla, 2023). In light of the fact that artificial intelligence and digital health applications, along with other digital technologies, are increasingly incorporated into our lives, the range of ethical issues they create and opportunities for human rights advocacy becomes multifaceted. The formulation of voluntary guidelines and ethical frameworks is one of the crucial responses to the ethical issues surrounding these technologies such as privacy, accountability, and discrimination as raised by Mittelstadt (2019). The frameworks seek to strike a balance between leveraging the technological benefits and mitigating the risks they pose to individual rights and societal values. Despite the progress emphasised as achieved by Taylor et al. (2017), the efforts are far from addressing the entire scope of ethical challenges, especially those affecting underprivileged groups. Noteworthy was the separate case of microworkers, major contributors to the digital economy but deprived of basic labour rights and protections as reported by Gray and Suri (2019). This observation exposed a glaring double standard in treating digital labour ethically and further establishes a need for robust, enforceable ethical frameworks that prioritise individuals' welfare and rights in the digital space.

Furthermore, the digital revolution provides opportunities for the promotion of human rights, such as the extension of access to information and freedom of expression and the stimulation of social and political mobilization (Krzywdzinski et al., 2023; Panda et al., 2023). Yet, the political feasibility of such promotion is uncertain, as powerful states, corporations, and other actors utilize digital technologies to monitor, control and even abuse human rights. From the standpoint of engineering and technology, adopting a human rights perspective means reframing the focus from micro-ethical issues that primarily concern individual conduct and decision-making – to a macro-ethical issue that deals with the structural and institutional impact of technology on society. Beaven and Laws (2004) argues that this broader vision is needed to focus on the performativity of actors as they contribute to and operate within the realms of social systems.

Furthermore, modern research, as exemplified by Latonero (2018) work, has broadened the dialogue in the field of technology development by calling for a rights-based perspective grounded in human dignity and social justice. The latter perspective, in turn, requires collaboration between a variety of experts, including technologists, ethicists, policymakers, and human rights supporters to ensure that digital innovations are developed for the common good and for the promotion, not restriction, of human rights. Scholarly works review issues and opportunities of the digital age for human rights and ethics (Levine et al., 2024). While the call for voluntary guidelines and ethical constructs is already promising, there is a need for more robust and enforceable measures that would ensure justice and equity for all in the digital space (Levine et al., 2024). As well, considering human rights matters in the design and deployment of digital innovations is essential to make sure they are beneficial to society. With the digital landscape continuing to evolve, continuous research inquiry from the perspective cultivated in an array of disciplines is instrumental in discerning ethical challenges and promoting the cause of human rights (Manroop et al., 2024).

The presented literature review offers a broad overview of the intersection between digital ethics, human rights, and the problems associated with it. In particular, the review identifies the major issues and indeed the solutions designed to settle them in the form of voluntarism guidelines and ethical frameworks. However, the review fails to critique the extent to which these measures are effective, and for which reasons. For instance, the existing power imbalance in the digital sphere could be discussed in greater detail. Finally, this review is mostly normative in its consideration of the issue. For instance, it does not delve too deep into the empirical evidence supporting the claims made therein. Hence, the accountability of the methodologies used to test human rights implications, such as a bibliometrics analysis, would provide the reader with a better understanding of the issue.

BIBLIOMETRIC ANALYSIS METHOD

The quality and relevance of research findings can be properly assessed only through a good understanding of the methodology behind the study. Moreover, according to Zhang, et al., (2021), bibliometric analysis as a method to study and measure the flow of specific terms or concepts in the literature of a certain scientific field remains an essential tool. Bibliometric analysis allows one to collect extensive metadata about the research and the factors associated with it, as mentioned by (Kaur et al., 2021). Consequently, this method sheds light on the field of digital ethics and human rights. Bibliometric analysis is a statistical description of the entire text body of works, trends, studies, the citations of which they contain or supply, the major concepts of the field, the keywords, the methodologies, the title and authorship information, and all the sources and citations of scientific literature. Recently, bibliometric method of reviewing literature is more and more utilized by scholars, and science mapping technique through platforms or databases which contain information about the relevant publications can be used by researchers to cover their wide area of study. Ahmi (2022) conducted his study based on a bibliometric analysis approach, paying special attention to the search for and analysis of targeted databases, particularly Scopus. Such an approach enabled her to obtain a significant number of publications and relevant target articles and increase her study's robustness and integrity. Importantly, the Scopus database gives users the opportunity to comprehensively explore a significant number of bibliographic areas.

Keywords and Search Strategy

Using appropriate keywords is a critical issue before conducting bibliometric analysis. In this context, to reach to my study goals, the researcher in this present study utilized publication title, abstract and keyword as the master keywords. The keyword should easily convey researcher understanding of possible variations, combinations and permutations for search result . Hence, the author restrict the article title, the abstract and the keyword field as the following query: TITLE-ABS-KEY digital AND ethic* AND human AND right* . The user can get good results regarding the keyword search in Title Keyword Abstract search string. Therefore, this article is based on an enormous body of published research in the Digital Ethics – Human Rights area has been retrieved through the online Scopus database. This high impact database was selected as the primary source of information for this research. Scopus is the largest abstract and citation database of technology, social science, business, and management literature with wide aspects of coverage the Digital Ethics and Human Rights databases. The database have only peer-review journals provided by high trusted and reputable academic publishers; Taylor and Francis Group, Springer, Elsevier, Inderscience, and Emerald. Figure 1 shows the search string and bibliometric workflow applied for the research to present a visual illustration of the process. Moreover, the PRISMA flow diagram presents the flow of the bibliometric analysis.

Analysis Tools

The researchers used Biblioshiny as an application programmed for the Bibliometrix R package to make an all-inclusive bibliometric analysis to sufficiently fulfilment the research objectives and clarify the research questions of this investigation. According to Aria and Cuccurullo, (2017), this specific application endeavoured to advance the achievement of scientific bibliometric investigation, which is deemed a complex process. The meticulous phases and thorough analyses that were transpired during this multifaceted investigation are summarized in Figure 2. In addition, the meticulous analysis presented in this scholarship was divided into two distinct extractions due to the following: descriptive analysis and network analysis. These strategies have been reallocated to the research questions posed in the previous section.

Figure 1. PRISMA flow diagram of the search strategy (Adapted from Page et al. (2021) and Morshidi et al. (2023))

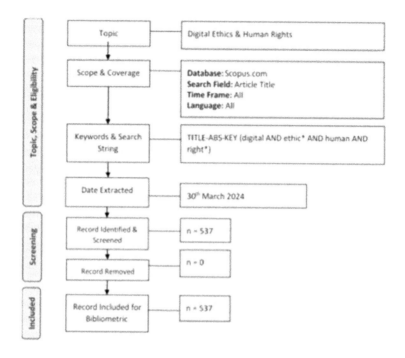

Figure 2. Detail steps for bibliometric analysis using Bibliometrix and Biblioshiny (Authors guided by Abdul Rahman et al.(2022))

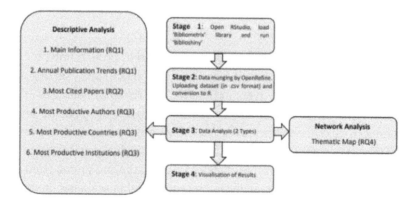

RESULTS

Main Information About Data

Table 1 shows a comprehensive overview provided by the descriptive data of a bibliometric dataset, spanning from 1966 to 2024, illuminates the extensive range and depth of documents within a specific field of study. This period of 58 years showcases sustained interest and ongoing research, supported by 436 diverse sources—including journals, books, and more—highlighting the interdisciplinary nature and the myriad platforms for research dissemination. With a total of 537 documents, reflecting a significant volume of research output, and an annual growth rate of 6.27%, the data points to a dynamic field with escalating scholarly engagement. The average document age of 5.54 years and an average of 14.7 citations per document underscore the relevance and impact of the research within the academic community. Moreover, the presence of 1690 Author's Keywords reveals the extensive thematic diversity covered, creating a rich semantic framework for analysis.

Table 1. Main information

Description	Results
MAIN INFORMATION ABOUT DATA	
Timespan	1966:2024
Sources (Journals, Books, etc)	436
Documents	537
Annual Growth Rate %	6.27
Document Average Age	5.54
Average citations per doc	14.7
References	0
DOCUMENT CONTENTS	
Keywords Plus (ID)	3223
Author's Keywords (DE)	1690
AUTHORS	
Authors	1692
Authors of single-authored docs	182
AUTHORS COLLABORATION	
Single-authored docs	197
Co-Authors per Doc	3.29
International co-authorships %	17.69
DOCUMENT TYPES	
article	302
article article	2
book	18
book chapter	64
conference paper	62
conference paper article	1
conference review	7
conference review article	1
editorial	5
editorial article	1
erratum	3
letter	2
note	14
review	51
short survey	4

The aforementioned comes in line with another prevalent theme, collaboration. The corpus verified 1692 authors, 182 with their single-authored documents, and 3.29 average co-authors per document, showcasing a collaborative trend in the field. The dataset reported a substantial proportion of individual research efforts, confirmed by 197 single-authored documents. In addition, the 17.69% rate of international co-authorships confirmed the global collaboration in the field, supporting a richer understanding through diverse perceptions and expertise. Finally, the compilation of articles and books, book chapters and conference papers and reviews epitomize the field variety of primary and sources through which findings are communicated to broader audiences. In conclusion, the corpus depicts an exciting rapidly developing field characterized by varied contributions, significant citation impact, and notable collaboration, both at national and international levels. The diversity of

A Bibliometric Analysis of Digital Ethics and Human Rights

the document types presented and an extended list of keywords indicate the fields, and topics complexity and interest across various fields opening opportunities for future research.

Annual Publication Trends

Figure 3. Growth and publication impact per year

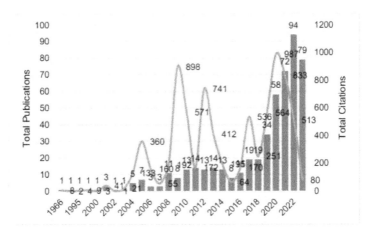

Figure 3 gives insight into how publications and citations have increased over the years regarding digital ethics and human rights. The orange line on the graph indicates how the level of publications has been increasing from 1966 to 2024 at an startling rate after 2006. This could indicate that people have been more interested and engaged in investigation regarding the topic more than before. It is important to note that in 2009, there were a highest 898 publications, possibly this was a year so many different achievements were made regarding digital ethics and human rights, or it received great global attention. Although some fluctuations were observed following this peak, the overall movement of the graph was upward, indicating sustained growth. On the other hand, the bar graph depicts the total number of citations each year's publications garnered, which also exhibits an upward trend. However, this trend is marked by several spikes that indicate years in which the publications had a significant impact, such as in 2008, when citations surged to 987. Interestingly, high citation numbers do not always align with the volume of publications, implying that the influence of certain publications may extend beyond their publication year. Recently, a downturn has been noted in 2023, with a decrease to 94 publications and 513 citations. This recent dip could be attributable to incomplete data collection for the year, a shift in research emphasis, or broader influences that have curtailed both research output and the ensuing academic discourse. Despite these annual inconsistencies, the field demonstrates a general trend of stability and consistent interest, as evidenced by the gradual increase in early year publications, which suggests the consolidation of digital ethics and human rights as a significant area of research.

As can be supposed from the observations, the fields of digital ethics and human rights are unarguably progressing and developing. The impact on the academic audience can be different, but overwhelmingly, they show stable and even enhanced growth, which demonstrates that these fields are becoming more and more important given that digital is gaining momentum in consideration of almost any aspect. Thus, considering the bibliometrics trends, the field is undoubtedly advancing and has all the rights to remain crucial in the consideration of the digital era.

Most Cited Papers

Table 2, on the other hand, paints a picture of the papers' scholarships' impact. Citation metrics provide a metaphorical acknowledgment node from one scholar to another. All woven together, an intellectual exchange web is formed and connects scholars across time. The Paper column links readers to the source of the publication, complete with the author's surname, year of publication, and the publication from. For example, the paper from Medical Teacher's 2009 period, such as Sandars's, is not only source of knowledge but also the touchstone for additional study. This particular paper has been cited 624 times across the field since publication. The Total Citations number is a straightforward way to measure how far a paper reached into the broader discussion. This is more than a personal achievement; this is the cumulative statement for a papers' reverberation through the field as a whole, and a high number suggests that it has served as a cornerstone for future works. Sandars's paper seems to be ubiquitous in its field, given that many of his peers cited it.

Table 2. Most cited papers

Paper	Total Citations	TC per Year
Sandars (2009)	624	39.00
Miller (2012)	427	32.85
Linnet Taylor (2017)	324	40.50
Schwendicke et al.(2020)	271	54.20
Coppola et al. (2019)	196	32.67
Gasser et al. (2020)	182	36.40
Illes et al. (2010)	173	11.53
Morley et al. (2020)	158	31.60
Hunt et al.(2006))	153	8.05
Michell et al. (2012)	151	11.62
Abascal & Nicolle (2005)	145	7.25
Ramsthaler et al. (2010)	132	8.80
Nemitz (2018)	131	18.71
Christenhusz et al. (2013)	126	10.50
Cain & Romanelli (2009)	102	6.38

The temporal dimension added to the impact of a paper by the metric, Total Citation per year, makes a wide presentation of an assumed scholar's impact on people's time. Thus, the metrics offer an analysis of both TC and TC per Year to a common impact on scholarly literature over a period hence presenting the best way to understanding papers' scholarly impact to their specific field. For instance, while the TC is an absolute indicator of the extent of the paper's contribution to the discipline, TC per Year (CI) tells us how long an academic paper has been developing and becoming integral to the field.

It is imperative to explore this aspect of scholars' academics, for some acquire fast acclaim while others decline in their scholarly utility. For instance, a TC of 624 and 54.20 CI per year on Taylor L. et al's 2017 paper in Big Data Society explains how the scholar's work gains prominence, and with such a CI per year, it ensures its pertinence to the latest conversation. Some aspects of the metrics' indicators will become important in fields that rapidly evolve and change trends, as such show the papers at the top at any time in the field. These indicators have diversified on the scholarly papers one offers to encompass the entire contribution to various fields of study. This is from the consideration that the publications provided range from psychological insights in Perspectives on Psychological Science, to the latest medical publications in The Lancet Digital Health. As such, the collection may offer a cross-sectional overview of scientific enquiry and intellectual diversity. The variance is in TC totals, from 102 to 624 and TC per year, from 8.05 to 54.20, shows the unique volumes fit which papers tend to gain scholarly influence. Some achieve acclaim fast while others grow at a steady pace. The volumes one paper gains in a unit time may depend on the value of the topic among specialized scholars or the entire scholarly community engaging in the field, and on the breadth and reputation of the publication to academics.

Most Productive Authors

Table 3. Most productive authors

Authors	Articles
Jia Liu	5
Effy Vayena	4
Tamar Ashuri	3
Johannes J Britz	3
Marcello Ienca	3
Camille Nebeker	3
Jian Zhang	3
Gaurav Acharya	2
Kristin MS Bezio	2

Table 3 presents a concise summary of publications authored by individual scholars in the field of digital ethics and human rights. The information outlined in the table offers essential insights into the contribution of specific researchers in the proposed field of study. Jia Liu is at the top of the list of authors who have published the highest number of articles in this field.

Specifically, the scholar has had a bibliography of five publications in the digital ethics and human rights discourse a move indicative of determination and focus. The second author in the list is Effy Vayena who has had four publications, showing that the scholar has had an immense contribution in the academic discussion of these topics. Meanwhile, Tamar Ashuri, Johannes J Britz, Marcello Ienca, Camille Nebeker, Jian Zhang have published three articles, and their level of research output is consistent and indicates that they are reputable figures in the field. Gaurav Acharya and Kristin MS Bezio each have two articles in the field and thus they seem to be emerging voices or specialized contributions. However, even with fewer publications, the two scholars have still dedicated their efforts to the assessments of important issues within the technology-ethics-society framework.

While this data gives a quantitative outlook on researchers' influence and busywork around the quest topic, they lack the extent, citation power, or quality of their sway, which are key indicators of assessing a better understanding of the role played in the field. Moreover, this said data is not exhaustive for an academically conformist assessment of authors' contribution to the field.

Most Productive Countries

Table 4. Most productive countries

Country	Freq
USA	271
UK	133
CHINA	62
ITALY	58
GERMANY	57
CANADA	53
AUSTRALIA	52
NETHERLANDS	50
INDIA	47
SWITZERLAND	46
SPAIN	29
SOUTH AFRICA	26
FRANCE	23
AUSTRIA	18
BRAZIL	18
NEW ZEALAND	18
SWEDEN	18
BELGIUM	17
IRELAND	15
UKRAINE	15
PORTUGAL	13
TURKEY	13
INDONESIA	12
IRAN	12
MALAYSIA	11

Table 4 provides a striking glance at the contribution made by countries to the field of digital ethics and human rights. The bibliometric data presented sheds light on the way knowledge is created and spread around the globe in the relevant to society and morality topic. The dataset may be described as a snapshot of human intellectual output, as it is, simultaneously profound. Unsurprisingly, the country that ranks number 1 is the United States, whose publication number reaches a tremendous 271. This figure may point both at the environment of intense intellectual work and the great amount of research investment specifically in the topic of digital ethics and human rights. In addition to the latter, such a prevalent position of the country, may be an indicator of attempts to guide global discussions on the morality of technology. Next, the narrative continues with the United Kingdom trajectory, as its publication number is slightly above 133. Not less importantly, it is being followed by a number of other regions, with China, Italy, and Germany around high fifties and sixties in publication count, respectively. The above may be understood as a testament to the expanding field of interest in the topic of ethical derivates of digital revolution and the truly global community of spirit.

The data set illustrates an ethnically inclusive international conversation. While established nations appear to have a larger contribution and would track the funding and extensive network of academic institutions, increased productivity is anticipated. However, the involvement of regularly less conspicuous countries, such as India, Brazil, and Malaysia, in global research indicates research is the more widely dispersed worldwide endeavour. This also notes that while these countries have yet to adequately publish additional information, the horizon in research and development in digital public issues and human rights is vastly wider into new areas, other human/age demographics. Numerous frequencies of publication prerequisites require a thorough examination of the reasons and determinants of variance. A country's economic condition, resource distribution for schooling and science, and goodwill of immensely performance-giving academic institutions can significantly influence research presents. Moreover, it is critical to understand the influences as a gateway to research existence that is both lively and equitable.

Table 5. Most productive source titles

Source Title	TP	NCA	NCP	TC	C/P	C/CP	h-index	g-index	Citable Year	m-index
ACM International Conference Proceeding Series	9	30	9	14	1.56	1.56	2	2	9	0.222
Lecture Notes in Computer Science (including subseries Lecture Notes in Artificial Intelligence and Lecture Notes in Bioinformatics)	7	8	3	16	2.29	5.33	2	4	8	0.250
CEUR Workshop Proceedings	5	24	0	1	0.20	0.00	1	1	5	0.200
Studies in Health Technology and Informatics	5	14	3	157	31.40	52.33	5	5	24	0.208
Technology in Society	3	9	3	46	15.33	15.33	3	3	6	0.500
Applied Sciences (Switzerland)	2	13	2	11	5.50	5.50	2	2	3	0.667
Lecture Notes in Networks and Systems	2	2	1	0	0.00	0.00	0	0	3	0.000
Computers and Education	2	3	2	59	29.50	29.50	2	2	11	0.182
Communications in Computer and Information Science	2	12	2	2	1.00	1.00	1	1	2	0.500
International Journal of Environmental Research and Public Health	2	12	2	12	6.00	6.00	2	2	4	0.500

It can be observed from the Table 5 that in academic context, Source Title is a major connect that leads a researcher towards the knowledge repositories within various publications. From ACM International Conference Proceeding Series to CEUR Workshop Proceedings, the titles in this row represent various forums that allow academic discussions. The whole list demonstrates a broader spectrum of perspectives on scholarship including setting-specific approach to conference procedures and community-based learning featured in workshops and discourse. Given the disparity in settings, each publication in the row has a contribution to make towards a collective understanding. When viewed in details, years of effort symbolized by Total Publications (TP) defines the scope of a scholarly work. For example, ACM, an annual series comprising nine publications is an example of a consistent exposure to new research. Nevertheless, the figure is superficial, measuring only the time frame without offering insight into how much a Source truly matters. Total Cited Articles (NCA) explains a publication's resonance with scholarly engagement. Themes treated herein have sparked debate, clarifications or personal opinions. CEUR, with 24 cited articles, has a rallying influence creating demand for further research.

In this context, the Number of Citing Publication represents the number of voices in the chorus, each conditioned by the publication, while the Total Citing represents the cumulative sound of the entirety of the voices. Studies in Health Technology and Informatics, for example, are heard in academic singing about as long as the work is cited 157 times. Indeed, in the time-space where Scholarship validates and disseminates knowledge, the impact of a work reverberates. Citations per publication reflect per work average impact, which is more telling. Since Scholars are conditioned to validate and circulate other works, the Citation per Publications suggests that the work is not only heard and recognized but also considered. When a work has a Citation per publication of 31.4 as Studies in Health Technology and Informatics does in this case, it discloses a consistent recurrence pattern of works being validated and included in other's work. Citation per citation Publication adds uncertain certainty to the interpretation. As a matter of fact, the value of Studies in

Health Technology and Informatics indicates that if its works are cited at all, they are described several times in the citing work. h-index and g-index could be seen as the heart and soul of the Scholar. They are a measure of the scholar's clout and demonstrate a solid- five articles that had gained attention. The g-index also of the value 5 is congruent with this measure.

The Citable Year introduces a temporal dimension by place indicating its achievements throughout the course of time. The 24-year citable years of Studies in Health Technology and Informatics indicate that it has been a long-resident and constantly-welcomed player in the field. Furthermore, the m-index, or the 0.222 metrics of influence of the ACM series, represents the "rate of influence over time", measuring the tempo and the intensity of the spread and lasting of the work in the academic world. Assuming the above metrics and indices, one can claim that Studies in Health Technology and Informatics is a robust source in the field, citing hundreds of publications with an exponentially expanding number of citations every year. The harmony between the h-index and the g-index shows that the works of are being cited, while the works of the ACM series have an unexpected result, although having a high publication base and a large number of cited works. The average citations per publication are still moderate, implying that the series may serve as a seed or an inciter instead of being a comprehensive and utmost authoritative source. The above comparison shows the multi-dimensional approach to the quantitative impact of scientific work, indicating that more than just quantity matters but also the stimulation of ideas and the ecosystems of competence development.

Table 6. Most productive institutions

Affiliation	Articles
UNIVERSITY OF CALIFORNIA	9
UNIVERSITY OF OXFORD	9
UNIVERSITY OF ZURICH	9
MONASH UNIVERSITY	8
DUKE UNIVERSITY	6
HARVARD UNIVERSITY	6
PENNSYLVANIA STATE UNIVERSITY	6
UNIVERSITY OF OTTAWA	6
ETHICS COMMISSION COLLEGE OF NURSING OF MADRID	5
STANFORD UNIVERSITY	5

Table 6 provides a snapshot of academic institutions leading in digital ethics and human rights research activities. The University of California, the University of Oxford, and the University of Zurich stand at the top with nine publications each, indicating a high level of involvement in these fundamental domains. The number of publications advanced by these institutions does not only imply that they are highly interested in this field but also that they have the financial and academic muscle to invest in producing such huge volumes. The institutions mentioned as the leaders in this field in these results, including Monash University and duke University, with eight and six publications, respectively present a global front in examining the ethical definitions and human rights risks associated with an increasingly digitalizing planet. Moreover, a general glance at the table illustrates that these interdisciplinary topics come from entities dedicated wholly to research, such as the Ethics Commission College of Nursing of Madrid, to internationally renowned research universities, such as Harvard and Stanford. Integrating different academic backgrounds from North America to Europe and the rest of the world depicts a global approach to contextualizing these discussions. It is the broad scope and possibilities which the cross-vision presents that powers this field, thus enhancing a shared dialogue and innovation in digital ethics and human rights. While the number of publications cannot exhaust the method used to estimate the output of each institution in this field, it is essential to look at influence and the outreaching of these scholarly works in terms of citations earlier publications and determining their impact on the academic and practical world.

Theme Evolution

Figure 4. The Sankey diagram of theme evolution

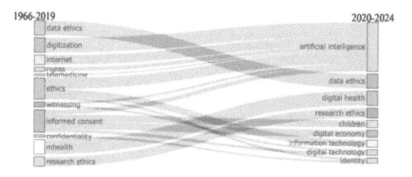

As illustrated by the Sankey Diagram, it captures the evolution of themes within the digital ethics and human rights field, as published in a range of occurrences sourced from two intervals; 1966-2019 and 2020-2024. The diagram was prepared in the order of the increasing or decreasing relative academic interest in each identified topic. The diagram aids in tracking the extent to which specific subjects demonstrated persistence in academic interest or their re-emergence in marked academic publications for each defined period of time. The left-hand column featuring the years of 1966-2019 records a longer list of related prior topics. This list of prior topics is captured in sequential order as follows: "data ethics", "digitization", "internet", "rights", "telemedicine", "ethics", "witnessing", "informed consent", "confidentiality", "mhealth", and "research ethics" are referred to as fundamental themes and subject matters of digital ethics and human rights. The arrows show which of these issues continued in stride in the enquiry into the realm of 2020-2024, shown in the right-hand column. The thickness of the lines represented the volume or strength of the connection between the issues of concern noted across each interval.

It is observed that digital ethics continue to be an imposing concern as illustrated using a thick line showing a continued "data ethics" theme. Similarly, this concern is manifested in the evolution from "digitization" and "internet" foundations to later "information technology", "digital technology", and digital economy, showing a transition into a more comprehensive digital ecological context. Further, the major concerns of "rights" and "ethics" can be encapsulated in the concerns of artificial intelligence, which is a recent concern, underlying recent focus due to the proliferation of artificial intelligence.

"Telemedicine", which was a distinct theme in the earlier period, now contributes to the broader term "digital health", which encompasses various aspects of health-related technology. Similarly, "research ethics" has maintained its relevance, perhaps reflecting continued concerns about the conduct of ethical research in an increasingly digital world.

The theme of "witnessing" has a less pronounced flow into the later period, suggesting that it may not be as central as it once was, or it has evolved into other themes. On the other hand, new themes such as "children" and "identity" emerge in the later period, which may represent new areas of concern in digital ethics, such as the impact of digital technologies on minors and the concept of identity in digital spaces. In short, the Sankey diagram indicates an evolving field of digital ethics and human rights that adapts to technological advancements and societal changes. This shows that certain themes remain relevant, while others gain prominence, highlighting the dynamic nature of this area of research.

Figure 5. Pre COVID 19 pandemic publication themes

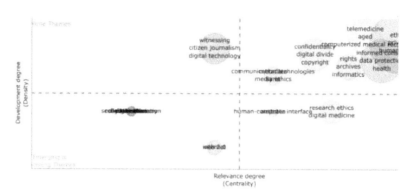

Figure 5 provides a dynamic representation of research themes in digital ethics and human rights, detailing their significance and evolution pre COVID 19 pandemic. Such visualizations are of great importance in tracing the thematic development of the field, particularly with regard to significant events such as the COVID-19 pandemic. In this study, centrality is displayed along one axis, demonstrating the prominence of a theme within the research community. High centrality indicates key topics, such as "computerized medical records" and "informatics," which are prominent in the field's literature and frequently associated with other important themes, thereby serving as the foundation of digital ethics and human rights research.

On the other hand, the density axis measures the level of development within a theme. Themes at a higher location on this axis, such as 'telemedicine' and 'ethical' on the x-axis, indicate well-studied subjects or topics with extensive literature available that attests to their maturity in the field. When centrality and density are plotted concurrently, the diagram splits into four distinct quadrants, each denoting a different level of theme maturity and centrality. The top right quadrant, 'Motor Themes,' encompass themes that are both central and mature, suggesting that these topics were dominant forces prior to the pandemic and were instrumental in defining the subject matter of digital ethics and human rights. Alternatively, Basic Themes on the lower right quadrant are tacky and are benchmark developments, indicating that they are both underlying and have further growth potential. In the top left quadrant, niche themes, such as 'Niche Themes,' arise as mature themes that are central but not extensible, marking specific research areas or practical topics. Finally, the last lower left quadrant is reserved for 'Emerging or Declining Themes,' signifying either new or declining issues ripe for revival post-pandemic. Before the pandemic, information management and rights research were firmly established, with increased data accessibility concerns. Telemedicine research portrays the strongest maturity in terms of leadership for the future of digital health research. Niche topics such as witnessing—a concept of responsiveness to digital media—demonstrate maturity, while emerging issues suggest topics of established concern that are reinvigorated or exposed anew by the pandemic. Ongoing and determined themes offer a perspective into the changing research priorities and societal demands in digital ethics and human rights in light of new circumstances in cutting a digital age.

Figure 6. Post COVID 19 pandemic publication themes

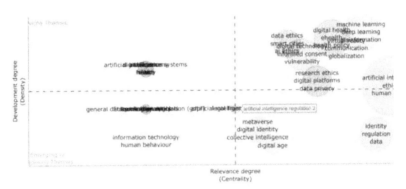

Figure 6 below displays a strategic map for positioning and processing research themes in the area of digital ethics and human rights after the COVID-19 pandemic. This figure visually illustrates the scholar's attention given to key and less common aspects and how it reorganizes after the onset of the pandemic. In this diagram, the most attention is given to the quadrants which ensures a more detailed examination of the themes. In the top-right quadrant are the motor themes, which are characterized by both high centrality and density. These are established and well-studied research lines and topics and have extensive literature such as machine learning, digital health, e-health, smart cities, and globalization. Examining the map, these aspects have been significantly affected by the pandemic, which has shown the importance of technological solutions for health and city development. Also, AI themes such as "artificial intelligence ethics" or "human rights" need to be re-examined to address our ethical challenges stemming from AI and its encounters with human rights.

The Basic Themes are fairly related in the lower-right quadrant. Basic Themes have a high centrality but not a lot of attention, with a lot of potential to explore such as "data ethics," "informed consent," and "data privacy" have been attracting more attention as the COVID-19 pandemic has increased human reliance on digital communication. On only the upper-left quadrant, Niche Themes, is fully developed but set aside or to the side aspects are seen, such as the themes discovered in the control directions. There is already a lot of research on the subject. Given the rapid development and stagnation of digital activity due to COVID-19, studying the impacts of AI on personal freedom could be likely. In the end, in the lower-left quadrant of Declining Themes, which might be lesser known. Other aspects can be inferred, such as the birth of new ethical considerations and discussions about the field of digital existence, including "metaverse", "digital identity" and "collective intelligence". Instead, a rare case may be found when the implications of AI will be studied under the influence of COVID-19-induced digital growth and the increased surveillance of people. The theme may still have a high essentiality and centrality but have attracted fewer scholars. The end of COVID-19 pandemic further led to the renaissance of the field to face the deployment of health data, the ubiquity of AI, and the dilemma of data governance. Therefore proved the ability of the field to adapt to a new world, notably a relevant one, given the role of ethical consideration in a world were innovation shapes daily life. Hence, the need for a comprehensive understanding of ethics and rights in constant dialogue with the evolution of the digital era shaping humanity's future.

DISCUSSION AND FUTURE RESEARCH AGENDA

The findings of the bibliometric analysis in Table 1 and Figure 3, in relation to RQ1, provide a comprehensive examination of the academic landscape of digital ethics and human rights between 1966 and 2024. This time frame, which spans almost six decades, reflects the long-standing history of the field and the many shifts and advances in technology and society throughout this period. The interdisciplinary nature of digital ethics and human rights is emphasized by the data, which includes multiple disciplines during each year. There is a collection of law, technology, sociology, and philosophy.

The large volume of research output: 537 documents and an annual growth rate of 6.27%, shows that the field of research is expanding rapidly and becoming more central to the digital age. The high average age of documents and a robust citation average indicate that these are also important contributions, sparking continued dialogue. A broad range of author keywords reveals the areas of focus, ensuring that the digital age continues to develop and adapt in the course of research directions.

The dataset collected includes a global network of 1692 authors, highlighting the collaborative nature of the research. Papers were also frequently written by a single author, underscoring the significant impact of individual contributions. These data, along with the rate of international co-authorship, remind us that this is a field with global implications and that the challenge is one that must be addressed from diverse cultural, legal, and regional perspectives.

The above data exhibit indicators of promising growth and collaboration; however, a critical assessment of the data must be undertaken, not least because of the recent decline in publications observed in 2023. One must think about whether this is merely a transient outcome or perhaps a deeper problem of saturation, disciplinary reconfiguration, or a lack of responsiveness to emerging issues and their integration into the discourse. Moreover, the initial surge in publications in 2009 followed by the increases and sharp decreases raises the questions of what is behind these fluctuations. Does the field react to global pressures or funding priorities and adjusts its publication course accordingly? Digital ethics and human rights is an increasingly desirable area to study and need more people as a result. In a world where technology continues to disrupt our freedoms, our behaviour, and our societal organization, this field's exploration is critical given the anticipated trajectories. It is the responsibility of scholars to ensure that research does not merely maintain the breakneck pace at which technology evolves, but that their work anticipates and guides it in a manner that does not violate human dignity and rights (Manroop et al., 2024; Wu, 2023). For future bibliometrics, it is critical to consider the broader sociotechnical scene and criticize the translation of research to keep the field dynamic and pertinent.

Table 2, answering Research Question 2, offers an angle of considering the impact of scholarly work through citation metrics. A complex web of academic discourse is shaped through paying attention to how many times have scholars from other sources siege each other's ideas. For example, using the same set of research papers and their authors, publication years and journal titles allow the reader to navigate the academic intellectual terrain. For example, the paper by Sandars (2009) was cited 624 times, suggests the paper remains a central piece to every scholar embarking on a similar research journey. In other words, such citation counts demonstrate not only a paper's breath of reach, but the foundational process of what a scholar does, which is based on other's insights. These counts also factor in TC per year, which adds another layer of the same argument.

A rising TC per year, such as that in Big Data Society by L. Taylor et al. (2017), shows that the paper reached an audience quickly and the topic remains extremely relevant currently. Hence this count is even more curious in rapidly changing discipline as it shows the scholarship at work in a fast-changing domain currently pivoted towards how to harness the power of data accessible online. Table 2 also lists what journals these counted papers were pulled from. There is a wide variety of those, from Perspectives on Psychological Science, and Lancet Digital Health. Thus, this corpus covers many fields of study, especially for its diversity. The range of total citations from 102-624 and from 8.05-54.20 per year on average shows different paths papers could take to be "famous". This could depend on how specific or approaching a more generalized topic received, the size of the field and how active field members communicate with one another, and journal's reach and prestige.

The third research question was identifying the top authors in the extensive area of digital ethics and human rights, and a bibliometrics analysis for this survey was prepared in Table 3. This table presented the objective evaluation of the various authors' scholarly contribution to the field, which was utilised as an assessment method. Liu Jia was presented as the more prolific author with five publications stressing the commitment of this researcher to the exploration of this increasingly vital area of study. Four publications were prepared by Effy Vayena and used to continue a strong academic presence which inevitably shaped the dialogue and debate within the field. Authors with the next highest number of publications, Tamar Ashuri, Johannes J Britz, Marcello Ienca, Camille Nebeker, and Jian Zhang, had three works. These outputs confirmed the regular publication activity in the sector that was most probably collected are used in different studies which informed and guided the fast-changing story of digital ethics and human rights. At the same time, there were authors with fewer works, such as Gaurav Acharya and Kristin MS Bezio, and this fact could be explained by used as a new voice or a contributing factor making the conversation even more diverse.

Table 4 expands the scope of this inquiry to the national level by showcasing an international landscape in which intellectual works on digital ethics and human rights are located and countries in which they originate. The US' large lead in publication volume suggests a lively intellectual climate and a deliberate prioritisation of the highlighted research agenda. The UK, China, Italy, and Germany also stand out, demonstrating an active and vibrant global research community. This international perspective underlines the reality that digital ethics and human rights issues do not stop at national borders and must be studied comprehensively and from a diverse perspective. While bibliometric indices are valuable sources of information to evaluate academic output, the dynamic and ever-changing nature of digital ethics and human rights demands a more inclusive and comprehensive assessment of the subjects' impact and significance, which goes beyond mere publication frequencies and citation counts.

Before the pandemic, from 1966 to 2019 on one side, the Sankey Diagram shows that before the onset of the pandemic, many themes were ranking high, which included important directions such as data ethics or even telemedicine as a part of the broader field of digital ethics and human rights. The arrows demonstrate how these themes have shifted to the pandemic era since 2020, and the thickness of the lines indicates the strength they have generated in terms of academic interest. One can see once the terms progress along the timeline that some themes remain crucial even during the pandemic and adjust to the new digital world.

The diagram indicates a convergence of topics, such as telemedicine, with broader notions of digital health, which may be seen as an adaptive response to the pandemic's influence on healthcare delivery. However, this diagram fails to capture the entirety of the situation. This suggests a changing landscape, but it does not address the qualitative aspects of the research impact. For instance, how have these thematic changes influenced policymaking or practice in digital ethics? Have emerging topics given rise to the development of new ethical frameworks, or are they merely theoretical trends that have yet to have a significant impact outside academia?

In essence, bibliometric data reveal the evolutionary development of digital ethics and human rights research, and yet they also invite a more profound examination of the practical outcomes of this scholarly endeavour. It is imperative to contemplate the extent to which the academic community has successfully transformed research into practical applications, particularly in the present era where digital technologies have become inextricably linked with human rights.

While looking to the future of research in this field, it is clear that the expansion beyond traditional bibliometric review is required (Morshidi et al., 2023). Although these evaluations give an overview of the amount of research and the progression of themes, they do not take into consideration how these are applied or the impact studies have. One such area is the creation of AI ethical guidelines and how that influences digital ethics and human rights. This can vary from drawing legislative conclusions or recommending policy changes or even a thorough review and subsequent analysis. The current priority in citation numbers and publication metrics ignore the contributions lowly cited works make in creating new fields or affecting specific areas. Publication numbers and citation rankings may indicate a paper's influence instead of its theoretical or practical value. The concerns regarding the insufficiencies of current methodologies in digital rights bibliometric analysis are warranted and necessitate thorough examination. One notable issue is the inclination to disregard synonyms and related terms in keyword searches. In academic literature, various terms and phrases may be employed to describe the same concept, and relying on a restricted set of keywords may result in a confined dataset. For instance, "digital privacy", "internet privacy", and "online privacy" might be used interchangeably by different authors, but a bibliometric analysis concentrating solely on "digital privacy" would overlook pertinent studies that employ alternative terminology. This shortcoming can lead to an incomplete comprehension of the field, as vital papers may be excluded from the analysis.

Future work should consider broader societal and technological changes, the potential for interdisciplinary approaches, and the impact of research on policy and industry when evaluating its impact. In addition, research should not only focus on the number of citations but also on the reasoning behind them and their contribution to the global conversation on digital ethics and human rights. To fully assess the impact of a researcher's work, it is important to consider both the quality of the work and the materials used, as well as the citations. For the future agenda, it is crucial to address the rapid ethical changes in technology, particularly in the post-pandemic world. Research should focus on the ethical implications of emerging technologies and their impact on human rights in a surveilled society, as well as adapting to the evolving global digital policies. Theoretical research findings should be practical and provide suggestions for policy implementation, framework creation, and ethical application of digital technology.

CONCLUSION

In summary, this chapter notes that digital ethics and human rights is an ever-changing and expanding field with diverse contributions and significant scholarly impacts. From pre-COVID-19 to post-COVID-19 themes, one notes that the pandemic has functioned both as a lens and a catalyst, bringing existing challenges into sharper focus while introducing new dimensions to ethical considerations in the digital arena. The bibliometric data between 1966 and 2024 reveals sustained interest and growing scholarly contributions indicates the level of interest and its varied dissemination platforms. However, the drop in publications and citations in 2023 requires critical evaluation of the field's responsiveness to global events, funder prioritization and changes in publication practices. In addition, the high levels of focus on themes such as data ethics, digital health, artificial intelligence ethics, and rights-ethical concern about digital technology before and after the pandemic indicates the urgencies in addressing the introductory and already manifesting themes regarding new technologies and digital human rights. The continued existence of foundational concepts and new and emerging areas points to the dynamism of this field and a need to sustain research to keep abreast with digital challenges ethically. As such, this bibliometric novelty calls for the necessity of a forward-looking research agenda that shall not only respond to the pace of technological change but also anticipate and influence direction positively towards human rights. Future digital ethics and human rights research should focus on closing the gap between theoretical exploration and practical implementation, ensuring that the insights translate into influential policy, legal, and developmental frameworks that are ethical. As one seeks to navigate through the digital complexities, this field is sustainability is essential to a just and equitable future.

REFERENCES

Abascal, J., & Nicolle, C. (2005). Moving towards inclusive design guidelines for socially and ethically aware HCI. *Interacting with Computers*, 17(5), 484–505. 10.1016/j.intcom.2005.03.002

Abdul Rahman, N. A., Ahmi, A., Jraisat, L., & Upadhyay, A. (2022). Examining the trend of humanitarian supply chain studies: Pre, during and post COVID-19 pandemic. *Journal of Humanitarian Logistics and Supply Chain Management*, 12(4), 594–617. 10.1108/JHLSCM-01-2022-0012

Ahmi, A. (2022). *Bibliometric Analysis using R for Non-Coders: A practical handbook in conducting bibliometric analysis studies using Biblioshiny for Bibliometrix R package.*

Ahn, M. J., & Chen, Y. C. (2022). Digital transformation toward AI-augmented public administration: The perception of government employees and the willingness to use AI in government. *Government Information Quarterly*, 39(2), 101664. 10.1016/j.giq.2021.101664

Aria, M., & Cuccurullo, C. (2017). bibliometrix: An R-tool for comprehensive science mapping analysis. *Journal of Informetrics*, 11(4), 959–975. 10.1016/j.joi.2017.08.007

Bankar, S., & Shukla, K. (2023). Performance Management and Artificial Intelligence: A Futuristic Conceptual Framework. In *Contemporary Studies of Risks in Emerging Technology, Part B* (pp. 341–360). Emerald Publishing Limited. 10.1108/978-1-80455-566-820231019

Beaven, Z., & Laws, C. (2004). Principles and applications in ticketing and reservations management. In Yeoman, I., Robertson, M., Ali-Knight, J., Drummond, S., & McMahon-Beattie, U. (Eds.), *Festival and Events Management* (pp. 183–201). Butterworth-Heinemann. 10.1016/B978-0-7506-5872-0.50017-X

Cain, J., & Romanelli, F. (2009). E-professionalism: A new paradigm for a digital age. *Currents in Pharmacy Teaching & Learning*, 1(2), 66–70. 10.1016/j.cptl.2009.10.001

Christenhusz, G. M., Devriendt, K., & Dierickx, K. (2013). Disclosing incidental findings in genetics contexts: A review of the empirical ethical research. *European Journal of Medical Genetics*, 56(10), 529–540. 10.1016/j.ejmg.2013.08.00624036277

Coppola, L., Cianflone, A., Grimaldi, A. M., Incoronato, M., Bevilacqua, P., Messina, F., Baselice, S., Soricelli, A., Mirabelli, P., & Salvatore, M. (2019). Biobanking in health care: Evolution and future directions. *Journal of Translational Medicine*, 17(1), 1–18. 10.1186/s12967-019-1922-331118074

Gasser, U., Ienca, M., Scheibner, J., Sleigh, J., & Vayena, E. (2020). Digital tools against COVID-19: Taxonomy, ethical challenges, and navigation aid. *The Lancet. Digital Health*, 2(8), e425–e434. 10.1016/S2589-7500(20)30137-032835200

Gray, M. L., & Suri, S. (2019). *Ghost work: How to stop Silicon Valley from building a new global underclass*. Eamon Dolan Books.

Hunt, P. A., Greaves, I., & Owens, W. A. (2006). Emergency thoracotomy in thoracic trauma - A Review. *Injury*, 37(1), 1–19. 10.1016/j.injury.2005.02.01416410079

Illes, J., Moser, M. A., McCormick, J. B., Racine, E., Blakeslee, S., Caplan, A., Hayden, E. C., Ingram, J., Lohwater, T., McKnight, P., Nicholson, C., Phillips, A., Sauvé, K. D., Snell, E., & Weiss, S. (2010). Neurotalk: Improving the communication of neuroscience research. *Nature Reviews. Neuroscience*, 11(1), 61–69. 10.1038/nrn277319953102

Kaur, M., Rekha, A., & Resmi, A. (2021). Research landscape of artificial intelligence in human resource management: A bibliometric overview. In Yadav, R., Yadav, R., & Gupta, S. B. (Eds.), *Artificial Intelligence and Speech Technology* (pp. 463–477)., 10.1201/9781003150664-51

Krzywdzinski, M., Gerst, D., & Butollo, F. (2023). Promoting human-centred AI in the workplace. Trade unions and their strategies for regulating the use of AI in Germany. *Transfer: European Review of Labour and Research*, 29(1), 53–70. 10.1177/10242589221142273

Latonero, M. (2018). *Governing Artificial Intelligence: Upholding human rights & dignity*. Data & Society.

Levine, M., Philpot, R., Nightingale, S. J., & Kordoni, A. (2024). Visual digital data, ethical challenges, and psychological science. *The American Psychologist*, 79(1), 109–122. 10.1037/amp000119238236219

Manroop, L., Malik, A., & Milner, M. (2024). The ethical implications of big data in human resource management. *Human Resource Management Review*, 34(2), 101012. 10.1016/j.hrmr.2024.101012

Michell, M. J., Iqbal, A., Wasan, R. K., Evans, D. R., Peacock, C., Lawinski, C. P., Douiri, A., Wilson, R., & Whelehan, P. (2012). A comparison of the accuracy of film-screen mammography, full-field digital mammography, and digital breast tomosynthesis. *Clinical Radiology*, 67(10), 976–981. 10.1016/j.crad.2012.03.00922625656

Miller, G. (2012). The Smartphone Psychology Manifesto. *Perspectives on Psychological Science*, 7(3), 221–237. 10.1177/17456916124421526168460

Mittelstadt, B. (2019). Principles Alone Cannot Guarantee Ethical AI. *Nature Machine Intelligence*, 1(11), 501–507. 10.1038/s42256-019-0114-4

Montasari, R. (2022). Artificial Intelligence and National Security. In *Artificial Intelligence and National Security*. 10.1007/978-3-031-06709-9

Morley, J., Cowls, J., Taddeo, M., & Floridi, L. (2020). Ethical guidelines for COVID 19. *Nature*, 582(7810), 29–31. 10.1038/d41586-020-01578-032467596

Morshidi, A. Bin, Satar, N. S. M., Azizan, A. A. D. A., Idris, R. Z., Idris, R., Radzi, M. S. M., Yusoff, S. M., & Sarjono, F. (2023). A Bibliometric Analysis of Artificial Intelligence and Human Resource Management Studies. In *Exploring the Intersection of AI and Human Resources Management* (pp. 85–117). IGI Global. 10.4018/979-8-3693-0039-8.ch006

Morshidi, A., b, K. Y. S., Yussof, K. M., Idris, R. Z., Idris, R., & Abas, A. (2022). Online work adjustment in the sustainable development context during COVID-19 pandemic: A systematic literature review. *International Journal of Work Innovation*, 3(3), 269–288. 10.1504/IJWI.2022.127668

Nemitz, P. (2018). Constitutional democracy and technology in the age of artificial intelligence. *Philosophical Transactions. Series A, Mathematical, Physical, and Engineering Sciences*, 376(2133), 20180089. Advance online publication. 10.1098/rsta.2018.008930323003

Page, M. J., McKenzie, J. E., Bossuyt, P. M., Boutron, I., Hoffmann, T. C., Mulrow, C. D., Shamseer, L., Tetzlaff, J. M., Akl, E. A., Brennan, S. E., Chou, R., Glanville, J., Grimshaw, J. M., Hróbjartsson, A., Lalu, M. M., Li, T., Loder, E. W., Mayo-Wilson, E., McDonald, S., & Moher, D. (2021). The PRISMA 2020 statement: An updated guideline for reporting systematic reviews. *International Journal of Surgery*, 88(March), 105906. Advance online publication. 10.1016/j.ijsu.2021.10590633789826

Panda, G., Dash, M. K., Samadhiya, A., Kumar, A., & Mulat-weldemeskel, E. (2023). Artificial intelligence as an enabler resiliency : past literature, present debate and future research directions. *International Journal of Industrial Engineering and Operations Management*. 10.1108/IJIEOM-05-2023-0047

Ramsthaler, F., Kettner, M., Gehl, A., & Verhoff, M. A. (2010). Digital forensic osteology: Morphological sexing of skeletal remains using volume-rendered cranial CT scans. *Forensic Science International*, 195(1–3), 148–152. 10.1016/j.forsciint.2009.12.01020074879

Sandars, J. (2009). The use of reflection in medical education: AMEE Guide No. 44. *Medical Teacher*, 31(8), 685–695. 10.1080/01421590903053741 19811204

Schwendicke, F., Samek, W., & Krois, J. (2020). Artificial Intelligence in Dentistry: Chances and Challenges. *Journal of Dental Research*, 99(7), 769–774. 10.1177/0022034520915714 32315260

Studiawan, H., & Sohel, F. (2021). Anomaly detection in a forensic timeline with deep autoencoders. *Journal of Information Security and Applications*, 63, 103002. 10.1016/j.jisa.2021.103002

Suleman, D., Rusiyati, S., Sabil, S., Hakim, L., Ariawan, J., Wianti, W., & Karlina, E. (2022). The impact of changes in the marketing era through digital marketing on purchase decisions. *International Journal of Data and Network Science*, 6(3), 805–812. 10.5267/j.ijdns.2022.3.001

Taylor, L. (2017). What is data justice? The case for connecting digital rights and freedoms globally. *Big Data & Society*, 4(2), 1–14. 10.1177/2053951717736335

Taylor, L., van der Sloot, B., & Floridi, L. (2017). Conclusion: what do we know about group privacy? In *Group Privacy: New Challenges of Data Technologies* (pp. 225–237). 10.1007/978-3-319-46608-8_12

Wu, Y. (2023). Data Governance and Human Rights: An Algorithm Discrimination Literature Review and Bibliometric Analysis. *Journal of Humanities. Arts and Social Science*, 7(1), 128–154. 10.26855/jhass.2023.01.018

Zhang, Y., Xu, S., Zhang, L., & Yang, M. (2021). Big data and human resource management research: An integrative review and new directions for future research. *Journal of Business Research*, 133(April), 34–50. 10.1016/j.jbusres.2021.04.019

Zhang, Y., Zhang, M., Li, J., Liu, G., Yang, M. M., & Liu, S. (2021). A bibliometric review of a decade of research: Big data in business research – Setting a research agenda. *Journal of Business Research, 131*(December), 374–390. 10.1016/j.jbusres.2020.11.004

KEY TERMS AND DEFINITIONS

AI Ethics: The branch of ethics concerned with the development and implementation of artificial intelligence in ways that promote fairness, accountability, transparency, and the avoidance of harm.

Algorithmic Discrimination: The bias that occurs when algorithmic systems reinforce or amplify social biases, leading to unfair treatment of individuals based on race, gender, age, or other characteristics.

Artificial Intelligence (AI): A branch of computer science that involves creating systems capable of performing tasks that normally require human intelligence, such as learning, reasoning, and problem-solving.

Bibliometric Analysis: A research method that uses quantitative measures to analyze the impact and development of scientific literature within a specific field, including citation patterns, collaboration networks, and thematic developments.

Contact Tracing: The process of identifying and notifying individuals who have been in close contact with someone diagnosed with a contagious disease, often utilizing digital tools and applications.

COVID-19 Pandemic: A global health crisis caused by the coronavirus SARS-CoV-2, significantly accelerating the adoption of digital technologies and exposing the ethical and human rights challenges associated with increased digital surveillance and data privacy issues.

Cyberbullying: The use of digital communication tools to harass, threaten, or intimidate individuals, often involving repeated and aggressive behavior.

Data Governance: The management of data availability, usability, integrity, and security in an organization, ensuring that data is handled in compliance with regulations and ethical standards.

Digital Divide: The gap between those who have access to digital technologies and the internet and those who do not, often due to socioeconomic, geographic, or demographic factors.

Digital Ethics: A branch of ethics that deals with the impact and implications of digital technology on society. It includes issues such as data privacy, cybersecurity, digital divide, and the ethical implications of artificial intelligence and algorithmic decision-making.

Digital Identity: The representation of an individual in digital environments, encompassing personal information, online activities, and digital interactions.

Digital Privacy: The right of individuals to control their personal information and how it is collected, stored, and shared in the digital environment.

Digital Surveillance: The systematic monitoring of people's digital activities and communications by governments, organizations, or other entities, raising concerns about privacy and civil liberties.

Digital Transformation: The integration of digital technologies into all aspects of business and society, fundamentally changing how organizations operate and deliver value to customers.

Human Rights in the Digital Era: The extension of established universal human rights to the digital realm. This includes rights to digital privacy, digital service access, freedom of expression, and protection from internet-related harms.

Informed Consent: The process of obtaining voluntary agreement from individuals to participate in research or use a service, based on a clear understanding of the potential risks and benefits.

Privacy-by-Design: A principle that promotes embedding privacy features and data protection measures into the design and architecture of information systems and technologies from the outset.

Smart Cities: Urban areas that use digital technologies and data analytics to improve the efficiency of services, enhance quality of life, and reduce environmental impact.

Surveillance: The monitoring of individuals' activities and behaviors, often by governments or corporations, through digital technologies for various purposes, including security, marketing, and public health.

Telehealth: The provision of healthcare services through digital communication technologies, including video consultations, remote monitoring, and digital health records.

Chapter 2
Artificial Intelligence:
A New Tool to Protect Human Rights or an Instrument to Subdue Society by the State

Abhishek Benedict Kumar
Symbiosis International University (Deemed), Pune, India

Aparajita Mohanty
https://orcid.org/0000-0003-2329-1798
Symbiosis Law School, Symbiosis International University (Deemed), Pune, India

ABSTRACT

With each technological revolution, mankind experiences a shift in the pace of history, leading to the emergence of a new age that fundamentally redefines human beliefs. Artificial intelligence (AI) is a technological subject that is causing a profound shift in human society towards a state characterised by robots and machines. AI encompasses several technologies such as machine learning, natural language processing, big data analytics, algorithms, and other related fields. Nevertheless, as human intelligence is marked by inherent bias in decision-making, AI products that rely on human-generated intelligence also display similar characteristics. The existence of bias and discrimination poses a threat to universal human rights, since these phenomena are deeply embedded in social institutions and have many technological origins. Indeed, because of its facilitation of prejudice, AI disproportionately affects the human rights of marginalized individuals and communities, hence giving birth to a new type of oppression rooted in technology.

DOI: 10.4018/979-8-3693-3334-1.ch002

1. INTRODUCTION

With each technological revolution, mankind experiences a shift in the pace of history, leading to the emergence of a new age that fundamentally redefines human beliefs. Artificial Intelligence (AI) is a technological subject that is causing a profound shift in human society towards a state characterized by robots and machines. AI encompasses several technologies such as machine learning, natural language processing, big data analytics, algorithms, and other related fields. Nevertheless, as human intelligence is marked by inherent bias in decision-making, AI products that rely on human-generated intelligence also display similar characteristics. The existence of bias and discrimination poses a threat to universal human rights, since these phenomena are deeply embedded in social institutions and have many technological origins. Indeed, because to its facilitation of prejudice, AI disproportionately affects the human rights of marginalized individuals and communities, hence giving birth to a new type of oppression rooted in technology (West & Allen, 2018).

Given its status as a general-purpose technology, artificial intelligence (AI) has a broad impact on all facets of society. The "inception report" on the broader societal impact of AI, published in 2017 by the European Economic and Social Committee, highlighted several key areas that have the most significant influence on society (Hunt, 2024/2014). These areas include safety, ethics laws and regulations, democracy, transparency, privacy, work, education, and equality. This implies that AI has a significant impact on the core principles of European society, including human rights, democracy, and the rule of law (Zhang & Lu, 2021). The AI High Level Expert Group on AI released the Ethics Guidelines for Trustworthy AI in 2019. Trustworthy AI is characterized as being resilient in terms of its socio-technical, ethical, and legal aspects, according to these principles (Kinder et al., 2023). The principles unequivocally acknowledge fundamental rights as the bedrock for AI ethics in relation to the ethical aspect of dependable AI. These concepts lack legal enforceability in isolation, although include elements directly drawn from established human rights. There has been an increasing need for new or stronger legally binding AI technologies in recent times (Valle-Cruz et al., 2020). In its Whitepaper on AI, the European Commission highlighted the need of AI adhering to the fundamental rights and regulations that safeguard them. The document also unveiled potential elements of a legal framework (Busuioc, 2020).

The chapter delves into the impact of AI on democracy, human rights, and the principles of legal governance. This document delineates the human rights that are now or are expected to be significantly impacted by artificial intelligence, as defined by the European Convention on Human Rights (ECHR), the Universal Declaration of Human Rights (UDHR), and its Protocols. It aims to provide many possible strategies that may be implemented simultaneously, if necessary (Truby,

Artificial Intelligence

2020). The text also examines the impact of the present system on human rights, democracy, and the rule of law, and proposes solutions that should be included into the existing framework. Owing to the fast advancement of technology and society, this research cannot be exhaustive, but it does emphasize the most notable consequences that are now recognized (Cheng-Tek Tai, 2020). On a daily basis, novel AI technologies are introduced, and AI is being used in the realm of human rights. It is easy to feel fearful or gloomy about the ongoing impact of AI on our surroundings (Ahmed & Bajema, 2019). The chapter also aims to elucidate the potential use of AI as a means to safeguard human rights and hold governments accountable for their transgressions against human rights. AI technologies provide a challenge to traditional beliefs about morality and the law, but they also offer promising opportunities for the advancement of human rights (Pelau et al., 2021). In order to ensure this outcome, we may evaluate the appropriate timing and manner in which AI should be used by examining case studies of existing implementations and exploring potential hazards and drawbacks. In recent years, the use of artificial intelligence systems has expanded exponentially, and in 2023, projects of a fresh type of such a system – "Large Language Models" – will take the lead. Several examples of such systems are Bard and ChatGPT (Ray, 2023). There is no discussion that while artificial intelligence technology – of which LLMs are just one example – is still in its infancy, it is easily apparent that the application of such systems will hugely affect most of us individually and professionally, as well as our communities' structure and functioning, probably most critically now. Of course, it is not true that algorithms are smarter than humans – algorithms are just faster and more cost-effective in completing a wide range of duties, from simple to very difficult ones, than most human intellects are. One of the fundamental challenges raised by the use of AI systems in public administration and court systems, as well as over certain indispensable services, granted by private persons, is how to ensure the continuation of respect for human rights, democracy, and the rule of law (Saltman, 2020). Showing the usage of AI technologies is currently a public conversation predominantly focused on the financial impacts of utilizing the technology (Dirican, 2015). The implementation of these kinds of systems can seriously threaten the main principle of democratic governance – the liberal democracy rule – represented upon such a core understanding: elections, the freedom of assembly and association, the right to hold views, as well as the capacity to get and relay information (Nemitz, 2018). Therefore, it is up to all governments worldwide to determine how to confront the challenges created by artificial intelligence systems and adopt the necessary legislation to foster the required innovation while securing these fundamental rights. We must consider how we can use AI systems to improve environmental command, increase social integration and stimulate economic activity without inadvertently establishing a dystopian, oligarchic world where algorithms rule over the society instead of law (hui et al., 2023). The development of artificial intelligence and its increased application in numerous aspects of social, economic and political life has increased the question of AI's influence on democracy. The concept and the practice of democracy involve a plurality of opinions, many of which come with subtle nuances. The corresponding political theory disputes appear to be quite fruitful and effective, pointing to variations in the normative, procedural, or structural aspects in how we understand and design democracy (Saetra, 2020). However, for the purpose of clarity and non - vagueness, these manifold questions must be reduced to a limited number of essential – sometimes-rooted democratic elements that are poised to be influenced by the use of AI Technology. Wherever the argument loses subtlety, a universally fitting theoretical framework makes up for it.

2. AI AS A TOOL IN PUBLIC SPHERE

The most well-known and well followed argument at the moment centres on artificial intelligence's effects on public life. It is also closely related to ongoing discussions about how the digital public sphere is structurally changing. The rise of social networks, which have torn down barriers between private and public interactions and amplified the personalization of news consumption, is already a result of the digital transition. These kinds of developments are sometimes attributed to the echo-chamber or filter-bubble effects, which are then blamed for the escalating political polarization that occurs in democracies across the globe.

Although empirical studies on echo chambers, filter bubbles, and societal polarization have persuasively demonstrated that the effects are greatly exaggerated and that numerous non-technology-related factors more adequately account for the democratic retreat, the proliferation of AI applications is frequently anticipated to resurrect the causal relationship between technological advancements and societal fragmentation that endangers democracy. Here, the supposition is that artificial intelligence (AI) would greatly expand the potential for monitoring and directing public conversation and/or accelerate the automated compartmentalization of will formation. Thus, the argument is that the strength of present AI applications is the capacity to observe and examine massive continuous flows of communication and information, to identify patterns, and to permit immediate and often opaque reactions. In a world of a surfeit of communication, automated content moderation is essential, and both commercial and political pressures ensure that digital tools are produced to supervise and intervene in communication flows. The possibilities for control are carved up amongst users, platforms, moderators, commercial actors and states, but all of these dynamics tend towards automation, albeit in highly asymmetrical

Artificial Intelligence

ways. Hence, AI is "baked into" the back end of all communication and becomes a more covert and yet powerful shaping influence.

The risk of this development is twofold. On the one hand, there can be malevolent actors who utilize these new possibilities for their vested interest. The endeavor to decipher and sway political discourses, like the Cambridge Analytica scandal, surely comes to mind. The other risk is that the relationship between a public and the private powers is changing. The private ones are more and more entangled in political questions, and their capacity to sway political processes diminished and grew more opaque for structural and technological issues (Muller, 2020). Furthermore, the remaking of the public sphere via private business models was pushed forward by the shifting economic rationality of the digital societies, like the development of the attention economy. The private powers strengthen and become less audience able to public actors. An AI application spurs this development, working on the wrong assumption that AI applications forge deathly dependencies and permit for opacity alike. As Shoshana Zuboff has already noted, the politics of surveillance capitalism is that it began to utilize the data accrued not only for including but also forecasting, converting, and controlling the citizens' operations. An AI application is a fundamental element in making this politics between a reality, as it enables this all (Attard-Frost et al., 2024). At the same time, it works like an insulation for the companies which develop and utilize it from the public audit due to the network effects on the one side and the opacity on the opposite. AI is relied on the accumulated data in massive sums and involves giving of the viable upfront costs. Still, while it is established, it is challenging to astound through the competitive markets. Although applications can be developed on multiple hands and to a lot of ends, the fundamental AI infrastructure is quite consolidated and challenging to refurbish. The leading actors are those capable of keeping a stranglehold on the most crucial resources, the general ones and the information, and benefiting from everyone's corporate user. Therefore, we can precursor that an AI application even more tightly ties the clasp of contemporary internet giants. The public powers are stimulated to employ AI applications more and more, and so they flourish even more reliantly which are apt to give the preeminent infrastructure, although it occurs commercially and technically in dark (Sergey Kamolov & Kirill Teteryatnikov, 2021). The developments outlined over the reduced political control and fortification of the private ones impacts each other, and it is to be inferred that countless of the deficiencies visible in today's digital masses will continue their escalation. Although, really, it is hard to assess whether these processes can be counterbalanced by a state input action, it is surprising how commonly it is also presented in association with varied networked communicational facts. A transformation is the expectancy that AI applications can also be innovated by the users who prevail its democratic risks, allowing for larger capacity for more influential and addressed public spaces. This is the expectancy of numerous of the more utopian strings of AI and public spheres literature, underlining that AI technologies can supply the potential to authorize the individuals to navigate complex and data-enriched ecosystems and coordinate for drafting action.

3. AI AND DEMOCRACY

The impact of AI on democracy starts with how public services are set up. The claim that we need to digitize politics often comes with the expectation that decision-making processes in public administration will become automated to a large extent (Salam et al., 2023). This goes from administration of welfare to tax systems to border control. More and more actors hope that in an ever more complex world a move to highly automated processes will make politics and political systems more efficient. Automation should take out mistakes and frustration, enable more finely grained and faster adjudicative tasks, require less resources to be spent on certain tasks, which can then be spent on other problems. So how can the way public services are organized be considered a challenge to democracy? A mostly unspoken premise of democracy is that it needs law and legal rules as the basic steering wheel of society. Written laws being applied by a judicially controlled executive is treated as a default democratic form (Ayşen et al., 2020). Procedural institutionalization is a necessary part of the idea where people govern themselves. The law and legal rules guarantee that democratic kinds of rule can be understood by its subjects and give the necessary ground to fight over them Federspiel et al. (2023). So, if the state starts moving at a slow pace the legal processes or supplementing them with automated and adaptive processes, then the workings and the legitimacy of democratic systems are challenged in contested ways that we should lit on with critical attention.

The idea that the people they will govern should choose their leaders is a cornerstone of democracy (Verma, 2024). Such self-rule is both a normative theory that defends the rulers' temporal power over the ruled and a pragmatic notion that decentralized decision-making is preferable to other more centralized forms of expert rule or normator decision-making (Shrestha et al., 2019). AI casts light on potential limits to self-rule in a variety of ways, impacting people's ability to achieve self-rule as well as the notion that distributed decision-making inside intricate social structures is preferable to expert rule.

The treaty will specifically address the potential threat that AI technology poses to democracy and democratic processes—namely, the ability of these systems to be exploited to influence or deceive people (Berryhill et al., 2019). This includes interfering directly or indirectly with the freedoms of expression, opinion formation and holding, assembly and association, and information sharing through the use of so-called "deep fakes," microtargeting, or other tactics. Legally enforceable requirements to offer appropriate protection against such acts will be included in the Framework Convention for its parties.

The "rule of law" is a well-established legal-philosophical concept that includes, among other things, the notions that everyone has access to impartial dispute resolution, that the law should be clear and widely known, that government and private actors are accountable under it, and that laws are enacted, administered, and enforced in an accessible, equitable, and efficient manner (The United Nations Educational, Scientific and Cultural Organization, 2017). It is clear that when developing and utilizing AI systems that may be used in delicate situations, like the creation of laws, public administration, and last but not least, the administration of justice through the legal system, this fundamental understanding of what makes a just and liberal, law-abiding society must be respected. Parties to this agreement will also have specific obligations outlined in the Framework Convention.

In May 2018, The Toronto Declaration: Safeguarding the Rights to Equality and Non-Discrimination in Machine Learning Systems was drafted under the leadership of Access Now and Amnesty International THE TORONTO DECLARATION (2018). The paper provided binding of international legal concepts as a foundation for the current focus on AI bias. The Toronto Declaration delineates the obligations of nations and private sector entities concerning the utilization of machine learning systems (THE TORONTO DECLARATION, 2018). These obligations encompass minimizing the impact of discrimination, promoting transparency, and furnishing efficacious redress to individuals who have suffered harm. The declaration's potential impact is still unclear because its proponents are still looking for supporters, especially from AI firms. Nevertheless, it is a noteworthy attempt to translate basic human rights for the AI domain (Council of Europe, 2019).

People's historical data representation determines how visible they are to AI. When it comes to identifying people who are members of underrepresented groups in the training data, AI struggles (Curzon et al., 2021). Minorities who are not often represented in data sets, for instance, will not be apparent to computer vision, and historically marginalized groups will not be connected to particular employment, increasing the possibility of prejudice in AI-assisted job operations. This general pattern has significant implications for democracy. For instance, because certain groups are often rendered invisible, their influence would be reduced in any AI-based depiction of the political system and in forecasts of its actions, preferences, and grievances (Leenes & Silvia De Conca, 2018).

Therefore, those who have already been granted the right to dissent may run the danger of experiencing additional discrimination and disenfranchisement when government services are implemented, of having their voices and preferences mediated by digital means used to shape policy agendas, or of being targeted more severely by the state security apparatus.

Based on this, we can infer that future AI-powered depictions of public opinion, the political system, and AI-assisted redistricting will be prejudiced against historically underrepresented groups. Certain groups may have more democratic impact than others depending on how visible they are to AI. For example, AI could improve resources for the privileged by increasing the visibility and accessibility of their voices, interests, attitudes, concerns, and grievances to decision-makers. When predicting political trends and the influence of policies, AI may take into account the preferences of visible groups while disregarding those of less visible groups.

The labor market may suffer as a result of AI. Although companies might theoretically invest in automation to free up workers' time for new projects and raise the value of their labor, it seems that companies primarily employ this to reduce labor expenses by replacing labor-intensive jobs with artificial intelligence (Frank, 2019). By replacing capital with labor, this reduces workers' income and negotiating power, which raises the possibility of economic inequality and erodes workers' collective bargaining strength. As a result, this may also reduce the political representation and power of workers (Carbonero et al., 2023).

Elections are essential to democracies because they give political factions a chance to rise to power within an established institutional framework, thereby channelling and managing political conflict. Democracy is a system of "organized uncertainty" if no faction perceives a real chance to seize power. Applications of artificial intelligence pose a danger to this sense of unpredictability around election results. There aren't many applications of AI in this industry, though.

The ability of data-driven methods to forecast the behavior of specific voters is constrained. While it is possible to anticipate the voting patterns of ardent party supporters with some degree of accuracy, at least in two-party systems, it is far more difficult to predict the voting patterns of those who are just marginally interested in politics (Charles et al., 2022). Voting is not always practiced, and when it is, the circumstances can be very different.

The majority of the time, modelers cannot forecast their vote choices, hence AI is not a good fit for the task of automatically predicting voting behavior. Thus, for the foreseeable future, electoral victories will remain unpredictable (Imik Tanyildizi & Tanyildizi, 2022). However, campaigns may create more pertinent data-driven models of elections, including a person's likelihood of casting a ballot or making a financial contribution, which could provide them an advantage over rivals. However, considering the widespread accessibility of AI-based technologies and the capacity of campaign organizations to learn from the triumphs and failures of others, any such advantage is probably short-lived (Ravi et al., 2022).

4. HOW AI CAN BE MISUSED

The public release of the High Commissioner's announcement comes as the Office of the High Commissioner for Human Rights releases a report assessing how AI systems affect the right people have to privacy (Türk, 2023). Also, the report considers other rights, as health, education, freedom of movement, of peaceful assembly and association, and of expression. The report also covers how artificial intelligence profiling and other machine-learning technologies technology affect these rights (Pizzi et al., 2020). Tim Engelhardt, a Human Rights Officer in the Rule of Law and Democracy Section, called the situation "dire." That is how he opened his presentation when releasing the report in Geneva. He also stated in his remarks that the situation has not been good and has been getting worse over time, noting that we are largely in the same position (Neuman, 2021). However, not all is lost. The report has shown that some states are trying to strengthen regulation; international commitments and accountability exist. Still, Engelhardt is most concerned about the short-term: on the contrary, the first critical actions must be taken immediately to prevent the capacity of consequences for people around the world. The sentiment was echoed by the OHCHR's Director of Thematic Engagement Peggy Hicks, who stated in the same that there is no "time buffer" before the harms associated with the countries' inaction get out of the League of Legends (ZHANG, 2021). The report points out "pockets" where states or companies rush to implement AI applications without proper due diligence, thereby allowing unjust treatment of people. For example, it is not difficult to imagine denial of social security benefits by mistake because of flawed AI tools or wrongful detentions due to inaccurate facial recognition software. Moreover, the document explains how AI systems work with vast datasets that consist of individual information, some of it shared and multiplied in an opaque manner. This shows how flawed, illiberal, or outdated data – and all these can be the case – present massive risks when utilized in AI systems. The High Commissioner also addressed the "significant accountability gaps in the lifecycle of data collection, storage sharing, and use" that tremendously impact people already (United Nation, 2024).

Apart from this AI systems in our new thumb paper urgency ballpark set up some more needs from that discussion well much of the conversation in AI and democracy is specifically on how generative AI gave you for pumping new types of content into the information space. This report takes a wider looking not just in jackpots but also in the facial recognition software in 2 second help make sense of government data from its systems that automate decision through this led to report addresses how BI systems are impacting human rights changing landscape for civic activism impacting partners of society organization so we've dedicated to traditional issues of democracy rights it shares what our participants saw this major royal environment in AI government and governance in the green corporate location strategy for civil society organizations thinking about how Democratic values in a world where AI systems play increasingly helped shape society and landscape computers and government artificial intelligence (Confederation of Indian Industry, 2023). It is one of the cross-cutting transnational challenges that are demanding out of each and innovation by Democratic activists' society in order to democracy community and society partners who need to do more to floss their stage sharing knowledge and experience across regions from about peace they will be trace awareness and deep and engage in its society at large embedding digital health library camping tape and crime innovation they're in love with people. Thanks to AI, according to the channel because it is time a strong and durable incentives to encounter the drivers of authoritarian objectives which can too easily prevailed when lovers pushing on I have a democratic principle sort of absent in the context of local democratic backsliding and rising apart your age of fluid the intentional abuse of their technologies presents grave Christmas democracy so children thought was to climb its technologies with the temptation to see them as a substitute for engaging with neighbors or constituents and that partnership of us drawing on the rest of our knowledge and experience within the book of community of Democrats is critical to steer us Roberts vendor better than you and character in today's conversation with that curly was done such outstanding work to tease out the ideas in this room they sharing takeaways from the paper in a conversation two of our workshop participants the Tamalia Party executive director of the Open Day Charter puts PN's Eduardo on the executive there are co-direct charities of the power flying Metro Bridge organs (de Almeida et al., 2021).

5. AI UNDER STATE MACHINERY

AI can also significantly increase government efficiency when applied by automating routine tasks and minimizing human error among its many benefits. Using AI, agencies may work smarter in a variety of ways. For example, the Department of Defence uses AI to evaluate and sort satellite reconnaissance to release experts to deal with more involved tasks. Additionally, the Internal Revenue Service uses AI to improve customer service by responding naturally to a limited amount of frequently asked inquiries (Wirtz et al., 2022). Finally, AI can help the momentous Data mining permits decision-makers to make improved judgments. AI can be extremely beneficial to government agencies since it can help them better determine and solve troubles. Most government agencies now have access to a wealth of information that can be utilized to forecast unruly future (Lt. Gen. Deependra Singh Hooda, 2023). AI can help sort out patterns and data that even experts overlook to give advice to the best action. The National Weather Service is taking advantage of AI to improve the accurateness of weather predictions and warn systems (SITNFlash, 2024). It's significantly increasing public safety by getting individuals out of safety harm's way during serious weather occurrences (Masterson, 2023).

While advantages of using AI for the purpose of government are apparent, there are several challenges to adopting these technologies. First, it is the weakness of financial resources. It is because the implementing of AI systems requires significant investments that should be unavailable to many government bodies (Singh et al., 2020). In addition, low technical expertise can be also an obstacle to implementing and maintaining AI systems in these organizations. This issue can be resolved through cooperation with external laboratories and academies or their dedicated training programs for government employees. The second obstacle is the complexity of the regulatory framework. In this case, also, government agencies should always remain compliant with the existing laws and regulations in terms of data security and data privacy. In addition, there are a number of ethical concerns in the sense of biases present in AI systems or the impact of its use in jobs (Dwivedi et al., 2021). Therefore, it is essential for government bodies to develop their own set of standards of the AI usage that should be transparent to the public. AI can be a real game changer for the variety of functions that are performed by government organizations at all levels. Through promoting the efficiency of the operations and the quality of the decision-making processes, improving the speed and accuracy of the data analysis, and potentially cutting the costs, it can help agencies to achieve their goals and their objectives better. However, there are also several challenges to face, such as the budget limitations, complexity of the regulations, and atom ethical questions (Nam et al., 2020). By following the best practices, such as a cooperation with the academia, transparency in communication with the public, and adherence to the common ethical rules, the government organizations can handle these challenges and use the potential of AI in the most effective way.

6. WAY FORWARD

Since 1951, AI has made significant advancements. In the first documented success of an AI computer program, Christopher Strachey's checkers program completed a full game on the Ferranti Mark I computer at the University of Manchester (Bleakley, 2020). Over the years, developments in machine learning and deep learning led to

IBM's Deep Blue defeating chess grandmaster Garry Kasparov in 1997, and IBM Watson winning Jeopardy in 2011.

Since then, generative AI has been at the forefront of the latest phase in AI's development, marked by OpenAI introducing its first GPT models in 2018. This led to the creation of OpenAI's GPT-4 and ChatGPT, sparking a rise in AI models capable of generating text, audio, images, and various other forms of content.

AI has also been utilized to assist in the sequencing of RNA for vaccines and in modelling human speech, technologies that depend on model- and algorithm-based machine learning and are increasingly emphasizing perception, reasoning, and generalization.

Business Automation

About 55 percent of organizations have embraced AI to different extents, signalling increased automation for numerous businesses soon. As chatbots and digital assistants become prominent, companies can utilize AI to manage basic customer conversations and address simple employee inquiries (Klumpp, 2017).

AI's capacity to analyse vast datasets and transform findings into user-friendly visuals can also expedite the decision-making process. Instead of manually sorting through data, company executives can leverage immediate insights to make well-informed decisions.

"Once developers grasp the potential of the technology and have a deep understanding of the domain, they begin to draw connections and ponder, 'Perhaps this could be tackled as an AI problem, or maybe that one as well,' rather than approaching it with a specific problem statement in mind," shared Mike Mendelson, a learner experience designer at NVIDIA.

Job Interruption

Business automation has evoked concerns about job losses. Research indicates that employees estimate nearly a third of their tasks could potentially be automated by AI. Despite advancements in AI implementation in various workplaces, its effects have been disparate across industries and professions (Shanmugasundaram & Tamilarasu, 2023). While occupations such as secretarial roles face the threat of automation, there has been an increased demand for roles such as machine learning specialists and information security analysts.

Workers in more skilled or creative roles are more likely to see their positions enhanced by AI rather than replaced. Whether through requiring employees to adapt to new technologies or assuming some of their responsibilities, on both at individual and organizational scale (Morandini et al., 2023).

Artificial Intelligence

Klara Nahrstedt, a computer science professor at the University of Illinois at Urbana-Champaign and director of the school's Coordinated Science Laboratory, emphasized the crucial need for substantial investment in education to retrain individuals for new roles for the success of AI in various domains (The Grainger College of Engineering Coordinated Science Laboratory, 2016).

Data Privacy

There is no doubt that his new era of AI systems has put privacy at stake. In year 2000s, when many companies were collecting data from people using internet-enabled devices there was at least certain checks and technological limitations that limited these companies from laying hands over unrestricted use of personal data. This is different in AI systems, which fail both in limitations and ethics. one is able to observe that there is a black box (AI) where we live and they are incapable of being under user control as to: (a) what data get scanned; how it been collected or why the thing was scamming; or so (c) what can be carried on receding us away from this pricing basis (Gravrock, 2022). Even today people who use any internet-based-products or services may simply already find it just about impossible to avoid wide-spectrum digital surveillance for some major life sectors - and indeed likely more difficult should AI capabilities be combined.

It further concerns the very threat of data usage and AI tools reaching evil hands that might be lethal. For instance, models of generative AI that are trained on internet scraped data may infringe some information's about people or their relationships with others. This information can greatly assist spear-phishing campaigns targeting individuals for identity theft and fraudulent activities (Miller, 2024). In fact, early signs are worrying - Criminals have already started using AI voice library to fraud with personal and corporate telephone lines. It maybe even with different kinds of data (for example a resume or photo) shared for another reason that they are reused to train non-transparent AI systems, which in turn impacts citizen rights.

Another area of the double edge nature of AI use is where we see employers are using predictive models in order to winnow down the pool of candidates who will be brought in for interviews. However, there have been several cases where AI tools utilized to assist in candidate filtration exhibited biases. Amazon suffered from this with an in-house AI hiring screening tool that actually put women at a disadvantage.

Companies need significant amounts of data to train the models driving generative AI tools, a process that has faced heightened scrutiny (Curzon et al., 2021). Concerns surrounding companies' data collection practices have prompted the Federal Trade Commission to launch an inquiry into whether OpenAI has adversely impacted consumers through its data gathering methods, potentially breaching European data protection regulations (Murdoch, 2021). In response, the Biden-Harris administration introduced an AI Bill of Rights that highlights data privacy as a fundamental principle (The White House, 2022). While this legislation may lack significant legal authority, it underscores the increasing emphasis on safeguarding data privacy and urging AI enterprises to be more transparent and careful in how they gather training data (Park, 2023).

Rules and Regulation

AI has the potential to change the way certain legal issues are viewed, depending on the outcomes of generative AI lawsuits in 2024. Notably, the debate over intellectual property has gained prominence due to copyright lawsuits brought against OpenAI by writers, musicians, and entities like The New York Times (The Economic Times, 2024). These legal battles have implications for how the U.S. legal system defines private and public ownership, with a negative ruling potentially causing significant setbacks for OpenAI and its rivals.

Ethical concerns related to generative AI have prompted increased pressure on the U.S. government to adopt a firmer stance. The Biden-Harris administration has maintained a moderate position through its latest executive order, which establishes general guidelines on data privacy, civil liberties, responsible AI, and other AI-related aspects (The White House, 2022). Nevertheless, the government may consider implementing stricter regulations depending on shifts in the political landscape.

Sustainability Climate Change and Environmental Issues

Artificial intelligence (AI) has the potential to accelerate sustainability across multiple industries by decreasing waste, expanding resource accessibility, and producing eco-friendly results. This article examines how artificial intelligence (AI) may support sustainability in manufacturing, transportation, utilities, healthcare, and construction activities, as well as in production and distribution systems (Shuford, 2024). It also looks at how AI may be used to deliver more specialized services and reduce hazards to people's health and infrastructure. This Chapter outlines the difficulties in applying AI for sustainability and offers workable fixes that institutions and decision-makers might use. All things considered, this chapter highlights how AI may promote sustainability and offers insightful information for creating AI-based solutions to urgent environmental issues.

On a larger scale, AI is positioned to significantly impact sustainability, climate change, and environmental issues. Optimists see AI as a tool to enhance supply chains' efficiency by enabling predictive maintenance and other measures to lower carbon emissions (Nishant et al., 2020).

Artificial Intelligence

Nevertheless, AI could also be considered a central contributor to climate change. The energy and resources necessary to develop and sustain AI models could lead to a substantial increase in carbon emissions, undermining sustainability efforts in the tech sector. Even when AI is utilized in climate-friendly technologies, the expenses associated with model construction and training could potentially worsen society's environmental predicament. Almost every major industry has already been influenced by modern AI (Bienvenido-Huertas et al., 2020).

Manufacturing

Manufacturing has embraced AI for decades. Utilizing AI-driven robotic arms and other industrial bots since the 1960s and 1970s, the industry has seamlessly integrated AI capabilities. These industrial robots often collaborate with human workers to complete specific tasks such as assembly and stacking, while predictive analysis sensors ensure optimal equipment performance. AI and machine learning can revolutionize manufacturing process efficiency, productivity, and sustainability. However, there are many challenges associated with using AI in manufacturing, including data acquisition and management, human factors, infrastructure issues, security risks, as well as issues of trust and implementation. For example, obtaining the data needed to train AI models is expensive for rare events or time-consuming for large unlabelled datasets. Additionally, AI models can open security vulnerabilities when integrated into control systems that run physical machinery. Furthermore, the manufacturing ecosystem may be averse to AI-based solutions, due to inertia or lack of understanding of its workings (Plathottam et al., 2023). However, AI can be extremely useful in manufacturing, notably in the areas of quality control, process optimization, and predictive maintenance. The adoption or rejection of AI in manufacturing must be based on the needs and capabilities of each individual case. In summary, this chapter updates developments, issues, and trends in ML/AI applied to manufacturing. This information is intended to help one better understand the range of AI/ML technologies available for solving manufacturing problems and to help companies prioritize and select the right ones (Nica et al., 2021). Through a review of industries, challenges, and relevant knowledge, one might identify opportunities where agents can focus research efforts to drive transformational impacts. AI/ML applied to manufacturing is still in its early stages, but the initial experience leaves no doubt about the promising costs and benefits that these technologies can offer, enabled by the availability of large volumes of data originated in autonomous manufacturing systems.

Healthcare

AI healthcare is already transforming the way people engage with medical professionals, despite seeming improbable. With its powerful big data analysis abilities, AI can expedite disease detection, enhance accuracy, accelerate drug development, and supervise patients using virtual nursing assistants (Dave & Patel, 2023).

Healthcare technologies have continued to develop at a rapid pace, reshaping the way medicine is practiced. Artificial Intelligence implemented in clinical practice is believed to supplement clinical decision-making providing diagnostic aid, and making usable suggestions or logical inferences from patient data characterized by ambiguity and complexity. Nonetheless, its functional execution as a nonhuman team's enabling member in healthcare requires the convenience of other user members: trust. Trust is generally considered as a consumer's faith in someone or something's reliability and trustworthiness (Tucci et al., 2021). Although it is possible to manipulate trust for AI, introduction, effective control and awareness through proper regulations and knowledge are good options to enhance the positive use of AI and most importantly trust on AI and its applications.

Finance

The accessibility of machine learning models is causing financial technologies to spread across all respective fields. Machine learning models have a decent chance of achieving a high accuracy but insufficient explainability. Furthermore, high-risk AI applications based on machine learning, according to the proposed regulations must be "trustworthy", and meet several mandatory requirements, including but not limited to Sustainability and Fairness (Giudici & Raffinetti, 2023). Till this day, there are no standardized metrics for obtaining an overall point of view on the AI applications in finance. Some of the known key factors that can be used to judge about the trustworthiness of the model is Accuracy of model.

Finance Banks, insurers, and financial institutions utilize AI across various applications such as fraud detection, audit processes, and customer loan evaluations. Traders leverage machine learning to assess vast amounts of data simultaneously, enabling quick risk assessment and informed investment decisions.

Education

Investing in development of comprehensive skills in primary and secondary education is crucial in addressing the underlying reasons for the gap in global skill levels. This will help in preparing the next generation of talent for lifelong learning and ensuring that future retraining and upskilling efforts are beneficial for individuals, businesses, and governments. This necessitates ongoing renewal activities by all stakeholders. Global education reform is already in progress, and we may witness more significant solutions to enhance efficiency in the near future (Udvaros & Forman, 2023). It is paramount that the new generation receives the highest quality education possible. An increasing number of experts, researchers, and policymakers are convinced that advancements in technology can greatly enhance the effectiveness of teaching methods and learning in the coming years. As this approach gains traction in the academic community, new research and scientific publications are emerging on this topic. The conventional education system as we presently know it is inadequate or faces challenges in delivering the education quality necessary for the continual and effective development of individual skills.

AI in education will revolutionize how people of all ages learn. Leveraging machine learning, natural language processing, and facial recognition, AI aids in digitalizing textbooks, identifying plagiarism, and interpreting students' emotions to pinpoint those who may be struggling or disinterested. AI customizes the learning experience to suit each student's unique requirements, both now and in the future (Kamalov et al., 2023).

Media

Artificial Intelligence is a transformative technology in the digital era and a crucial mindset for businesses, particularly those in the media sector dealing with a diverse range of digital content products and advertising opportunities. The potential disruptive power and benefits of AI in the media industry compared to tangible goods sectors remain uncertain due to its distinct market characteristics. This assessment delves into the practical applications of AI in the media sector, its role within the value chain, and the obstacles faced when integrating cognitive technologies into this field. The analysis reveals that AI's impact in media spans across eight key areas: audience content recommendations and discovery, audience engagement, enhanced audience experiences, message optimization, content management, content creation, audience insights, and operational automation (Chan-Olmsted, 2019). The industry encounters notable challenges in striking a balance between effectiveness and efficiency, and between human and AI decision-making processes.

Journalism is also tapping into the power of AI, and will continue reaping its benefits. An instance of this can be observed in The Associated Press' utilization of Automated Insights, a tool that generates thousands of earnings report stories annually (de-Lima-Santos & Ceron, 2021). However, with the emergence of generative AI writing tools like ChatGPT in the market, concerns about their application in journalism are arising.

Customer Service

In the realm of digital environments, chatbots serve as customer service representatives aiding consumers in decision-making processes. This research, grounded in the computers-are-social-actors perspective, explores the perceived distinctions in communication quality and privacy concerns among various service agents and how these factors influence consumers' willingness to adopt such services (Pizzi, Scarpi, et al., 2020). The study also probes whether these perceived variances might hinge on the users' need for human interaction. Through a series of five scenario-driven experiments that gathered data and assessed hypotheses, it was revealed that: different kinds of service agents have a direct impact on consumers' adoption tendencies; the perceived communication quality and privacy risks act as mediators between service agent types and adoption intentions; and the impacts of service agent types on perceived accuracy, communicative skills, and privacy risks are influenced by the desire for human interaction (Song et al., 2022). These findings shed light on the judicious implementation of human-computer interaction in e-commerce.

Most individuals fear receiving a robocall, but AI in customer service has the potential to equip the industry with data-driven tools that offer meaningful insights to both customers and service providers (Fernandes & Oliveira, 2021). The customer service sector leverages AI tools such as chatbots and virtual assistants to enhance interactions.

Transportation

Transportation systems have developed into advanced systems to accommodate the growing population and traffic in response to sophisticated mobility. Conventional methods may not be sufficient when dealing with a large amount of data generated from in-vehicle sensors and network devices (Machin et al., 2018). In response, Artificial Intelligence algorithms have been applied across all portions and areas of transportation to alleviate these difficulties. The paper investigates the influence of an array of AI methods on Information Transport Systems. The approaches are split into three fields: Control of Vehicles, Control and Prediction of Traffic, and Predicting Accidents and Roads Safety. The results indicate that an ensemble of numerous AI methods is most successful in analyzing the large quantities of data produced in transportation data.

AI Can Create Inequality

Predictions by AI in the social sector regard human behaviors such as labor, housing, education, and criminal justice. Based on these predictions, then according to these, the access of a person's resources and opportunities might vary. Automating the classification of human behavior is intrinsically difficult and error-prone, subject to heterogeneous definition of behaviors noisy, non-representative data (Zuiderwijk et al., 2021). In social domains, the greatest challenge with AI includes defining the target behavior, having access to unbiased data and not necessarily observing all of the outcomes.

At work, AI has the power both to reinforce racially and sexually invidious patterns of discrimination while also amplifying them. Algorithms can end up with the same bias in their design or training pixelating preconceived notions tying together social groups, to produce different results on account of them belonging to oppressed identities. Because AI reinforces class inequality by consolidating power among those with capital, the capitalists. AI tools could enhance employer captures, decrease worker rights and autonomy, strengthen managerial control and maintain existing power dynamics. The current anti-discrimination laws are not enough to manage the problems related to biased AI. While human rights laws were designed to combat this kind of intentional discrimination, algorithmic bias is often subtle and systemic in nature (Kim, 2021). Existing laws are not great at protecting worker data, and the erosion of union power in recent decades has meant that workers have few bargaining chips to negotiate terms concerning how AI tools are used (Capraro et al., 2024).

7. CONCLUSION

All significant advancements in technology result in a variety of both favorable and unfavorable outcomes. Ensuring the successful progression of artificial intelligence is not only one of the most pivotal issues of our era but also arguably one of the most critical inquiries in the annals of humankind (Mckinsey & Company, 2024). This necessitates governmental resources, funding, attention, and involvement. Currently, nearly all resources allocated to AI are focused on accelerating the advancement of this technology. In contrast, efforts aimed at enhancing the safety of AI systems struggle to secure the necessary resources. Researcher Toby Ord approximated that between $10 to $50 million was dedicated to addressing the alignment problem in

2020. On the other hand, corporate AI investments in the same year exceeded this amount by over 2000 times, totalling $153 billion.

This doesn't just apply to the AI alignment issue. Efforts to address the full spectrum of adverse social outcomes from AI are inadequately supported compared to the substantial investments in enhancing the capacity and deployment of AI systems (Roser, 2022).

It's frustrating and worrisome for society that AI safety efforts are largely overlooked, with insufficient public funding dedicated to this critical research area. However, for individuals, this neglect presents a significant opportunity to truly make a positive impact by focusing on this issue now. Despite the relatively small size of the AI safety field, there are valuable resources available to guide those interested in contributing concretely to addressing this challenge. Good resources are accessible for those looking to engage in this work.

We aspire for increased commitment from individuals towards this cause, however, collective action is essential. A technology reshaping our society must capture the attention of everyone. It is crucial for our society to reflect on the societal implications of AI, educate ourselves about the technology, and grasp the significance of the situation.

When our kids reflect on this day, I anticipate they will struggle to grasp the limited attention and resources we invested in the advancement of secure AI. I am hopeful that this mindset shifts in the future, and we prioritize allocating more resources towards ensuring that advanced AI is developed in a manner that serves both our current and future generations well.

If we do not cultivate a comprehensive understanding, then it will be a small elite of financiers and developers of this technology who will shape how one of the most powerful technologies in human history will revolutionize our world. Entrusting the development of artificial intelligence solely to private corporations also means granting them control over what lies ahead for the future of humanity.

REFERENCES

Ahmed, S., & Bajema, N. (2019). *Artificial Intelligence, China, Russia, and the Global Order Technological, Political, Global, and Creative Perspectives.* https://www.airuniversity.af.edu/Portals/10/AUPress/Books/B_0161_WRIGHT_ARTIFICIAL_INTELLIGENCE_CHINA_RUSSIA_AND_THE_GLOBAL_ORDER.PDF

Attard-Frost, B., Brandusescu, A., & Lyons, K. (2024). The governance of artificial intelligence in Canada: Findings and opportunities from a review of 84 AI governance initiatives. *Government Information Quarterly*, 41(2), 101929–101929. 10.1016/j.giq.2024.101929

Ayşen, A., Gül, Y., Dilek, P., & Elmer. (2020). *Digital transformation in media & society.* https://westminsterresearch.westminster.ac.uk/download/cf99c72689c33eacd4e7cc8ad5680d1c0e939e2a2b9d3ffff86b6693c213a815/4794770/2B799F5AB7A14F628F90BFC458B38029.pdf#page=165

Berryhill, J., Kok Heang, K., Clogher, R., & Mcbride, K. (2019). Hello, World: Artificial intelligence and its use in the public sector. *OECD Working Papers on Public Governance, 36*(19934351). 10.1787/19934351

Bharadiya, J. (2023). Artificial Intelligence in Transportation Systems A Critical Review. *American Journal of Computing and Engineering*, 6(1), 34–45. 10.47672/ajce.1487

Bienvenido-Huertas, D., Farinha, F., Oliveira, M. J., Silva, E. M. J., & Lança, R. (2020). Comparison of artificial intelligence algorithms to estimate sustainability indicators. *Sustainable Cities and Society*, 63, 102430. 10.1016/j.scs.2020.102430

Bleakley, C. (2020). Artificial Intelligence Emerges. *Poems That Solve Puzzles*, 75–92. 10.1093/oso/9780198853732.003.0005

Busuioc, M. (2020). Accountable Artificial Intelligence: Holding Algorithms to Account. *Public Administration Review*, 81(5), 825–836. Advance online publication. 10.1111/puar.1329334690372

Carbonero, F., Davies, J., Ernst, E., Fossen, F. M., Samaan, D., & Sorgner, A. (2023). The impact of artificial intelligence on labor markets in developing countries: A new method with an illustration for Lao PDR and urban Viet Nam. *Journal of Evolutionary Economics*, 33(3), 707–736. Advance online publication. 10.1007/s00191-023-00809-736811092

Chan-Olmsted, S. M. (2019). A Review of Artificial Intelligence Adoptions in the Media Industry. *International Journal on Media Management*, 21(3-4), 193–215. 10.1080/14241277.2019.1695619

Charles, V., Rana, N. P., & Carter, L. (2022). Artificial Intelligence for data-driven decision-making and governance in public affairs. *Government Information Quarterly*, 39(4), 101742. 10.1016/j.giq.2022.101742

Cheng-Tek Tai, M. (2020). The impact of artificial intelligence on human society and bioethics. *Tzu Chi Medical Journal, 32*(4), 339–343. 10.4103/tcmj.tcmj_71_20

Confederation of Indian Industry. (2023, May 17). *Artificial Intelligence in Governance*. CII Blog. https://ciiblog.in/artificial-intelligence-in-governance/

Council of Europe. (2019). *Artificial intelligence and data protection*. Council of Europe Publishing; Council of Europe. https://edoc.coe.int/en/artificial-intelligence/8254-artificial-intelligence-and-data-protection.html

Curzon, J., Kosa, T. A., Akalu, R., & El-Khatib, K. (2021). Privacy and Artificial Intelligence. *IEEE Transactions on Artificial Intelligence*, 2(2), 1–1. 10.1109/TAI.2021.3088084

Dave, M., & Patel, N. (2023). Artificial intelligence in healthcare and education. *British Dental Journal*, 234(10), 761–764. 10.1038/s41415-023-5845-237237212

de Almeida, P. G. R., dos Santos, C. D., & Farias, J. S. (2021). Artificial Intelligence Regulation: A Framework for Governance. *Ethics and Information Technology*, 23(3), 505–525. 10.1007/s10676-021-09593-z

de-Lima-Santos, M.-F., & Ceron, W. (2021). Artificial Intelligence in News Media: Current Perceptions and Future Outlook. *Journalism and Media*, 3(1), 13–26. 10.3390/journalmedia3010002

Dirican, C. (2015). The Impacts of Robotics, Artificial Intelligence On Business and Economics. *Procedia - Social and Behavioral Sciences, 195*, 564–573. 10.1016/j.sbspro.2015.06.134

Dwivedi, Y. K., Hughes, L., Ismagilova, E., Aarts, G., Coombs, C., Crick, T., Duan, Y., Dwivedi, R., Edwards, J., Eirug, A., Galanos, V., Ilavarasan, P. V., Janssen, M., Jones, P., Kar, A. K., Kizgin, H., Kronemann, B., Lal, B., Lucini, B., & Williams, M. D. (2021). Artificial Intelligence (AI): Multidisciplinary Perspectives on Emerging challenges, opportunities, and Agenda for research, Practice and Policy. *International Journal of Information Management*, 57(101994). Advance online publication. 10.1016/j.ijinfomgt.2019.08.002

Federspiel, F., Mitchell, R., Asokan, A., Umana, C., & McCoy, D. (2023). Threats by artificial intelligence to human health and human existence. *BMJ Global Health*, 8(5), e010435. 10.1136/bmjgh-2022-01043537160371

Fernandes, T., & Oliveira, E. (2021). Understanding consumers' acceptance of automated technologies in service encounters: Drivers of digital voice assistants adoption. *Journal of Business Research*, 122, 180–191. 10.1016/j.jbusres.2020.08.058

Frank, M. R., Autor, D., Bessen, J. E., Brynjolfsson, E., Cebrian, M., Deming, D. J., Feldman, M., Groh, M., Lobo, J., Moro, E., Wang, D., Youn, H., & Rahwan, I. (2019). Toward Understanding the Impact of Artificial Intelligence on Labor. *Proceedings of the National Academy of Sciences of the United States of America*, 116(14), 6531–6539. 10.1073/pnas.1900949116 30910965

Gen, H. (2023). *Implementing Artificial Intelligence in the Indian Military*. Www.delhipolicygroup.org. https://www.delhipolicygroup.org/publication/policy-briefs/implementing-artificial-intelligence-in-the-indian-military.html

Giudici, P., & Raffinetti, E. (2023). SAFE Artificial Intelligence in finance. *Finance Research Letters*, 56, 104088. 10.1016/j.frl.2023.104088

Hui, Alamri, & Toghraie. (2023). Greening Smart Cities: an Investigation of the Integration of Urban Natural Resources and Smart City Technologies for Promoting Environmental Sustainability. *Sustainable Cities and Society*, 99, 104985–104985. 10.1016/j.scs.2023.104985

Hunt, E. B. (2014). Artificial Intelligence. In *Google Books*. Academic Press. https://books.google.co.in/books?hl=en&lr=&id=9y2jBQAAQBAJ&oi=fnd&pg=PP1&dq=artificial+intelligence&ots=uQSx0laUyD&sig=nHVwQiR88VrgtdCsKvev3yZwB74#v=onepage&q=artificial%20intelligence&f=false

Imik Tanyildizi, N., & Tanyildizi, H. (2022). Estimation of voting behavior in election using support vector machine, extreme learning machine and deep learning. *Neural Computing & Applications*, 34(20), 17329–17342. Advance online publication. 10.1007/s00521-022-07395-y

Kamalov, F., Calonge, D. S., & Gurrib, I. (2023). New Era of Artificial Intelligence in Education: Towards a Sustainable Multifaceted Revolution. *Sustainability (Basel)*, 15(16), 12451. 10.3390/su151612451

Kinder, T., Stenvall, J., Koskimies, E., Webb, H., & Janenova, S. (2023). Local public services and the ethical deployment of artificial intelligence. *Government Information Quarterly*, 40(4), 101865. 10.1016/j.giq.2023.101865

Klumpp, M. (2017). Automation and artificial intelligence in business logistics systems: Human reactions and collaboration requirements. *International Journal of Logistics*, 21(3), 224–242. 10.1080/13675567.2017.1384451

Leenes, R., & De Conca. (2018). Artificial intelligence and privacy—AI enters the house through the Cloud. *Edward Elgar Publishing EBooks*. 10.4337/9781786439055.00022

Machin, M., Sanguesa, J. A., Garrido, P., & Martinez, F. J. (2018). *On the use of artificial intelligence techniques in intelligent transportation systems | IEEE Conference Publication | IEEE Xplore*. Ieeexplore.ieee.org. https://ieeexplore.ieee.org/abstract/document/8369029

Masterson, V. (2023, December 14). *AI beats top weather forecasting computers*. World Economic Forum. https://www.weforum.org/agenda/2023/12/ai-weather-forecasting-climate-crisis/

Mckinsey & Company. (2024, April 30). *What's the future of AI? | McKinsey*. Www.mckinsey.comMckinsey & Company. https://www.mckinsey.com/featured-insights/mckinsey-explainers/whats-the-future-of-ai

Miller, K. (2024, March 18). *Privacy in an AI Era: How Do We Protect Our Personal Information?* https://hai.stanford.edu/news/privacy-ai-era-how-do-we-protect-our-personal-information

Morandini, S., Fraboni, F., De Angelis, M., Puzzo, G., Giusino, D., & Pietrantoni, L. (2023). The Impact of Artificial Intelligence on Workers' Skills: Upskilling and Reskilling in Organisations. *Informing Science*, 26(1), 39–68. 10.28945/5078

Muller, C. (2020). *The Impact of Artificial Intelligence on Human Rights, Democracy and the Rule of Law*. https://allai.nl/wp-content/uploads/2020/06/The-Impact-of-AI-on-Human-Rights-Democracy-and-the-Rule-of-Law-draft.pdf

Murdoch, B. (2021). Privacy and Artificial Intelligence: Challenges for Protecting Health Information in a New Era. *BMC Medical Ethics*, 22(1), 122. Advance online publication. 10.1186/s12910-021-00687-334525993

Nam, K., Dutt, C. S., Chathoth, P., Daghfous, A., & Khan, M. S. (2020). The adoption of artificial intelligence and robotics in the hotel industry: Prospects and challenges. *Electronic Markets*, 31(3), 553–574. https://link.springer.com/article/10.1007/s12525-020-00442-3. 10.1007/s12525-020-00442-3

Nemitz, P. (2018). Constitutional democracy and technology in the age of artificial intelligence. *Philosophical Transactions. Series A, Mathematical, Physical, and Engineering Sciences*, 376(2133), 20180089. 10.1098/rsta.2018.008930323003

Neuman, S. (2021, September 16). *The U.N. Warns That AI Can Pose A Threat To Human Rights*. NPR.org. https://www.npr.org/2021/09/16/1037902314/the-u-n-warns-that-ai-can-pose-a-threat-to-human-rights

Artificial Intelligence

Nica, E., Stanciu, C., & Ionu Stan, C. (2021). Internet of Things-based Real-Time Production Logistics, Sustainable Industrial Value Creation, and Artificial Intelligence-driven Big Data Analytics in Cyber-Physical Smart Manufacturing Systems. *Economics, Management, and Financial Markets, 16*(1), 52–62. https://www.ceeol.com/search/article-detail?id=939242

Nishant, R., Kennedy, M., & Corbett, J. (2020). Artificial intelligence for sustainability: Challenges, opportunities, and a research agenda. *International Journal of Information Management*, 53(53), 102104. 10.1016/j.ijinfomgt.2020.102104

Park, E. (2023, February 8). *The AI Bill of Rights: A Step in the Right Direction.* Social Science Research Network. https://papers.ssrn.com/sol3/papers.cfm?abstract_id=4351423

Pelau, C., Ene, I., & Ionut Pop, M. (2021). The Impact of Artificial Intelligence on Consumers' Identity and Human Skills. *Amfiteatru Economic*, 23(56), 33–45. 10.24818/EA/2021/56/33

Pislaru, M., Vlad, C. S., Ivascu, L., & Mircea, I. I. (2024). Citizen-Centric Governance: Enhancing Citizen Engagement through Artificial Intelligence Tools. *Sustainability (Basel)*, 16(7), 2686. 10.3390/su16072686

Pizzi, G., Scarpi, D., & Pantano, E. (2020). Artificial intelligence and the new forms of interaction: Who has the control when interacting with a chatbot? *Journal of Business Research*, 129, 878–890. Advance online publication. 10.1016/j.jbusres.2020.11.006

Pizzi, M., Romanoff, M., & Engelhardt, T. (2020). AI for humanitarian action: Human rights and ethics. *International Review of the Red Cross*, 102(913), 145–180. 10.1017/S1816383121000011

Plathottam, S. J., Rzonca, A., Lakhnori, R., & Iloeje, C. O. (2023). A review of artificial intelligence applications in manufacturing operations. *Journal of Advanced Manufacturing and Processing*, 5(3), e10159. Advance online publication. 10.1002/amp2.10159

Ravi, S. K., Chaturvedi, S., Rastogi, N., Akhtar, N., & Perwej, Y. (2022). A Framework for Voting Behavior Prediction Using Spatial Data. *International Journal of Innovative Research in Computer Science & Technology*, 10(2), 19–28. 10.55524/ijircst.2022.10.2.4

Ray, P. P. (2023). ChatGPT: A Comprehensive Review on background, applications, Key challenges, bias, ethics, Limitations and Future Scope. *Internet of Things and Cyber-Physical Systems*, 3(1), 121–154. 10.1016/j.iotcps.2023.04.003

Roser, M. (2022, December 15). *Artificial Intelligence Is Transforming Our World — It Is on All of Us to Make Sure That It Goes Well*. Our World in Data. https://ourworldindata.org/ai-impact

Sætra, H. S. (2020). A shallow defence of a technocracy of artificial intelligence: Examining the political harms of algorithmic governance in the domain of government. *Technology in Society*, 62, 101283. 10.1016/j.techsoc.2020.10128332536737

Salam, R., & Sinurat, M. (2023). Implementation of Artificial Intelligence in Governance: Potentials and Challenges. *Influence: International Journal of Science Review*, 5(1), 243–255. https://influence-journal.com/index.php/influence/article/view/122

Saltman, K. (2020). Artificial intelligence and the technological turn of public education privatization: In defence of democratic education. *London Review of Education*, 18(2), 196–208. 10.14324/LRE.18.2.04

Sergey & Teteryatnikov. (2021). Artificial Intelligence in Public Governance. *Springer EBooks*, 127–135. 10.1007/978-3-030-63974-7_9

Shanmugasundaram, M., & Tamilarasu, A. (2023). The impact of digital technology, social media, and artificial intelligence on cognitive functions: A review. *Frontiers in Cognition*, 2, 1203077. Advance online publication. 10.3389/fcogn.2023.1203077

Shrestha, Y. R., Ben-Menahem, S. M., & von Krogh, G. (2019). Organizational Decision-Making Structures in the Age of Artificial Intelligence. *California Management Review*, 61(4), 66–83. 10.1177/0008125619862257

Shuford, J. (2024). Interdisciplinary Perspectives: Fusing Artificial Intelligence with Environmental Science for Sustainable Solutions. *Journal of Artificial Intelligence General Science*, 1(1), 1–12. 10.60087/jaigs.v1i1.p12

Singh, R. P., Hom, G. L., Abramoff, M. D., Campbell, J. P., & Chiang, M. F. (2020). Current Challenges and Barriers to Real-World Artificial Intelligence Adoption for the Healthcare System, Provider, and the Patient. *Translational Vision Science & Technology*, 9(2), 45–45. 10.1167/tvst.9.2.4532879755

SITNFlash. (2024, March 4). *A Sky Full of Data: Weather forecasting in the age of AI*. Science in the News. https://sitn.hms.harvard.edu/flash/2024/ai_weather_forecasting/

Song, M., Xing, X., Duan, Y., Cohen, J., & Mou, J. (2022). Will artificial intelligence replace human customer service? The impact of communication quality and privacy risks on adoption intention. *Journal of Retailing and Consumer Services*, 66, 102900. 10.1016/j.jretconser.2021.102900

The Economic Times. (2024, April 30). US newspapers sue OpenAI for copyright infringement over AI training. *The Economic Times*. https://economictimes.indiatimes.com/tech/technology/us-newspapers-sue-openai-for-copyright-infringement-over-ai-training/articleshow/109737033.cms?from=mdr

The Grainger College of Engineering Coordinated Science Laboratory. (2016, September 6). Klara Nahrstedt. *Archives*.library.illinois.edu. https://csl.illinois.edu/directory/faculty/klara

The Toronto Declaration. (2018). https://www.torontodeclaration.org/

The United Nations Educational, Scientific and Cultural Organization. (2017). *AI and the Rule of Law: Capacity Building for Judicial Systems | UNESCO*. Www.unesco.org. https://www.unesco.org/en/artificial-intelligence/rule-law/mooc-judges

The White House. (2022, October). *Blueprint for an AI Bill of Rights*. The White House; The White House. https://www.whitehouse.gov/ostp/ai-bill-of-rights/

Truby, J. (2020). Governing Artificial Intelligence to benefit the UN Sustainable Development Goals. *Sustainable Development (Bradford)*, 28(4), 946–959. Advance online publication. 10.1002/sd.2048

Tucci, V., Saary, J., & Doyle, T. E. (2021). Factors influencing trust in medical artificial intelligence for healthcare professionals: A narrative review. *Journal of Medical Artificial Intelligence*, 0. Advance online publication. 10.21037/jmai-21-25

Türk, V. (2023, July 12). *Artificial intelligence must be grounded in human rights, says High Commissioner*. OHCHR; Uniter Nations. https://www.ohchr.org/en/statements/2023/07/artificial-intelligence-must-be-grounded-human-rights-says-high-commissioner

Udvaros, J., & Forman, N. (2023). Artificial intelligence and Education 4.0. *INTED2023 Proceedings*, 6309–6317. 10.21125/inted.2023.1670

United Nation. (2024, April 11). *UN: AI resolution must prioritise human rights and sustainable development*. https://www.article19.org/resources/un-ai-resolution-must-prioritise-human-rights-and-sustainable-development/

Valle-Cruz, D., Criado, J. I., Sandoval-Almazán, R., & Ruvalcaba-Gomez, E. A. (2020). Assessing the public policy-cycle framework in the age of artificial intelligence: From agenda-setting to policy evaluation. *Government Information Quarterly*, 37(4), 101509. 10.1016/j.giq.2020.101509 von Gravrock, E. (2022, March 31). *Artificial intelligence design must prioritize data privacy*. World Economic Forum; World Economic Forum. https://www.weforum.org/agenda/2022/03/designing-artificial-intelligence-for-privacy/

West, D., & Allen, J. (2018, April 24). *How Artificial Intelligence Is Transforming the World*. Brookings; The Brookings Institution. https://www.brookings.edu/articles/how-artificial-intelligence-is-transforming-the-world/

Wirtz, B. W., Weyerer, J. C., & Kehl, I. (2022). Governance of artificial intelligence: A risk and guideline-based integrative framework. *Government Information Quarterly*, 39(4), 101685. 10.1016/j.giq.2022.101685

Zhang, J. (2021). *AI: can it be regulated?* FIFDH. https://fifdh.org/en/festival/program/2024/forum/ai-can-it-be-regulated/

Zhang, C., & Lu, Y. (2021). Study on Artificial Intelligence: The State of the Art and Future Prospects. *Journal of Industrial Information Integration*, 23(23), 100224. 10.1016/j.jii.2021.100224

Chapter 3
Interplay of Artificial Intelligence and Recruitment:
The Gender Bias Effect

Shikha Saloni
University School of Business, Chandigarh University, India

Neema Gupta
University School of Business, Chandigarh University, India

Ambuj Kumar Agarwal
https://orcid.org/0000-0001-5181-5750
Sharda University, India

Raj Gaurang Tiwari
Chitkara University, India

Vishal Jain
https://orcid.org/0000-0003-1126-7424
Sharda University, India

ABSTRACT

In this chapter, the authors focus on different ways in which AI is incorporated in the process of recruitment. Along with the above stated objective, they also explore the forms of AI based recruitment, the benefits of AI based recruitment, and the challenges that might be encountered during the process with an emphasis on gender bias. In the findings, they aim to describe the gender bias in professional functions in businesses. On the other hand, they hope to gain insight into potential

DOI: 10.4018/979-8-3693-3334-1.ch003

Copyright © 2024, IGI Global. Copying or distributing in print or electronic forms without written permission of IGI Global is prohibited.

gender discrepancies between operational and leadership positions, as well as between departments. The findings of this chapter will benefit researchers, academics, and managers in analyzing gender-related practices and policies. Organizations can become more aware of their gendered practices, which affect the recruitment procedure and the varied roles and responsibilities assigned to men and women, by giving voice to the prejudices that generate gender biases. Along with this, they provide the implications, limitations, and future scope of the study.

INTRODUCTION

Work has changed dramatically as a result of technological advancement since the starting to fourth industrial era revolution. The recent-most industrial revolution brought the most turbulent innovations such as big data and artificial intelligence (Zhang and Chen, 2023). There has been an exponential increase in the use of digital technology which leads brings us to an existing possibility of developing a comparatively higher level Artificial Intelligence (Beneduce, 2020). AI is considered reliable and a dependable tool as it is able to process and analyze data thoroughly and quicker and accurately than what humans can (Chen, 2022). It is able to process and analyze vast amount of data which is beyond human capabilities (Shaw, 2019). The incorporation of AI and modern technology has transformed the work altogether and enhanced Human resource management which led to improved outcomes (Hmoud& Laszlo, 2019).Pan (2016) discovered that AI positively impacts the personnel management within businesses. To be able to advance HRM, it was concluded that more investment would be required in AI technology application. Davenport et al, (2020) discussed AI inferences in human resource management, particularly in the recruitment process. The purpose of the study was to determine whether AI is replacing humans in certain Indian software companies' hiring processes. According to the research, it appears that AI replaces workers in a good way. AI is professed to have great potential in the preliminary phases of the recruitment process, despite the fact that interaction with humans is preferable and recommended during the interview and wage package negotiating stages. By incorporating AI, the hiring process can become more efficient and expedient while eliminating bias. By making it possible to comprehend talent more thoroughly than rivals, AI deployment may give an edge over competitors and raise a company's level of competitiveness (Johansson and Herranen, 2019).Via algorithms, AI is able to receive data as well as commands input. Miasato and Silva (2019) contend that algorithms cannot eradicate bigotry on their own, despite the fact that AI developers think their algorithmic hiring practices simplify the recruiting process and reduce prejudice. The very first information that artificial intelligence gets shapes the decisions that it makes. In the event that the fundamental data is biased, the ensuing algorithms may reinforce partiality, prejudice, or unfair treatment, which may lead to widespread inequalities (Bornstein, 2018).

Objective of the Chapter

- To put light on the various ways AI is incorporated in the process of recruitment in organizations.
- To explore the forms of AI based recruitment.
- To explore the benefits of AI based recruitment.
- To put forward the challenges of AI based recruitment.
- To dive deeper into the gender biases encountered in AI based recruitment.

AI-BASED RECRUITMENT

The four main phases of the recruiting process are searching, screening, interviewing, and selection (Bogen and Rieke, 2018). The recruitment procedure comprise of an array of events. AI technology can impact how each step is carried out, and each phase consists of different activities. The goal of the searching phase is to create a web content navigation system. By using social networking and recruitment sites, it filters dormant job applicants online and analyzes their profiles in accordance with the job descriptions that have already been established. According to Hmoud and Laszlo (2019), the engine that searches matches candidates' profiles with job posts and job advertisements through a web-based search that takes into account the significance of the searched information. Employers use AI technology to help them score prospects and measure their abilities during the process of screening (Bogen and Rieke, 2018). Screening involves examining and evaluating candidates' qualifications. In order to better align the resumes with the position of employment description, they are screened. Applicants may be ranked by the algorithm based on how relevant the qualification metrics are. The interview phase comes next. The choosing process is likely to be the most personalized at this point, making it unlikely for artificial intelligence to completely replace it. Nevertheless, recruiters may perform video interviews and analyze candidates' facial expressions, speech tones, and emotions with the help of certain AI techniques (Ahmed, 2018). The wrapping-up phase is the selection phase, which encompasses the ultimate hiring decision made by the employer. At this juncture, AI systems can provide compensation and incentives to businesses and predict the likelihood that applicants will break policies at work. (Rieke and Bogen, 2018).

There is a troubling component to AI tools for hiring that should not be disregarded, emphasizing the necessity of addressing these issues via managerial or technical approaches (Raub, 2018). Algorithms and AI have a tendency to lead to discriminatory hiring practices and unaccountable instances of bias, but there is also growing evidence that these technologies are more unbiased than previously thought. Organizations should carefully choose their strategies, encourage the use of dependable algorithms, and advocate for maximum ethnic and gender diversity in high-tech businesses in order to fully retrieve the advantages of AI in recruitment. According to Njoto (2020), the information set is examined through the use of agents that represent different genders and attributes, including racial background, sexual orientation, and political views. A lengthy record of purposeful and unintentional gender and racial discrimination has meant that these biases are often present in the algorithms. Algorithmic bias is the error that occurs when AI interprets algorithmic input with preconceived notions that are repeated in the system (Jackson, 2021).

ADVANTAGES OF AI-BASED RECRUITMENT

The Quality of Recruitment

Some major corporations think that unconscious bias has an impact on the caliber of hiring, according to research by Beattie et al. (2012). Employers might need to bring on employees with greater expertise in order to prevent losses (Newell, 2015). To eliminate unintentional human bias that influences the employment process, artificial intelligence has entered the recruitment sector to automate the recruiting and selection procedure (Raub, 2018). One among the concepts driving the advancement of AI for applicant selection is to raise the bar for the interview process, regardless of what the interviewer's school of thought and stances are (Miasato and Silva, 2019). When the description of any job and targeted advertisements matches a potential employee's abilities and skills to the job performance, artificial intelligence technologies can begin their work. Then, based on every the applicant's profile, they may determine which one would be best suited for the job (Johnson et al. 2020). Furthermore, hiring managers can evaluate a greater number of applicants who otherwise would not be able to due to automated resume screening technologies (Beneduce, 2020). The appointment of candidates has become more impersonal due to advancements in AI technology, as it is now based on data that is publicly readily accessible and shared with the organization.

Recruitment Efficiency

For each vacancy, HR departments may receive a large number of applications. The most costly and daunting recruiting procedure is traditional screening and selection, which basically relies on human interference to assess information about applicants (Hmoud and Laszlo, 2019). AI possess the ability to accelerate the hiring process, along with enhancing the experience of candidate as well as minimizing the expenses (Johansson and Herranen, 2019). It can provide candidates with employment information more quickly, enabling them to decide very initially in the process of recruitment regardingthe kinds of positions that interests them. The number of prospects recruiters must choose from later can be decreased by using artificial intelligence to filter out a large number of indifferent candidates and exclude them from the applicant pool. With artificial intelligence's assistance, it is even feasible to find reserved individuals, giving you more time to focus on the ideal fit. Artificial intelligence is capable of classifying candidates automatically according to the job description, in addition to automating the large-scale, rapid examination of hundreds of resumes. Additionally, as per Raveendra (et al, 2020) it is easier to provide the candidate with the final results following the hiring choice.

Transactional Workload

The use of AI in recruitment has ushered in a "new era in human resources," as it has the potential to substitute for human recruiters' repetitive activities and alter the industry's long-standing practices (Upadhyay and Khandelwal, 2018). In terms of lessening administration and routine work, the majority of experts think AI is advantageous for recruiters (Johansson and Herranen, 2019). Time consuming and tiring tasks like hiring, screening and interviewing will be taken care of with the help of AI which renders the managers to be free and be able to focus on strategic decisions (Upadhyay and Khandelwal, 2018).

CHALLENGES

Even if the benefits of applying AI are acknowledged, there are still many obstacles to overcome before AI can be effectively used in human resources. They cover a wide range of topics, from real-world problems like prejudices in the data that is used to guide AI systems to concerns about justice, ethics, and the law (Tambe et al., 2019).

- First, artificial intelligence (AI) systems may favor a specific set of candidates due to biases in algorithms developed using past data. For the aforementioned reason, Amazon experienced an issue with their employment algorithm in 2018. The system was developed using historical work performance data, which was predominately composed of male, white employees who performed better than average. Male applicants in the same demographic category received higher marks from the AI algorithm as a result. Because it was unable to make the algorithm gender-neutral, the organization subsequently stopped using it for hiring (Meyer, 2018).
- Second, owing to the property rights, AI robots and algorithms are typically not available for general public, since AI robots and algorithms are typically unavailable to the public due to property rights, there lies an emerging apprehension regarding the ethics of AI-powered recruitment tools and whether people's perceptions of these tools' usage impact the amount of trust of people in the companies that use them (Figueroaetal., 2022a; 2022b).
- Finally, in certain areas, employment choices and final applicant selection still require human decision-making due to legal issues (under the General Data Protection Regulation in the EU, Laurim et al., 2021).

AI is anticipated to greatly boost recruiters' efficiency and provide them free time to take over more tactical and human-centered work, notwithstanding these reservations (Upadhyay & Khandelwal, 2018). Empirical research hasn't, however, fully investigated AI technology from the standpoint of those working in human resources (HR) in terms of advantages, effectiveness, and contact experiences. According to Laurim et al. (2021) some of the research only included participants from Bangladesh or Germany, or had very small sample sizes (3–5 recruiters) or distribution of nations. Although most research on AI in recruitment has concentrated on AI technologies (e.g., augmentation, automation, machine vision, and natural language processing) and how they affect the hiring and selection process (Hemalatha et al., 2021; Kong et al., 2021), how recruiters use AI in hiring (Albert, 2019; Johansson &Herranen, 2019), or ethical and trust issues (Figueroa-Armijos et al., 2022a, 2022b; Mujtaba&Mahapatra, 2019), little is known about how recruiters view AI-based tools in hiring and selection.

UNDERSTANDING AND ADDRESSING GENDER BIAS IN THE WORKPLACE

Gender equality in any organization is crucial for fostering a productive, innovative, and harmonious work environment (Filho et al., 2022). Ensuring gender equality means providing equal opportunities for all genders in hiring, promotion, and leadership roles, and creating a workplace free from discrimination and bias. (Kossek & Buzzanell, 2018). This approach not only aligns with the principles of fairness and justice but also has significant practical benefits for organizations. Research consistently shows that diverse teams, which include a balanced representation of genders, tend to perform better, make more informed decisions, and are more innovative. A diverse workforce brings a variety of perspectives and experiences, which can enhance creativity and problem-solving abilities, leading to better overall performance and competitiveness (Kundu & Mor, 2017). From a theoretical standpoint, gender equality can be explained through several key frameworks. One such framework is the Social Role Theory, which posits that societal roles and expectations significantly influence individuals' behavior and opportunities. By challenging and changing these traditional roles, organizations can create a more inclusive environment where all employees can thrive. Another relevant theory is the Human Capital Theory, which emphasizes the importance of investing in all employees' education, training, and development (Almendarez, 2013). By ensuring that both men and women have equal access to these resources, organizations can maximize their talent pool and enhance their overall effectiveness.

The concept of intersectionality, introduced by (Carbado et al. (2013) is also essential in understanding gender equality. This theory highlights how different aspects of identity, such as race, gender, and socioeconomic status, intersect and create unique experiences of discrimination or privilege. Organizations that recognize and address these intersecting identities can create more comprehensive and effective diversity and inclusion strategies. Additionally, the Equity Theory, developed by John Stacey Adams (1965), suggests that employees are motivated by fairness in the workplace. When individuals perceive that they are being treated equitably, they are more likely to be satisfied, committed, and productive. Gender equality initiatives can help ensure that all employees feel valued and fairly treated, which can lead to higher morale and lower turnover rates. All in all gender equality is not just a moral imperative but a strategic advantage for organizations. By embracing diverse talents and fostering an inclusive environment, organizations can drive innovation, improve decision-making, and achieve better business outcomes. Theoretical perspectives like Social Role Theory, Human Capital Theory, intersectionality, and Equity Theory provide valuable insights into why gender equality should be respected and actively promoted within any organization.

GENDER BIAS IN AI BASED RECRUITMENT

Gender prejudice may arise in the systems and software developers and data miners during the algorithmic training phase of AI based recruitment systems because of the gender disparities that are encountered initially. Previous researches have indicated the same and stated that there is indeed a lack of gender diversity in AI development teams. (Johnson, 2019; Clifton et al., 2020; Lee, 2018; Martinez & Fernandez, 2020). Any department of an organization that encounters a lack of gender diversity results in a lack in diversity of mindsets as well when it comes to the developing teams that create the AI based decision making systems. One illustration of lack of gender diversity is among the AI developers and STEM workers which is a male dominated, homogenous IT industry (Wang, 2020; Lee, 2018; Johnson, 2019).Consequently, gender-biased results could be a consequence of this, as it would strengthen the power and dominance of one gender (male) over algorithms and choices. The facial recognition programs used in the United States in 2015, for example, struggled to govern diversity (Daugherty et al., 2018; Otterloo, 2019). Additionally, Lambrecht & Tucker (2019) looked into how more males than women view career prospects in STEM advertisements, which ultimately leads to less women applying for STEM positions.

DETRIMENTAL EFFECTS OF GENDER BIAS

Gender bias in the workplace can lead to numerous problems, including reduced employee morale and productivity (Rotimi et al., 2023). It fosters an environment of inequality, where certain individuals are overlooked for promotions, raises, and key assignments, based solely on gender. This not only stifles talent and potential but also diminishes overall organizational effectiveness and innovation. Gender bias can lead to higher turnover rates, as affected employees may feel undervalued and seek opportunities elsewhere (Rothausen et al., 2015). Additionally, it can harm the organization's reputation, making it difficult to attract and retain top talent. Ultimately, gender bias undermines the principles of fairness and equality, essential for a thriving workplace.

REALTIME INSTANCES OF GENDER BIAS IN AI BASED RECRUITMENT

1. Amazon's AI Recruiting Tool

In 2018, Amazon scrapped an AI recruiting tool after discovering it was biased against women. The tool was found to favor male candidates over female ones, penalizing resumes that included words like "women's" as in "women's chess club captain." The AI was trained on resumes submitted to Amazon over a 10-year period, which were predominantly from men. As a result, the algorithm learned to prefer male candidates. This incident highlights how training data that reflects historical gender imbalances can lead to biased AI outcomes. Amazon's experience underscores the importance of using balanced and representative training data and continuously monitoring AI systems for bias.

2. HireVue's Video Interviewing Platform

HireVue, a company that uses AI to analyze video interviews, faced criticism for potentially reinforcing gender bias. The AI algorithms used facial recognition and voice analysis to rate candidates on traits such as confidence and enthusiasm, which can be gender-biased. Women might be unfairly assessed due to differences in communication styles and facial expressions that the AI interprets as less confident or enthusiastic. This bias could result in fewer women being recommended for further consideration. Addressing this requires refining the algorithms to account for diverse communication styles and implementing regular audits to ensure fair evaluations.

3. Unilever's AI-Powered Recruitment

Unilever used an AI-driven hiring process, which included game-based assessments and video interviews analyzed by AI. While the approach aimed to reduce bias, it still faced challenges related to ensuring fairness across different gender groups. Although Unilever's AI system was designed to evaluate candidates based on data-driven insights, any underlying biases in the training data or the AI's interpretive models could still result in gender-biased outcomes. To mitigate this, Unilever needed to ensure that their AI models were regularly tested for bias and calibrated to treat all candidates fairly, regardless of gender.

APPROACHES TO ALLEVIATE GENDER BIAS

AI Technology-Related Approaches

In order to counteract unsuitable and/or unjust datasets, Clifton et al. (2020) suggested to gathering inputs from every aspect of excluded, gender-diverse segments of community and including multi-dimensional datasets into the designing of the algorithms used for making decisions. Analogously, Hayes et al. (2020) contend that equitable population representation in data sets will lead to equitable AI results.

Management Approaches for Fair AI

It was found in the previous researches (Hayes et al., 2020) that there are certain cases where the ability of AI based decision are fallacious when the algorithms self learn from their own pernicious feedback. These fallacious events happen in case of unethical data practices, when there is misreporting of data or any other misconducts related to data collection and preparation. To address such issues, researchers have put forward the idea of testing and auditing the AI based decision making systems in the design and implementation phase itself (Miron et al., 2020; Johnson, 2019). Researchers that specialize in compliance management, such as Martinez and Fernandez (2020) may be involved in this. AI specialists, for instance, can employ interpretation tools to identify possible issues and difficulties as well as conduct routine evaluation and certification of AI-based systems for making decisions (Wu et al., 2019).

Regulatory Approaches and AI Governance

The necessity and importance of developing AI legislation for interacting with humans are underscored by the latest suggestion from the European Union (2021). According to Marabelli et al. (2021) the implementation of regulations and guidelines pertaining to AI-powered decision-making systems guarantees enhanced efficacy in the final results.

Community and Societal-Focused Approaches

In order to promote diversity and inclusivity, social interventions including improving professional education and training on gender diversity in the community should be supported (Hayes et al., 2020; Prates et al., 2019). Regarding the sharing of any private data for any such decision-making processes, public policies to safeguard users' privacy are also suggested as a potential strategy to boost trust in

AI-based decision-making systems (Clifton et al., 2020). Cirillo et al. (2020) suggest that governments or legislators create a "ecosystem of trust" to make sure that systems abide by the fundamental laws that safeguard consumer and human rights, especially when it comes to AI-based decision-making systems.

IMPLICATIONS

Companies must get ready for the AI revolution as HRM is becoming more and more digitalized. This includes reevaluating organizational structures and manager roles [Institute of Employment Studies (IES), 2018]. One such strategy is to define and improve the role of hiring professionals in the process of recruitment. Currently, recruiting algorithms are invaluable for searching through the vast amount of internet data about possible applicants; however, they are primarily designed to carry out routine as well as mundane duties (Upadhyay and Khandelwal, 2018) and are not supposedly not intended to adopt the role of a recruiter. Alternatively, hiring algorithms can reorganize the work that is performed by human recruiters whilst also improving the forecast of the candidate's future performance, even though the ultimate decision regarding the hiring will still be in the hands of manual recruiter (Agrawal et al., 2019). Employers may have more time to engage with prospects and concentrate on strategic matters if this role is implemented. Employing data science approaches in HR presents a number of difficulties because HR concerns are complicated phenomena (Cappelli et al., 2018). Among these difficulties, Cappelli et al. note that complicated tasks are linked with other jobs, making it impossible to fully isolate individual success from collective performance. As a result, they believe that it is not quite clear what a "good employee" is. Concerns about justice and ethics are another difficulty mentioned by these writers.

When a candidate is rejected and someone else with identical qualities is accepted, it could be challenging for an HR professional to convey to them the AI-based factors that led to that choice regarding their future. Additionally, candidates and staff members may attempt to fool the AI or exhibit other hostile behaviors that could have an impact on the organizational outcomes. Recruiters need to be sufficiently informed about the capabilities and limitations of the technology their employer uses. They also need to have their concerns about their jobs and the possibility of AI replacing them all addressed if they are to be comfortable using digital hiring algorithms and depending on the results. Like other professional endeavors, it remains to be seen how this may change over time.

LIMITATIONS AND FUTURE SCOPE

In AI-based recruitment research, gender bias is still in its infancy. Numerous limitations of this study present interesting directions for future research. Diverse scenarios provide a great means of examining the attitudes, behaviors, feelings, and perceptions of actual circumstances (Taylor 2005). The experimental approach, however, precluded first-hand experience from real application scenarios. Therefore, the responses of candidates in actual application scenarios may be the subject of future research. Applications will grow more accustomed to these new selection methods as a result of the growing use of AI systems in HRM, which could eventually lead to a rise in the acceptability of AI in general. The experiment could be repeated in later research to see if the negative affective reactions go away with time. According to Jatoba et al. (2022) research ought to be done on the best ways to retrain workers in intuitive and empathetic skills in order to keep employment. In addition, we will look at how developing technology will affect the future of the employment market, including workers' adaptability and resilience. Investigating the psychological issues raised by these new technologies and the ways in which different HR procedures could lessen the psychological issues that these technologies cause for workers while also providing insightful information.

CONCLUSION

A common misperception regarding AI is that it can perform tasks more efficiently than humans. It does in many instances. However, AI ought to be created to function differently. Since humans make mistakes, monitoring, correcting, anticipating, and occasionally punishing these blunders has been a cornerstone of the social contract for ages. However, there has always been a certain degree of understanding for human errors, given the fact that humans are not perfect and are bound to make mistakes which will result in some biasness. But that's different in case of algorithms, and so, we tend to be less tolerant with algorithmic errors. Humans making mistake is a very normal thing, it is just how humans are regardless of what the supporters of rational approach to economic theory have been advocating. This implies that information gathered about people and their behaviors is inherently skewed. Thus, concerns pertaining to biased data sets must be addressed in addition to the problem of biased algorithms and their unfair results. Therefore, we are supposed to understand that it is our responsibility, in HR technology as well as in any other AI system, to ensure that we not only understand these problems but also develop the appropriate laws and technologies that will be fairly applied by lawmakers, regulatory bodies, businesses, and individuals alike. Lastly, it's critical to remember that, despite its effectiveness, we are continuously refining both the safeguard algorithms that account for biases and the primary algorithm described in this article. Our algorithm, as it stands at this point in its development, is a tool to help human recruiters make decisions rather than the decision maker. While matching a resume to a job offer, the final say still belongs to a person because AI is unable to account for human variables at this time. However, the speed and volume of CVs the algorithm can evaluate and choose from to assist the recruiter give it a competitive edge.

REFERENCES

Agrawal, A., Gans, J. S., & Goldfarb, A. (2019). Exploring the impact of artificial intelligence: Prediction versus judgment. *Information Economics and Policy*, 47, 1–6. 10.1016/j.infoecopol.2019.05.001

Ahmed, O. (2018). Artificial intelligence in HR. *Int J Res Anal Rev*, 5(4), 971–978.

Alam, M. S., Dhar, S. S., & Munira, K. S. (2020). HR Professionals' intention to adopt and use of artificial intelligence in recruiting talents. *Business Perspective Review,* 2(2), 15–30.

Almendarez, L. (2013). Human Capital Theory: Implications for educational development in Belize and the Caribbean. *Caribbean Quarterly*, 59(3–4), 21–33. 10.1080/00086495.2013.11672495

Beattie, G., & Johnson, P. J. P. P. (2012). Possible unconscious bias in recruitment and promotion and the need to promote equality. *Perspectives*, 16(1), 7–13. 10.1080/13603108.2011.611833

Beneduce, G. (2020). *Artificial intelligence in recruitment: just because it's biased, does it mean it's bad?* NOVA—School of Business and Economics.

Black & van Esch. (2020). AI-enabled recruiting: what is it and how should a manager use it? *Bus Horiz, 63*(2), 215–226. https://doi.org/.2019.12.00110.1016/j.bushor

Bogen M, Rieke A (2018) Help wanted: an examination of hiring algorithms, equity, and bias. Academic Press.

Bornstein, S. (2018). Antidiscriminatory algorithms. *Alabama Law Review*, 70, 519.

Cain, G. G. (1986). The economic analysis of labor market discrimination: A survey. *Handbook Labor Econ*, 1, 693–785. 10.1016/S1573-4463(86)01016-7

Carbado, D. W., Crenshaw, K. W., Mays, V. M., & Tomlinson, B. (2013). INTERSECTIONALITY: Mapping the movements of a theory. *Du Bois Review*, 10(2), 303–312. 10.1017/S1742058X1300034925285150

Chen, Z. (2023). Collaboration among recruiters and artificial intelligence: Removing human prejudices in employment. *Cognition Technology and Work*, 25(1), 135–149. 10.1007/s10111-022-00716-036187287

Cirillo, D., Catuara-Solarz, S., Morey, C., Guney, E., Subirats, L., Mellino, S., Gigante, A., Valencia, A., Rementeria, M. J., Chadha, A. S., & Nikolaos, M. (2020). Sex and gender differences and biases in artificial intelligence for biomedicine and healthcare. *Digital Medicine*, *8*(3). https://www.nature.com/articles/s41746-020-0288-5

Clifton, J., Glasmeier, A., & Gray, M. (2020). When machines think for us: The consequences for work and place. *Cambridge Journal of Regions, Economy and Society*, 13(1), 3–23. 10.1093/cjres/rsaa004

Davenport, T., Guha, A., Grewal, D., & Bressgott, T. (2020, October). How artificial intelligence will change the future of marketing. *Journal of the Academy of Marketing Science*, 48(1), 24–42. 10.1007/s11747-019-00696-0

Feuerriegel, S., Dolata, M., & Schwabe, G. (2020). Fair AI: Challenges and opportunities. *Business & Information Systems Engineering*, 62(4), 379–384. 10.1007/s12599-020-00650-3

Figueroa-Armijos, M., Clark, B. B., & da Motta Veiga, S. P. (2022). Ethical perceptions of AI in hiring and organizational trust: The role of performance expectancy and social influence. *Journal of Business Ethics*, 1–19.

Filho, W. L., Kovaleva, M., Tsani, S., îrcă, D., Shiel, C., Dinis, M. P., Nicolau, M., Sima, M., Fritzen, B., Salvia, A. L., Minhas, A., Kozlova, V., Doni, F., Spiteri, J., Gupta, T., Wakunuma, K., Sharma, M., Barbir, J., Shulla, K., & Tripathi, S. (2022). Promoting gender equality across the sustainable development goals. *Environment, Development and Sustainability*, 25(12), 14177–14198. 10.1007/s10668-022-02656-1

Hayes, P., Poel, I. V. D., & Steen, M. (2020). Algorithms and values in justice and security. *AI & Society*, 35(3), 533–555. 10.1007/s00146-019-00932-9

Hemalatha, A., Kumari, P. B., Nawaz, N., & Gajenderan, V. (2021). Impact of artificial intelligence on recruitment and selection of information technology companies. In *2021 international conference on artificial intelligence and smart systems (ICAIS)* (pp. 60–66). IEEE. 10.1109/ICAIS50930.2021.9396036

Hmoud, B., & Laszlo, V. (2019). Will artificial intelligence take over human resources recruitment and selection? *NetwIntell Stud*, 7(13), 21–30.

Jackson, M. C. (2021). Artificial intelligence & algorithmic bias: The issues with technology reflecting history & humans. *J Bus Technol Law*, 16, 299.

Jatobá, M. N., Ferreira, J. J., Fernandes, P. O., & Teixeira, J. P. (2023). Intelligent human resources for the adoption of artificial intelligence: A systematic literature review. *Journal of Organizational Change Management*, 36(7), 1099–1124. 10.1108/JOCM-03-2022-0075

Johansson, J., & Herranen, S. (2019) The application of artificial intelligence (AI) in human resource management: current state of AI and its impact on the traditional recruitment process. Bachelor thesis, Jonkoping University.

Johnson, K. N. (2019). Automating the risk of bias. *George Washington Law Review*, 87(6). https://www.gwlr.org/wp-content/uploads/2020/01/87-Geo.-Wash.-L.-Rev.-1214.pdf

Johnson, R. D., Stone, D. L., & Lukaszewski, K. M. (2020). The benefits of eHRM and AI for talent acquisition. *J Tour Futur*, 7(1), 40–52. 10.1108/JTF-02-2020-0013

Kong, Y., Xie, C., Wang, J., Jones, H., & Ding, H. (2021). AI-assisted recruiting technologies: Tools, challenges, and opportunities. In *The 39th ACM international conference on design of communication* (pp. 359–361). ACM.

Kossek, E. E., & Buzzanell, P. M. (2018). Women's career equality and leadership in organizations: Creating an evidence-based positive change. *Human Resource Management*, 57(4), 813–822. 10.1002/hrm.21936

Kulik, J. A., Mahler, H. I., & Moore, P. J. (1996). Social comparison and affiliation under threat: Effects on recovery from major surgery. *Journal of Personality and Social Psychology*, 71(5), 967–979. 10.1037/0022-3514.71.5.9678939044

Kundu, S. C., & Mor, A. (2017). Workforce diversity and organizational performance: A study of IT industry in India. *Employee Relations*, 39(2), 160–183. 10.1108/ER-06-2015-0114

Kyriazanos, D. M., Thanos, K. G., & Thomopoulos, S. C. A. (2019). Automated decisions making in airports checkpoints: Bias detection toward smarter security and fairness. *IEEE Security and Privacy*, 17(2), 8–16. 10.1109/MSEC.2018.2888777

Lambrecht, A., & Tucker, C. (2019). Algorithmic bias? An empirical study of apparent gender-biased discrimination in the display of STEM career ads. *Management Science*, 65(7), 2947–3448. 10.1287/mnsc.2018.3093

Laurim, V., Arpaci, S., Prommegger, B., & Krcmar, H. (2021). Computer, whom should I hire?–acceptance criteria for artificial intelligence in the recruitment process. In *Proceedings of the 54th Hawaii international conference on system sciences* (p. 5495). 10.24251/HICSS.2021.668

Lee, N. T. (2018). Detecting racial bias in algorithms and machine learning. *Journal of Information, Communication, and Ethics in Society*, 16(3), 252–260. 10.1108/JICES-06-2018-0056

Marabelli, M., Newell, S., & Handunge, V. (2021). The lifecycle of algorithmic decision-making systems: Organizational choices and ethical challenges. *The Journal of Strategic Information Systems*, 30(3), 101683. Advance online publication. 10.1016/j.jsis.2021.101683

Martinez, C. F., & Fernandez, A. (2020). AI and recruiting software: Ethical and legal implications. *Paladyn : Journal of Behavioral Robotics*, 11(1), 199–216. Advance online publication. 10.1515/pjbr-2020-0030

Meyer, D. (2018). Amazon reportedly killed an AI recruitment system because it couldn't stop the tool from discriminating against women. *Fortune*. fortune.com/2018/10/10/amazon-ai-recruitment-bias-women-sexist/.

Miasato, A., & Silva, F. R. (2019). Artificial intelligence as an instrument of discrimination in workforce recruitment. ActaUnivSapientiae. *Legal Studies*, 8(2), 191–212.

Miron, M., Tolan, S., Gomez, E., & Castillo, C. (2020). Evaluating causes of algorithmic bias in juvenile criminal recidivism. *Artificial Intelligence and Law*, 29(2), 111–147. 10.1007/s10506-020-09268-y

Mujtaba, D. F., & Mahapatra, N. R. (2019, November). Ethical considerations in AI-based recruitment. In *2019 IEEE International Symposium on Technology and Society (ISTAS)* (pp. 1-7). IEEE.

Newell, S. (2015). *Recruitment and selection. Managing human resources: personnel management in transition*. Blackwell Publishing.

Njoto, S. (2020) Research paper gendered bots? Bias in the use of artificial intelligence in recruitment.

Oullier, O., & Basso, F. (2010). Embodied economics: How bodily information shapes the social coordination dynamics of decision-making. *Philosophical Transactions of the Royal Society of London. Series B, Biological Sciences*, 365(1538), 291–301. 10.1098/rstb.2009.016820026467

Pan, Y. (2016). Heading toward artificial intelligence 2.0. *Engineering (Beijing)*, 2(4), 409–413. 10.1016/J.ENG.2016.04.018

Prates, M., Avelar, P., & Lamb, L. C. (2018). Assessing gender bias in machine translation – A case study with google translate. *Neural Computing & Applications*, 32(10), 6363–6381. 10.1007/s00521-019-04144-6

Raub, M. (2018). Bots, bias and big data: Artificial intelligence, algorithmic bias and disparate impact liability in hiring practices. *Arkansas Law Review*, 71, 529.

Raveendra, P., Satish, Y., & Singh, P. (2020). Changing landscape of recruitment industry: A study on the impact of artificial intelligence on eliminating hiring bias from recruitment and selection process. *Journal of Computational and Theoretical Nanoscience*, 17(9), 4404–4407. 10.1166/jctn.2020.9086

Rothausen, T. J., Henderson, K. E., Arnold, J. K., & Malshe, A. (2015). Should I stay or should I go? Identity and Well-Being in Sensemaking about Retention and Turnover. *Journal of Management*, 43(7), 2357–2385. 10.1177/0149206315569312

Rotimi, F. E., Brauner, M., Burfoot, M., Naismith, N., Silva, C. C., & Mohaghegh, M. (2023). Work environment challenge and the wellbeing of women in construction industry in New Zealand – The mediating role of work morale. *Engineering, Construction, and Architectural Management*. Advance online publication. 10.1108/ECAM-02-2023-0152

Shaw, J. (2019). Artificial intelligence and ethics. *Perspect: Policy Pract High Educ*, 30, 1–11.

Tambe, P., Cappelli, P., & Yakubovich, V. (2019). Artificial intelligence in human resources management: Challenges and a path forward. *California Management Review, 61*(4), 15–42. Upadhyay AK, Khandelwal K (2018) Applying artificial intelligence: implications for recruitment. *Strategic HR Review*, 17(5), 255–258.

Tambe, P., Cappelli, P., & Yakubovich, V. (2019). Artificial intelligence in human resources management: Challenges and a path forward. *California Management Review*, 61(4), 15–42. 10.1177/0008125619867910

Taylor, P. J. (2005). Leading world cities: Empirical evaluations of urban nodes in multiple networks. *Urban Studies (Edinburgh, Scotland)*, 42(9), 1593–1608. 10.1080/00420980500185504

W. (2019). Mitigating gender bias in Natural Language Processing: A literature review. *Proceeding of 57th Annual Meeting of Association for Computational Linguistics*. 10.18653/v1/P19-1159

Wilson, H. J., & Daugherty, P. R. (2018). Collaborative intelligence: Humans and AI are joining forces. *Harvard Business Review*, 96(4), 114–123.

Wu, W., Huang, T., & Gong, K. (2019). Ethical principles and governance technology development of AI in China. *Engineering (Beijing)*, 6(3), 302–309. 10.1016/j.eng.2019.12.015

Zhang, J., & Chen, Z. (2023). Exploring human resource management digital transformation in the Digital Age. *Journal of the Knowledge Economy*. Advance online publication. 10.1007/s13132-023-01214-y

Chapter 4
Socially Responsible Application of Artificial Intelligence in Human Resources Management

Ana Marija Gričnik
https://orcid.org/0009-0009-5981-2234
Faculty of Economics and Business, University of Maribor, Slovenia

Matjaž Mulej
Faculty of Economics and Business, University of Maribor, Slovenia

Simona Šarotar Žižek
Faculty of Economics and Business, University of Maribor, Slovenia

ABSTRACT

Humankind faces growing artificial intelligence (AI) and AI-based applications, influencing almost every activity, including human resource management (HRM), revolutionizing humans' work nature and content, workers, workplaces, HRM processes, etc. AI can support various HRM practices, such as candidate selection, employee training, data analysis, evaluation, etc. If organizations appropriately utilize AI, they can enhance productivity and their general/individual work performance, streamline processes, and increase efficiency, ultimately improving employee engagement and well-being. Hence, organizations can use AI to stay ahead of their competitors and help develop an innovative sustainable socially responsible society (ISSRS) to overcome crises. AI should only be used as a tool and not to replace humans, which is essential for a creative, efficient, satisfying, and successful work environment.

DOI: 10.4018/979-8-3693-3334-1.ch004

1. INTRODUCTION

Social responsibility is individuals' responsibility for influence on society (ISO, 2010, in: ISO 26000); it must not be limited to corporations and their free will, which is focused on in ISO 26000 (ISO, 2010). Artificial Intelligence (AI) can be defined as "making a machine behave in ways that would be called intelligent, - if a human were so behaving" (McCarthy et al., 1955). Despite being defined in 1955, its prominence has only recently increased due to the worldwide technological revolution (Dwivedi et al., 2021). AI's influence makes social responsibility grow, but not unavoidably (more in Mulej et al., 2024 a, b, c, d, e; Šarotar Žižek et al., 2023 a, b, c). AI should become socially responsible AI (SRAI), especially in HRM, focusing on human attributes and influencing activities.

AI is causing a vital shift that extends beyond technical spheres and has far-reaching sociotechnical implications. In addition to healthcare, education, commerce, and finance, AI also impacts everyday activities. Therefore, AI presents both opportunities and risks. The question of whether AI deserves our trust has resurfaced frequently recently in various spheres, including academia, industry, healthcare, and services (Cheng, Varshney, & Liu, 2021).

HRM activities include various repetitive and time-consuming activities, which are susceptible to human influence, biases, and subjectivity. Because of these, HRM is considered a promising area for applying AI (Rodgers et al., 2023; Tambe et al., 2019). AI has significant promise for enhancing HRM functions in organizations. However, it is important to remember that the automation of HRM processes comes with potential risks and limitations. Identifying these is crucial to facilitate the proper use of AI in HRM (Bujold et al., 2023).

The use of AI is becoming more prevalent across HRM functions, including but not limited to sourcing and selecting job applicants, talent acquisition and management, performance evaluation, workforce planning, health and wellbeing, and compensation (Bujold et al., 2023; Bankins, 2021).

Although the application of AI in HR may offer several advantages, it can also cause negative consequences, if it is not SRAI, implemented carefully and deliberately. This raises ethical concerns over AI's use in HRM, which handles sometimes sensitive facets of individuals' employment lifecycles. Despite this, most research at the intersection of HRM and technology continues to examine the potential applications of AI instead of looking into its vital ethical considerations and how to efficiently involve human workers in the use of AI (Bankins, 2021). SRAI is missing.

For many HRM professionals, even mentioning AI generates a worrisome feeling of apprehension, an unsettling sense, or even fear of the unknown. On the other hand, some people have an overly optimistic perspective on AI's potential and capabilities, creating unrealistic ideas comparable to those shown in science fiction movies. The dynamic between fear and fascination comes from a fundamental lack of understanding. However, in today's world, when AI is widely discussed, HRM professionals must have an increased awareness of potential issues and their consequences on each stakeholder group. This will enable them to proactively use AI to improve HRM procedures (Chang & Ke, 2023), being SRAI, or tending to be SRAI, at least, or even innovative sustainable socially responsible society's (ISSRS) HRM. (More about ISSRS see in Mulej et al, 2024 a, b, c, d, e.)

Organizations are becoming increasingly aware of the importance of digital transformation to stay competitive in the rapidly evolving global business environment (Li, 2020). This has changed the business style and pushed firms to discover how to capture value and opportunities effectively (Mithas et al., 2013). Therefore, no one can predict what the future holds for the HRM profession and how HRM practices will change. Nevertheless, thinking about the future helps us to prepare for it better. Thinking about the future may lead to new ideas and enhance HRM procedures (Ulrich, 1997).

The main aim of the research briefed in this chapter is to explore and present a literature review of existing empirical literature and research on SRAI - the socially responsible application of AI in HRM. In light of this, we contribute to the existing research on AI and HRM by extending it further to examine the socially responsible use of AI in this context, as the vast majority of literature only examines how AI can be applied in HRM's functions. The main goal of the research briefed in this chapter is to address the existing void in the literature, while offering direction to HRM practitioners/experts regarding the implementation of HRM enabled by SRAI. Both the sustainability and corporate social responsibility (CSR) efforts of organizations would benefit from this, possibly leading to ISSRS, not CSR only.

2. HUMAN RESOURCE MANAGEMENT AND ARTIFICIAL INTELLIGENCE

In modern times, both individuals and society have become progressively dependent on AI technology. AI has played and will continue to play a pivotal role in innumerable aspects of life, livelihood, and liberty. In light of this, AI can drive humanity towards a prosperous future. On the other hand, significant hazards of oppression and catastrophe are also associated with AI (Cheng et al., 2021).

Every organization has three key resources: physical, economic, and workforce. Physical resources include materials, machinery, equipment, tools, buildings, etc. Economic resources include cash, debt, and credit, whereas human resources constitute the organization's employees and partners. It is well-known that employees are invaluable assets that serve a crucial function within a business organization.

For example, essential components of a business strategy may include technologies, marketable products and services, and efficient procedures. This success depends on the workforce's capacity to effectively create and implement workplace plans, procedures, programs, and strategies (Boxall & Purcell, 2011), and norms, and values, culture, ethics. expressed (VCEN) in prevailing habits (For more see e.g., Šarotar Žižek et al., 2023 a, b, c.)

We begin by exploring Human resource management and its value for organizations.

2.1 Human Resource Management

Organizations deeply rely on people. They must recruit and develop competent, skilled, and engaged employees, manage their performance, reward them based on their contributions, develop, and maintain positive employment relationships, and ensure their well-being. Essentially, this is the basis of HRM (Boxall, 2014).

It is a common misconception that human resources are defined only as those employed in a specific organization. HR should be viewed as humans' intrinsic resources and can apply to many different tasks throughout our lives (Boxall, 2013). However, organizations can't survive without human resources; they require the services individuals can only provide using physical, symbolic, and financial resources (Penrose, 1959). The skills and abilities of human beings are essential for developing, growing, and renewing organizations and inevitably contribute to making organizations somewhat idiosyncratic. Consequently, HRM plays a vital role in organizations (Boxall & Purcell, 2011).

Every organization serves some purpose in the wider society. The workforce is one of the most critical assets in business management, and like other resources, it can be difficult to manage and retain human resources in today's environment. As employees are social beings, they cannot function like machines. Therefore, the employers must ensure the well-being of these valuable human resources (Majumder et al., 2023), along with their values, i.e., values and knowledge management (more in Šarotar Žižek et al., 2024).

To understand HRM and its crucial aspects well, it is vital to examine its various functions inside the organization. HRM functions can be classified into three broad categories: i) managerial, ii) operative, and iii) advisory, as shown in Figure 1.

Figure 1. HRM functions (Majumder et al., (2023))

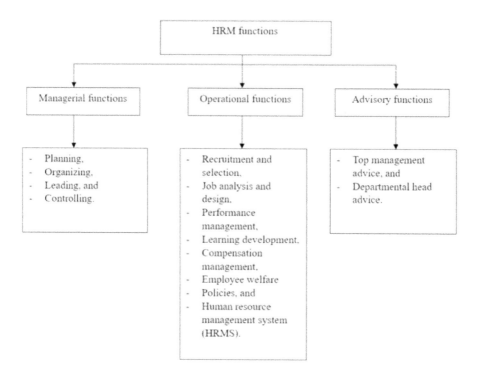

N.B. Peers' advice can also be helpful.

How HRM operates will differ based on the nature of the organization, the employees it employs, and the environment in which it operates. This is explained by contingency theory, which believes that HRM methods must be customized to the environment and circumstances of the organization. This is "best fit," instead of "best practice." The latter makes the argument that there is a set of universal best HRM practices that are best in every scenario, problematic. Contingency theory emphasizes two needs: first, to ensure that any proposed or existing HR(M) practice is compatible with the organization's needs and characteristics, and second, to prevent the simple replication of best practice, regardless of its relevance (Boxall, 2014), and adaptation.

Businesses can only achieve success with the efforts of their employees. By effectively managing employees, organizations can increase profitability and build brand loyalty among customers and other partners. Therefore, HRM is a key to an organization's growth and successful business strategies (Rao & Teegen, 2009). HRM is a strategic approach to employee recruitment, growth, and development, as well as the overall health and well-being of the employees. It includes several managerial actions that connect the organization to its employees (Noe et al., 2007). HRM is responsible for every managerial decision related to the workforce of an organization. It focuses on making successful human resource decisions and enhancing their performance to meet organization's goals. It balances employee needs with satisfaction with the organization's capabilities and profitability (Majumder et al., 2023).

The HR(M) system helps organizations manage employees to achieve the desired outcomes. The range and characteristics of HRM activities are determined by the organization's setting (including its internal and external environment – natural, social, and economic) and its people management philosophy, strategies, and policies (Boxall, 2014). A strong HRM system can boost company performance by establishing powerful and focused organizational cultures that influence and drive employee behavior and efforts toward organizational goals (Bowen & Ostroff, 2004) and partners' behavior by responding to employees', managers' and owners' behavior.

A HR(M) system encompasses (Boxall, 2014):

- *HR policies,* which define the functions of HR and set guidelines for the application and implementation of specific HR aspects.
- *HR practices* which encompass the HR activities associated with employee development and management, as well as employment relationship management. These parts are essential to the HR architecture and provide the basis of the HR(M) system. HR values, stakeholder interests, and CSR considerations influence them.
- *HR strategies,* which define the direction that HRM should take in each of its major areas of activity.

Now that we have covered the fundamental aspects of HRM, our next step involves researching its connection with organizational performance.

2.2 Human Resource Management and Organizational Performance

Organizational and business performance, both in the short and long term, is influenced by internal factors such as the quality of employees at each level, the organization's capacity to innovate and compete, its cost-effectiveness, and the efficiency of its operational strategies and actions. On the other hand, the external factor consists of domestic and international competition, labor market conditions, governmental financial and business policies and legislation, international pressures and influences, and economic trends. However, the success of organizations mainly relies on the quality, commitment, passion, skills, and expertise of their employees (including managers). Therefore, HRM offers additional benefits and helps achieve a "sustainable and sustained competitive advantage" by strategically developing the organization's unique, difficult-to-imitate, and difficult-to-replace human resources (Boxall, 2014). A significant aspect of HRM is its foundational assumption that improved organizational efficiency is achieved through employees within the organization, mostly. Thus, if HR policies and practices are appropriate, HRM will improve organizational performance (Guest, 1997).

Organizational performance may be measured using both financial and non-financial indicators. In light of this, Boxall (2014) states the following financial and non-financial measures:

Financial measures:

- *Shareholder value* – monetary benefits that an organization provides to its shareholders.
- *Profit (in an organization)* – the surplus of sales over costs.
- *Financial performance (in a not-for-profit organization)* – expenditure control; the surplus or deficit of income over expenditures.
- *Sales turnover* – the economic value of the business's sales.
- *Returns on capital employed* – the percentage of the organization's financial assets that are earned as profits.
- *Earnings per share* – earnings in proportion to the quantity of shares.
- *Productivity* – added value generated by the organization, divided by the number of employees (the difference between the value of sales and the cost of labor and items bought is added value).
- *Cost per unit of output* – production costs as a percentage of the total number of units produced.

Non-financial measures

- *Quality and execution of corporate strategy* – (the degree to which goals have been accomplished).
- *Effectiveness of human capital management* – the capacity to attract, retain, and develop talented (and committed, socially responsible, we add) people.
- *Quality of product or service.*
- *Levels of customer service.*
- *Innovation and research leadership.*
- *Market share.*
- *Management expertise.*
- *Reputation as a business and as an employer.*

- *Exercise of CSR, particularly about environmental concerns.*
- *Ratio of number of employees to output* - such as units produced per individual.

HRM contributes to organizational performance in three distinct ways: (1) HR(M) practices may impact employee characteristics such as commitment, motivation, engagement, and skills; (2) if employees have these qualities, it is likely that organizational performance, in terms of quality, productivity, added value, and the delivery of outstanding client service, will improve; and (3) improving these aspects of organizational performance will improve financial outcomes. This may be defined as the HR value chain. These propositions emphasize the presence of an intermediary factor that lies between HRM and financial performance. This factor includes the HRM outcomes affected by HR practices in the form of employee characteristics. Therefore, HRM does not directly impact organizational performance (Boxall, 2014). This is shown in Figure 2.

Figure 2. Impact of HRM on organizational performance (Boxall, 2014)

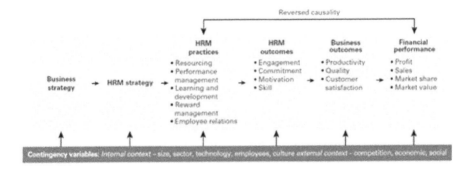

Bowell (2014) combined the results of several studies on the link between HRM and organizational performance. His findings are presented in Table 1.

Table 1. Research on the relationship between HRM and organizational performance

Researcher(s)	Methodology	Outcomes
Birdi et al., (2008)	The University of Sheffield Institute of Work Psychology undertook a 22-year longitudinal study of 308 companies to examine how HRM and operational practices affect company productivity.	Empowerment, especially job enrichment, increased employee value by 7%. About 6% was acquired by extensive training. Quality management, collaboration, and just in time have little to no effect.

continued on following page

Table 1. Continued

Researcher(s)	Methodology	Outcomes
Guest et al., (2000)	An analysis of the 1998 WERS study, which surveyed 28,000 workers in 2,000 firms.	Employee attitudes and workplace performance have a strong connection with HRM.
Patterson et al., (1997)	The study examined HR practices, organizational culture, and business performance.	HR practices explained 19% and 18% of productivity and profitability variances. Two HR strategies stood out: employee skill development and flexible, responsible, and varied job design.
Purcell et al., (2003)	Bath University longitudinal research of 12 organizations on how people management affects organizational success.	Most successful companies have a "big idea." A strong vision and integrated values define them. Their focus was maintaining optimal performance and flexibility.
Thompson (2002)	A UK study on high-performance work practices in aerospace firms covering teamwork, appraisal, job rotation, broad-banded grade structures, and business information sharing.	The quantity and depth of HR practices divide successful from unsuccessful firms.
West et al., (2002)	A UK study of 61 hospitals collected HR strategy, policy, and procedures data from CEOs, HR directors, and mortality rates.	Research linked some HR practices to lower mortality rates. The overall system runs better if HR strategies prioritize effort and skills, foster employee skill development, and promote teamwork, collaboration, creativity, and synergy.

Source: adapted from Boxall (2014)

The next subsection is dedicated to Artificial Intelligence, since it is supposed to support HRM efforts.

2.3 Artificial Intelligence

Over the past several years, many significant developments and innovations have occurred in information technology (IT), with AI emerging as a particularly important one. In practical terms, digital technologies such as machine learning (ML) and AI have become integrated into various aspects of daily work in the workplace, resulting in an enormous transformation and development in business management. In addition, they significantly influence the management of employees' workflow, also known as the HRM. AI is highly beneficial for various business functions that reduce employees' stress, pressure, and burden (Ghosh et al., 2023).

AI is a relatively recent technique that emerged due to technical advancements in information technology (IT) (Ghosh et al., 2023). It is the capacity of a human-made system made up of algorithms and software to recognize, understand, provide insights, and learn from the data sources used to achieve particular predetermined tasks and goals (Chowdhury et al., 2022). It refers to a technology used for carrying out tasks requiring a particular degree of intelligence to be completed. Furthermore,

AI may also refer to executing a specific work by integrating human endeavors and a trained technological pattern. The distinctive components of AI, such as its ability to do fast calculations, use advanced algorithms, and analyze vast amounts of high-quality data, set it apart from other ordinary types of software systems (Ghosh et al., 2023). The definition mainly covers the two key aspects of this new technology: (1) it is made (artificial), and (2) it has some intelligence, or the capacity to learn from data in a similar way to how people learn from their experiences in life (Chowdhury et al., 2022).

Although the development of AI is not exactly a new endeavor, its success was only widely acknowledged once machine learning made a significant breakthrough and changed the history of AI. Machine learning encompasses various techniques, such as unsupervised, supervised, and reinforcement learning, that enable machines to acquire knowledge and produce outcomes. This is accomplished through four primary stages: 1) data processing, 2) model development, 3) assessment of results, and 4) practical application and implications (Chang & Ke, 2023).

For businesses embracing a data-centric and digital culture, AI has emerged as the key source for business model innovation, process transformation, disruption, and competitive advantage (Ransbotham et al., 2020). AI has not only revolutionized the process of generating and using information for decision-making (Mikaelf et al., 2017). Furthermore, it has significantly transformed business ways, having an essential effect on trade and management practices across various industries that provide increasingly competitive and sustainable products or services (Kuo & Smith, 2018).

Users' degree of control or interaction over AI systems is essential in fostering responsible application and eventual adoption since humans need it to maintain their position as decision-makers (van der Broek et al., 2019; van den Broek et al., 2021). These findings align with the notion of accountability, which asserts that individuals should retain responsibility and be held accountable for their judgments, even when assisted by AI systems (Bujold et al., 2023). SRAI would help, because it might provide more holism and prevent more mistakes, therefore.

In today's modern society, AI is present in almost every aspect of life. Besides that, it is arguable that the AI revolution, which is expected to occur shortly, is already here (Makridakis, 2017). Presently, AI systems are being widely adopted by governments and businesses, both for proactive benefit acquisition and to avoid falling behind as others do so (De Sousa et al., 2019). At the same time, the sustainability of all human activity is becoming an increasingly important concern for businesses, civil society, politicians, and regulators (Walker et al., 2019). It is now expected that all major businesses have a comprehensive understanding of their impact on the environment and society and that they publicly disclose and document their involvement in such matters. This is accomplished through a wide range of criteria, frameworks, and measurements associated with what is known as ESG (environmental, social, and governance), which is now often replacing the older term CSR (corporate social responsibility) (Verbin, 2020).

On a global scale, the increasing economic significance of AI may lead to an increase in inequalities due to the uneven distribution of education and computing worldwide. People who are already well-off and educated may be the primary beneficiaries of the great wealth technology driven by AI may create, while others will be left worse off due to job displacement. Moreover, the preexisting biases present in the data used to train AI algorithms could worsen, eventually resulting in increased levels of discrimination (Vinuesa et al., 2020).

After examining AI, in the following subsection we focus on its application in HRM.

3. APPLICATION OF ARTIFICIAL INTELLIGENCE IN HUMAN RESOURCE MANAGEMENT

HRM has been experiencing a significant revolution due to the widespread implementation of AI systems, which are assisting in the resolution of numerous issues and, as a result, are slowly being integrated into the execution of various HR functions in a timely and effective manner. Once carried out entirely and exclusively by people, these HR functions have been reshaped with the adoption of AI (Ghosh et al., 2023). The technical progress in HRM may be traced back to the industrial revolution. However, it has mostly affected either physical or mental services. Recent developments, however, are increasingly offering HRM alternatives in functions that traditionally require human engagement and communication. As a result, organizational structures and the nature of work are revolutionized (Colbert et al., 2016; Luo et al., 2019). These intelligent technologies have crucially transformed HRM's traditional tasks, providing significant advantages and possibilities for HRM. However, they also present major issues, such as the risk of outdated job-specific skills (Go & Sundar, 2019).

Using intelligent machines, AI appeals to organizations since it allows HR managers to keep their attention on the needs of employees (Ravipolu, 2017). The use of AI in HRM continues to rise due to its capacity to create value for consumers, employees, and companies. However, recent research has revealed that organizations have yet to experience the anticipated benefits of using AI despite dedicating considerable energy, money, and time to it (Chowdhury et al., 2022).

Decisions made by AI have the potential to be discriminatory and biased since they are directly influenced by the data upon which they are based (Tambe et al., 2019). For example, in the context of talent acquisition, AI has the potential to perpetuate and exacerbate preexisting disparities within the dataset. This might lead to an unclear probability of granting job interview invitations to individuals from underrepresented groups (Köchling et al., 2021). In addition to that, scanning emails for data causes significant privacy concerns regarding AI in HRM (Eckhaus, 2018). Furthermore, there was a slight, yet positive correlation noticed between the level of automation in the job application process and the privacy concerns expressed by candidates (Langer et al., 2020). Therefore, applying AI to HRM raises a significant concern regarding its appropriateness. Instances have already emerged where the use of AI in HRM has led to outcomes that exhibit racial and gender biases in recruiting processes, infringements against employee data security, and compromised fairness and justice in employee-related matters (Scholz, 2019; Tambe et al., 2019; Robert et al., 2020). AI was not SRAI enough.

According to Bujold et al. (2023), irresponsible use of AI in HRM is characterized by various principles: a) no responsible principle applied, b) bias and discrimination, c) perceived justice and trust, d) privacy, e) explain-ability and transparency, f) human role. More SRAI is needed for success.

Regarding HRM tasks, most organizations have implemented AI to enhance employee performance efficiencies and achieve organizational goals across various functional areas (Ghosh et al., 2023). In order to fully benefit from AI adoption, organizations must look over technical resources and prioritize the development of non-technical ones, such as organizational culture, human skills and competencies, governance strategy, innovation mindset, and AI-employee integration (Chowdhury et al., 2022).

As organizations eagerly adopt new technologies, the relationships between humans and machines in the workplace are experiencing significant changes. To answer the question: "How can HR teams and managers help ensure these relationships are reshaped so that organizations and their employees win now and in the future? "Oracle & Future Workplace (2019) conducted a research. Their study surveyed over 8000 HR leaders, managers, and employees from 10 countries regarding their views and behavior towards AI. Their findings uncovered some unexpected developments (Oracle & Future Workplace, 2019):

- Approximately 50% of the survey respondents reported that they are currently using some AI in their work. This represents a significant increase compared to the 32% who stated the same in their last survey conducted in 2018.
- Just as the adoption rate of modern technologies is increasing, so are people's perceptions of them and how they engage with them. Although some individuals continue to voice concerns regarding the potential of job losses resulting from greater use of AI in the workplace, research indicates that most respondents are enthusiastic about these technologies.

- The relationship between technology and employees is warming up; 53% of employees are optimistic and enthusiastic about having robot co-workers.
- The relationship between employees and managers is changing. At the same time, employees' trust in AI-enabled technologies is increasing; 64% of surveyed respondents said they would trust a robot more than their manager.
- 36% of respondents believe machines are better than managers at delivering unbiased information.
- Aside from data security and privacy concerns, the most significant barriers to greater adoption of AI technology are concerns regarding its complexity. Security concerns prevent 71% from using AI in the workplace.

On this basis, now follows a subsection focused on the advantages of AI in HRM.

3.1 Advantages of Artificial Intelligence in Human Resource Management

In today's increasingly competitive environment, data utilization has increasingly become essential in any organization's strategic planning and everyday functioning. Data analytics are essential to HRM strategic decisions, influencing employee recruitment, selection, development, motivation, and many other areas. AI systems can quickly identify patterns and recognize correlations that would be difficult and time-consuming for ordinary people to discover. This predictive intelligence enables HR departments to implement a more proactive and strategic approach in their planning and programs. Additionally, AI technology empowers HR managers and leaders with a tool to effectively attract, retain, and motivate skilled employees, thereby contributing to the overall growth and success of the organization (Ravipolu, 2017).

The primary advantages of using AI technology in HRM functions include improved employee selection, enhanced integration of analytics, and the implementation of time- and cost-saving automation systems. This enables the development of a fair, free, and unbiased decision-making system in recruiting (Ravipolu, 2017). AI is crucial in performing many HRM functions, benefiting managers, employees, and organizations. HRM professionals have benefited from implementing this technology by reducing huge workloads and administrative and routine tasks. Because of this, both time and money have been saved (Ghosh et al., 2023).

Ravipolu (2017) states the following benefits of using AI in HRM functions:

- **Customized training and development:** Employees undergo training to develop and enhance their performance and productivity. Employees need personalized training to enhance their talents, which must be regularly updated. In light of this, the AI tool provides personalized training for individual

employees, improving their performance and promoting their career development and growth.
- **Orientation of new employees:** Orientation plays an integral part in introducing new employees to the organizational culture. Additionally, it helps familiarize employees with the company's policies and processes. However, employers dislike spending time on orientation. AI helps with employee orientation and provides a platform for resolving questions or uncertainties.
- **Saving time:** Using AI tools to perform repetitive tasks saves time. Recruiters, for instance, have to spend much time reviewing candidates' resumes. That is also a repetitive task. Therefore, AI enables the optimization of recruiters' time.
- **Quality hiring:** AI enhances the quality of new employees by providing recruiters with vast amounts of data and unbiased filtering and screening processes.
- **Reducing turnover:** AI consistently provides employees with up-to-date data and responses to their questions. This leads to employee satisfaction and fosters their engaged participation. Furthermore, it contributes to lower turnover as satisfied employees remain committed to working for the organization.
- **Productive workforce:** As a consequence of AI, hiring is qualitative. Furthermore, AI also contributes to employee development and training. This results in improved efficiency and productivity of the employees.
- **No bias in recruitment:** The absence of individuals participating in the recruitment process ensures that screening and selection are increasingly free of any potential bias.
- **Quality candidates:** AI aids in comprehending candidates' qualities, competencies, abilities, and knowledge; only the most qualified are hired.

After discussing the advantages of AI in HRM, a subsection focused on its disadvantages and issues follows.

3.2 Disadvantages and Issues of Artificial Intelligence in Human Resource Management

Although AI holds great potential to revolutionize the world, it seems as we are not ready to entirely embrace it. On the one hand, it is challenging for many businesses to collect data in order to properly train a high-quality AI algorithm; on the other hand, the lack of transparency, potential harm, and unspecified social responsibilities when AI malfunctions have led to a lack of trust in many applications. Similar to any novel intervention, AI offers both opportunities and issues (Miller, 2019).

Despite the potential application of AI techniques in various HRM functions (e.g., recruitment, training, analysis, and retention), most organizations have yet to adopt these systems primarily because of related expenses or costs (Matsa & Gullamajji, 2019).

The use of AI for evaluating and predicting human behavior carries an inherent risk due to underlying bias in the data. Using AI to automate the targeting of online job advertising has been argued to present many discriminatory issues. These issues are primarily connected to the biases that have been present in the selection procedures that human recruiters carried out in the past. Considering this, it is vital to modify the data preparation process and specifically adapt the AI-based algorithms used for selection processes to prevent biases of this kind (Dalenberg, 2018).

The application of AI in HRM presents multiple issues that organizations may encounter while using it (Ghosh et al., 2023):

- An important issue arises from the incorrect and improper implementation of AI approaches into the process and a lack of trust in the system. HR professionals must have proper knowledge and adaptability regarding AI technology to implement AI for HRM functions within the organization. If HRM professionals are not experienced and skilled in applying AI software and other tools in their approach, adopting these technologies could result in crucial failure.
- Another issue that arises during the implementation of AI systems in HRM functions relates to the mindset of both employers and employees. The belief that AI systems will replace the function of HRM in organizations contributes to HRM professionals having a negative mindset. As a result, they fail to recognize the interdependence of human resources and intelligent machines pursuing better outcomes and increased productivity. The delayed growth of a positive mindset among professionals may be due to their poor understanding and knowledge of modern technology's application and operation process. As a result, the adoption of AI may be significantly delayed.
- A severe issue also results from eliminating any bias and prejudice. While the AI system is supposed to be unbiased, it examines past data for learning and provides insights afterward; past information may present a particular trend in unconscious bias. The case of unconscious bias has been clarified as follows: During data analysis, an AI system might discover a recent increase in attrition or conflict among e.g. male employees. Consequently, the AI system may develop a bias against men and, in recruitment, choose to review a resume pool only of female candidates. As such, in the context of HRM

functions, the utilization of AI techniques might also be prone to errors and biases, just like humans. Furthermore, it is argued that an AI system cannot be unbiased as there are instances where a candidate's suitability for a job is determined not just by their work experience, skills, and knowledge but also by their enthusiasm or attitude toward the job in question. It is possible that the AI system will not recognize a candidate's motivation and attitude and will rely only on his or her qualifications and experiences, leaving VCEN aside too much.
- Protecting privacy and secrecy is an important issue and obstacle within the AI system. The HR information is confidential and must be maintained highly securely.

Having covered the fundamental principles of AI and HRM, we will focus on the intricacies of team dynamics, due to AI application.

3.3 Integration of Artificial Intelligence into Various Functions of Human Resource Management

HR(M) professionals are moving toward the digital revolution by optimizing HR operations through AI while employing other techniques, such as big data analysis and cloud management (Ghosh et al., 2023).

There is a notable imbalance in the level of attention given to implementing AI throughout different HR functions. There are multiple explanations for this occurrence. For example, the use of AI systems in talent acquisition may be increasingly prevalent due to its reputation for being a time-consuming and repetitive process and its sensitivity to human bias. Furthermore, the data accessible for AI systems encompasses both existing and potential candidates, resulting in a significantly larger quantity of data (Eubanks, 2022).

Recruitment

The recruitment process entails extensive tasks, including advertising job openings, assessing resumes of suitable candidates, conducting preliminary and final interviews, and ultimately selecting the most qualified candidates. This process is monotonous and time-consuming, burdening HR personnel and recruiters with overwhelming tasks (Ghosh et al., 2023). Selecting the most qualified and remarkable candidate for a specific job position is critical and valuable to any organization's success. Therefore, all organizations must work carefully to identify and select the most suitable, highly talented individuals with the desired work potential (Raub, 2018). We would add to talent the ambition, habit of working hard and with commitment, including permanent learning and interdisciplinary teamwork. (More in Šarotar-Žižek, Mulej et. al., 2023 a, b, c.) Organizations can use AI technology to simplify and optimize the recruitment process. This involves using robots, machines, AI software, and similar technologies to assist with candidates' screening, selection, and talent evaluation (Ghosh et al., 2023). Issues of VCEN seem to be rarely covered in literature we detected.

Recently, HR departments have increasingly adopted AI due to its proven effectiveness and importance in recruitment. AI technology benefits recruitment due to its ability to assess available options very quickly and analyze massive amounts of data (Agrawal et al., 2018). Additionally, the utilization of AI in candidate interviews has proven its advantageous nature, as software-driven (AI) interviewers do not include any emotional elements when evaluating the personal, mental, physical, or other external characteristics of candidates, as human interviewers might do. It is frequently seen that during in-person interviews and human-administered evaluations, interviewers tend to make irrational or unscientific decisions regarding candidate selection based on their personal preferences, opinions, and emotions. As a result, the selection process is often somewhat biased, and the opportunity to select the most qualified, skilled, or experienced candidate is gone (Raub, 2018).

The utilization of AI technology in recruiting processes offers significant advantages for business (and other) organizations. It saves time and effectively reduces costs, enhancing cost-effectiveness (Upadhyay & Khandelwal, 2018). Modern software (AI) for recruitment offers numerous advantages beyond the mere automation of administrative or work-related tasks. In addition, the AI system also gives HR professionals more detailed information quickly, reducing operating expenses (Leong, 2018). AI technologies are frequently used to automate the process of screening resumes. Namely, these tools offer recommendations to evaluate potential employees' experience, talents, and qualifications, providing comprehensive insights that assist when making recruiting decisions (Ravipolu, 2017).

Ravipolu (2017) suggests the following use of AI in the recruitment process:

- **Initial screening of candidates:** HR managers use AI tools and chat boxes to conduct the initial candidate screenings. These tools are often used for questioning candidates, and HR professionals then evaluate the responses to gain insight into the diverse attributes of the candidates.
- **Involvement of candidates:** Very often, most job applicants do not receive a response from the organization. The AI tool assists in delivering personalized messages and information to each candidate.
- **Re-involvement of candidates:** The vast majority of candidate records are often lost once the vacancy's requirements are achieved. Nevertheless, the

candidate's record is utilized and constantly updated. The candidate's extra qualifications or roles are also updated, with AI, especially with SRAI.

- **Post-offer acceptance:** Once a candidate is selected, he or she is on notice. If there is no contact between the employee and the employer during this notice period, the employee may leave the organization and seek employment elsewhere. AI can help in resolving this issue. With consistent communication, AI helps maintain the relationship with employees. This saves a substantial amount of time and money that would have been spent on recruiting new employees while also helping in employee retention.
- **Talent mapping:** AI aids in understanding the needs and skills of candidates. This helps the recruiters in career planning and finding suitable job positions for the candidates.
- **Let us add VCEN mapping**: It seems to be found less crucial than it is covered in e.g., ISO 26000 that quotes seven attributes: accountability, transparency, ethics, respect for stakeholders, for the rule of law, for international norms and for human rights (ISO 26000 by ISO, 2010).

The recruitment process assesses applicants' fit for a specific position by evaluating their educational qualifications, experience, and other information in their resume. HR professionals universally agree that this process remains or tends to be the most time-consuming and tedious activity linked to recruitment procedures. The traditional approach required significant effort and time investment, making it cost-intensive. Namely, AI-powered software has enabled organizations to identify the most suitable candidates by analyzing the information provided. AI screens and selects the best candidate from a pool of candidates for a specific job position (Ghosh et al., 2023).

Talent Acquisition

The task of talent acquisition entails implementing integrated systems and strategies specifically designed to enhance the recruitment process by identifying and selecting talented etc., see above) and skilled individuals. Furthermore, it also involves retaining and developing these employees' skills to meet current and future organizational requirements. Most employers and HR professionals dedicate a significant amount of their productive time to this process, which includes attracting highly skilled individuals, organizing resumes, tracking and assessing applications, arranging and conducting interviews, and providing information to candidates regarding their position and status. With AI techniques implemented to perform the above tasks, this vast, time-consuming work has been drastically reduced. Namely, talent acquisition (AI) software can quickly scan, analyze, and evaluate candidates' qualities and eliminate undeserving applicants. Implementing AI reduces an estimated 75% of the labor required for this recruitment process (Ghosh et al., 2023).

Training and Development of Employees

In today's constantly evolving technological environment, employees must be aware of the patterns behind occurring changes and prioritize learning and developing their professional skills. Every essential employee must receive an equal opportunity and the necessary resources to participate in periodic or repetitive training programs that enhance their job-related knowledge, skills, and performance (Ghosh et al., 2023). VCEN are impacted, too, most often indirectly (More in Šarotar-Žižek and Mulej, ed., 2023).

Implementing AI technologies into HRM functions has not only enhanced the content, course, schedule, and delivery methods of employee training, but it has also significantly contributed to the motivation of employees, thereby increasing their overall engagement and participation in such initiatives. Furthermore, compared with traditional methods, AI-driven training programs have shown greater effectiveness in enhancing employees' knowledge and skills, leading to increased performance. AI techniques and software have transformed employee training by providing online courses, digital classrooms, video conferencing, virtual meetings, seminars, and machine learning processes. These developments have significantly reduced the workload and responsibilities of HRM professionals compared to the complex traditional training methods. AI has successfully addressed many common challenges for HRM professionals in organizing and delivering employee training. It has enabled them to save more time and effort by offering a more efficient method of gaining knowledge and skills, thus promoting the development of employees' skills (Ghosh et al., 2023).

How and for what purposes the skills are attained, applied and controlled, depends on VCEN, of course.

Implementing AI technology ensures that every employee receives a personalized training experience. Besides that, AI also determines its position and level based on the results of periodic evaluations conducted at the end of each training program. Another important benefit of integrating AI techniques into an organization's training and development programs is that it speeds up and simplifies career tracking of the employees through collecting and aggregating data on each employee from the employer, as well as feedback forms and appraisals submitted by the employees themselves. Through data analysis, the AI technique aids in creating career paths and goals that are achievable yet challenging for each employee. AI-driven training and development programs assist employees during the training sessions and provide them with real-time data and information that enables them to make adequate suggestions in real time, even during the training session. Furthermore, the use of AI linked with an online internal learning system supports the development of employees within an organization by providing them with insights into the pros and cons of the machinery they will be in charge of operating effectively after the training session. Developing interactive training modules and plans ensures employees know crucial technical aspects and safety protocols (Schermerhorn et al., 2012).

AI promotes proper and quick decisions. AI can predict and inform relevant authorities on areas where employees need help and guidance by analyzing employee data. Moreover, AI algorithms may assist in creating and organizing learning programs to meet the diverse needs of employees with varying work backgrounds, cultural origins, personalities, educational qualifications, experiences, hobbies, and behavior patterns. Likewise, different training and learning experiences must be adjusted to suit a particular employee. This is beneficial since not all training programs are suitable for all employees, as some employees might require greater emphasis on specific areas than others, and vice versa. Therefore, it is quite logical, useful, and both time and cost-efficient to offer individualized training for each employee; HR managers are not required to manage multiple training programs, which would be a waste of resources that could be applied to other work and projects (Bhatia, 2018).

Employee Performance and Management

The process of performance evaluation is an essential task for HRM professionals. If conducted properly, this evaluation might reveal the company's overall position, as well as the probability and potential for achieving the set and desired goals of the organization. When evaluating the performance of employees and managers of a company using traditional or orthodox methods, the process often turns out to be time-consuming and lengthy, with unsatisfactory results. This may be because the HRM professionals assigned to the task were/are unable to gather all the necessary information or input related to a large number of employees in various positions within the organization. Namely, it was noticed that performance evaluations were often incomplete and incorrect due to insufficient information about all employees and the managers' negligence, unwillingness, and limited time and energy to process and analyze a large amount of information about numerous employees working in different departments of large organizations (Ghosh et al., 2023). As a result, significant employee contributions were either left out or inadequately collected, resulting in inaccurate performance evaluations. This might result in employee demotivation and significantly reduce their productivity. Therefore, by implementing AI software, it becomes easier and less difficult to gather and analyze data and information and do real-time performance evaluations, facilitating responsiveness and potential interactions with employees. Besides that, implementing AI performance evaluation systems may result in the exclusion of managers' subjective opinions and attitudes regarding their employees. Therefore, prevalent human biases related to gender, race, religion, place, and other factors of VCEN may be eliminated from the evaluation process (Pawar, 2019).

Now that we have explored AI, HRM, and their connection, our attention turns to socially responsible AI.

4. SOCIALLY RESPONSIBLE USE OF ARTIFICIAL INTELLIGENCE

Socially responsible artificial intelligence (SRAI) is a new concept that incorporates three interrelated concepts vital to sustainability: CSR[1], ESG, and UN sustainable development goals (SDGs) (Chang & Ke, 2023).

AI's related social responsibility encompasses efforts to tackle both technological and societal challenges. It refers to a process driven by human values/VCEN and includes inclusiveness, fairness, transparency, accountability, reliability and safety, privacy and security, and other values as the guiding principles; developing socially responsible AI Algorithms is the method. The primary goal is to meet the social expectations of generating shared value, which means improving AI's abilities and the benefits it provides to society (Cheng et al., 2021). The term "socially" is used to emphasize the societal perspective of AI responsibility, as opposed to the individual view of AI responsibility (Cheng & Liu, 2023).

Understanding SRAI is particularly crucial in HRM because AI-enabled PA (people analytics) is a high-stakes application of AI whose results could change people's lives (Rudin, 2019). PA is an emerging HRM innovation emphasizing decision science based on evidence and facts rather than gut instincts (Ghatak, 2022). It has been widely used and researched in HRM to help organizations with employee sentiment analysis, career planning, expertise estimation, performance management and evaluation, job analysis, applicant sorting, turnover prediction and retention, competence analytics, onboarding and training, coaching, and many other HRM activities (Chang & Ke, 2023).

Nevertheless, despite the increasing needs, SRAI is relatively new and poorly understood in HRM. Furthermore, the current body of literature is overly optimistic and fails to address major risks and ethical issues (Tursunbayeva et al., 2018; Chowdhury et al., 2022). Given the uncertainties of AI's implications, SRAI is essential in directing and ensuring responsible AI creation and implementations. This involves maximizing AI applications' benefits, promoting the greater good, and reducing the social effects and potential damages. Therefore, to truly understand SRAI, it is vital first to understand sustainability and its synergy with AI (Chang & Ke, 2023).

SRAI needs to be defined in a manner that encompasses the full spectrum of responsibilities (Cheng et al., 2021). In the context of Caroll's Pyramid of CSR (1991) Cheng et al., (2021), suggest four kinds of social responsibilities that constitute SRAI, as shown in Figure 3.

Figure 3. The pyramid of SRAI; adapted from the Pyramid of CSR (Adapted from Cheng et al., (2021); Carroll (1991))

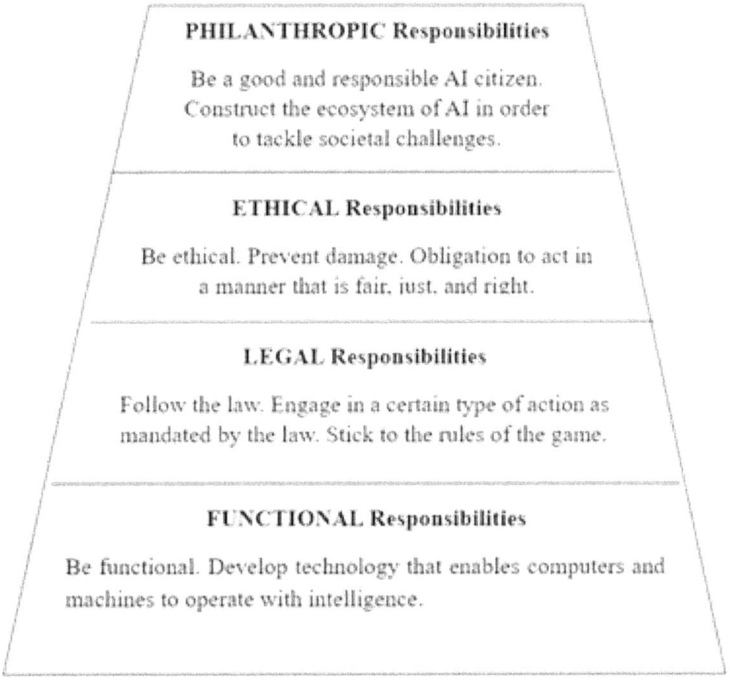

PHILANTHROPIC Responsibilities
Be a good and responsible AI citizen. Construct the ecosystem of AI in order to tackle societal challenges.

ETHICAL Responsibilities
Be ethical. Prevent damage. Obligation to act in a manner that is fair, just, and right.

LEGAL Responsibilities
Follow the law. Engage in a certain type of action as mandated by the law. Stick to the rules of the game.

FUNCTIONAL Responsibilities
Be functional. Develop technology that enables computers and machines to operate with intelligence.

Adapted from: Cheng et al., (2021); Carroll (1991)

N.B. to Figure 3: Legal and other relations are not socially responsible, if abuse is allowed.

While these responsibilities have always been there, until recently, the functional ones were prioritized. Every category of responsibility necessitates careful consideration. The pyramid represents all four aspects of SRAI, starting with the fundamental idea that the functional competence of AI provides the basis for everything else. Although these four aspects appear as separate concepts, they are not mutually exclusive. They are in a constant but dynamic state of tension with one another (Cheng et al., 2021).

In comparison to other similar concepts, SRAI holds a societal perspective on AI. It encompasses existing concepts and takes into account the fundamental responsibilities of AI systems, which include their ethical, legal, and functional requirements, as well as their philanthropic responsibilities to benefit society. Although SRAI is closely intertwined with other similar other concepts, there are significant differences between them (Cheng & Liu, 2023). The latter is presented in Table 2.

Table 2. Definitions of concepts related to SRAI

Concepts	Definitions
Ethical AI	AI systems that follow fair, right, and just principles; prevent any harm (Cheng & Liu, 2023).
Trustworthy AI	AI systems that are technically robust, ethically compliant, and lawful. The establishment of trust is essential in the development, deployment, and utilization of AI (Thiebes et al., 2021).
Fair AI	AI systems that are free of any bias or favoritism toward a group or an individual based on their inherent or acquired traits (Mehrabi et al., 2021).
Safe AI	AI systems that are used in a manner that does not cause any harm to humanity (Feige, 2019).
Dependable AI	AI systems that prioritize security, verifiability, reliability, and explainability (Singh et al., 2021).
Human-centered AI	AI systems that are providing an effective human-robot experience while continuously improving due to human input (Cognizant, 2024).

Adapted from: Cheng & Liu (2023).

Furthermore, Chang & Ke (2023) offered a comprehensive framework for SRAI, which takes the shape of an inverted pyramid model with five hierarchical levels: economic, legal, ethical, philanthropic, and environmental, as seen in Figure 4.

The social aspects of Caroll's CSR pyramid are represented by the first four layers, starting from the bottom of the framework. In addition to these four levels, authors incorporate environmental respect, drawing inspiration from the environmental concerns of ESG and SDGs that CSR does not adequately address. By incorporating the fifth layer, this framework expands beyond Caroll's CSR model, thereby offering a more comprehensive approach encompassing all crucial aspects of CSR, ESG, and SDGs. Furthermore, the inverted pyramid structure represents the five layers comprising the stakeholders' scope and impacts, as well as the outcomes concerning time. For instance, similar to Caroll's CSR model, the economic layer (SDG: 1, 2, 3, 8, 9) is the fundamental level that encompasses a smaller number of stakeholders and focuses only on the lower level of stakeholders' concerns and requirements. Fulfilling economic responsibility is considered essential, but it only offers short-term benefits and is inadequate for ensuring the sustainability of an organization.

Socially Responsible Application of Artificial Intelligence

Figure 4. The inverted pyramid of SRAI framework (Chang & Ke, 2023)

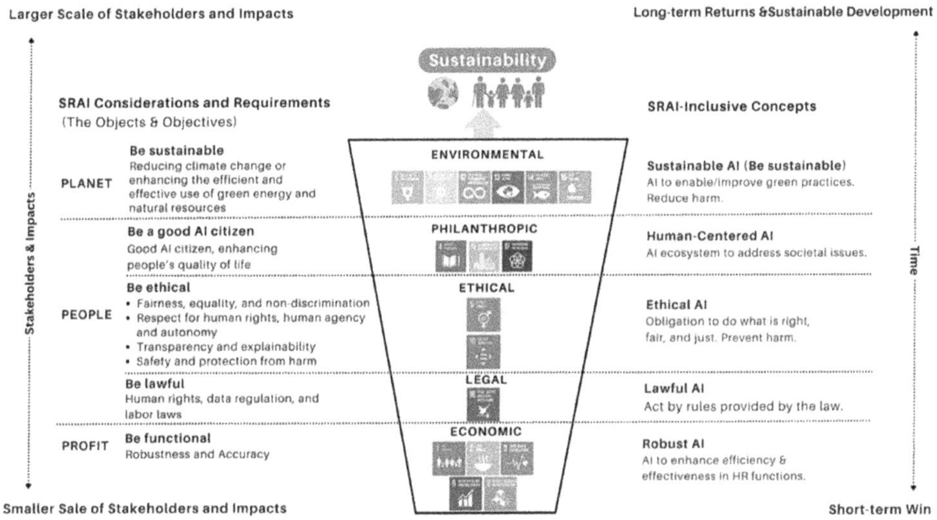

CSR, ESG, and SDGs collectively align effortlessly to achieve sustainability due to their shared economic, social, and environmental elements. Moreover, through the integration of AI into the CSR/ESG/SDGs profile, we establish the SRAI framework, which encompasses five essential needs and considerations for sustainability: 5) environmental, 1) economic, 2) legal, 3) ethical, and 4) philanthropic. This inclusive definition of SRAI guides our evaluation, ensuring an in-depth analysis of AI's influence on social responsibility and sustainable development. SRAI, in particular, is a system accountable for the effects of its choices and actions on society and the environment. It also contributes to sustainable development by satisfying stakeholders' economic, legal, ethical, philanthropic, and environmental considerations and demands (Chang & Ke, 2023).

In light of this Chang & Ke (2023) conceptualized SRAI by adapting AI to the proposed framework with five fundamental aspects:

- *The economic aspect* refers to the functional responsibility of AI to maximize operational efficiency and effectiveness to maximize gains in terms of both productivity and profitability (Jobin et al., 2019).
- *The philanthropic aspect* is the anticipation that AI will act as a responsible member of society, actively contributing to resolving social issues while enhancing people's quality of life (Jobin et al., 2019). In particular, AI systems need to function in a manner that is in line with society's philanthropic and

charitable aspirations to improve individuals' quality of life (Cheng et al., 2021).
- *The legal aspect* necessitates that AI functions in accordance with legal laws and regulations (Jobin et al., 2019), which establish the parameters of acceptable and unacceptable behavior in our society (Cheng et al., 2021).

    The ethical aspect encompasses upholding human rights, ensuring transparency, promoting non-discrimination and fairness, safeguarding privacy and safety, and preventing damage (non-maleficence) (Jobin et al., 2019). At its most fundamental level, ethical obligations entail the duty to act morally correctly, just, and fair while also taking measures to avoid or minimize any negative impacts on stakeholders, such as users and the environment. In order to meet its ethical responsibilities, AI systems must operate in a manner that aligns with social expectations and ethical standards. These principles should not be compromised to fulfill the functional duties of AI (Cheng et al., 2021).
- *The environmental aspect* acknowledges the vital function that AI can have in tackling environmental issues, such as climate change, and helping or enhancing eco-friendly practices, such as optimizing the consumption of energy, facilitating the establishment and supervision of renewable energy sources, assisting in sustainable agriculture, enhancing waste management, and achieving smart modes of transportation (Sætra, 2021).

HRM professionals are vital in promoting ethical and socially responsible organizational practices. They can benefit by gaining a better understanding of SRAI and implementing activities empowered by it (Chang & Ke, 2023). To achieve this, HRM professionals must undergo AI competence training to fully understand the functioning of AI and familiarize themselves with relevant rules and SRAI concerns. This will enable them to carefully execute, supervise, and evaluate AI-enabled HRM activities (Stahl et al., 2023). These activities involve promoting ethical hiring, developing objective performance evaluation algorithms, providing training, monitoring data governance, analyzing organizational factors, evaluating ethical and socially responsible practices, and ultimately contributing to organizations' long-term growth, reputation, and success (Chang & Ke, 2023). These requirements direct humans VCEN and hence select information and related actions (See more in Šarotar-Žižek and Mulej, ed., 2023 a, b, c).

In order to operationalize the SRAI framework within the realm of HRM, Table 3 provides an overview of SRAI and recommendations for HRM professionals to address the identified SRAI requirements and considerations of various stakeholders. These recommendations direct HRM professionals as they carry out SRAI-enabled activities that promote sustainability (Chang & Ke, 2023).

Table 3. SRAI considerations and recommendations for HRM activities[2]

SRAI Considerations	AI-Enabled HRM Activities	Recommendations for HRM Professionals
1. **Economic level**: precision and robustness. 2. **Legal level**: human rights, labor laws, and data regulation.	**Learning and development**: applying AI to adapt employee training based on individual learning preferences, performance, and career goals (Naim, 2023).	• To ensure precise and reliable personalized training recommendations, reliable and validated AI models and algorithms must be prioritized during learning and development. • Regularly audit AI algorithms to identify and remove biases that could lead to unequal treatment or bias in personalized training recommendations. • Allow employees to control their educational path and ensure AI-generated recommendations don't impact their autonomy. • Use explainable AI models instead of black-box AI and provide employees explanations of AI-driven learning recommendations. This will increase process transparency and system confidence.
	Coaching: by using AI for evaluating an employee's strengths, weaknesses, and ambitions, personalized coaching plans can be created that adapt to their specific needs and objectives (Graßmann & Schermuly, 2021).	• Validated AI algorithms and models may precisely assess employee strengths, errors, and goals, setting the basis for personalized coaching plans. • Regularly assess coaching AI algorithms for biases to provide fair and unbiased coaching for all employees. • Offer staff detailed explanations of AI-generated coaching plans to build trust and transparency. • Employee data used in AI-driven coaching must be easily accessible, diverse, and properly represented to avoid biased or distorted coaching recommendations.

continued on following page

Table 3. Continued

SRAI Considerations	AI-Enabled HRM Activities	Recommendations for HRM Professionals
	Knowledge management: implementing AI to capture, analyze, and optimize organizational knowledge in order to improve productivity and decision-making (Jarrahi et al., 2023).	• Accurate and resilient AI algorithms and models should be used to efficiently capture, analyze, and optimize organizational data. • Give employees comprehensive explanations of how AI algorithms manage values and knowledge. This will help to develop their transparency and understanding of AI-driven insights. • Ensure organizational knowledge is accessible and properly shown in the AI system; examine various perspectives to eliminate bias in values and knowledge management outcomes. • Secure the AI system and organizational knowledge with thorough cybersecurity protocols, thereby ensuring the preservation of sensitive data's confidentiality.
3. **Ethical level**: data security, respect for human rights, equality, and non-discrimination, data privacy, transparency and explainability, data bias and misinformation, human agency and autonomy, fairness, data accessibility and representation. 4. **Philanthropic level**: improving the overall well-being of individuals. 5. **Environmental level**: mitigating climate change, optimizing the utilization of renewable energy and natural resources in an efficient and effective manner.	*Career development (CD)* **Career planning**: using AI to do an analysis of an employee's values, skills, interests, experiences, and career objectives in order to make recommendations for personalized career paths that are aligned with the employee's aspirations and strengths (Guo et al., 2022).	• Use precise and robust AI algorithms to analyze and recommend career planning, guiding individuals to suitable careers. • Respecting employees' goals and allowing them control over their career choices empowers them and gives them autonomy. • Employ strong data privacy procedures to protect employee career planning data and comply with data privacy laws.

continued on following page

Socially Responsible Application of Artificial Intelligence

Table 3. Continued

SRAI Considerations	AI-Enabled HRM Activities	Recommendations for HRM Professionals
	Employee wellbeing: applying AI to keep track of employee health data and provide recommendations for personalized wellness programs and initiatives (Jones et al., 2019)	• Human rights, data protection, and labor laws must be followed when collecting and evaluating wellness data to protect employee rights and privacy. • Allow employees to manage their wellbeing data and participate in wellness initiatives while protecting their privacy. • Use explainable AI to show employees how AI algorithms examine wellness data and make personalized recommendations to foster trust. • Implement strict cybersecurity and data protection safeguards to maintain confidentiality and security, and protect the AI system and employee wellness data from unauthorized access. Ethically and responsibly apply AI into employee wellbeing initiatives to improve employees' quality of life and wellbeing.
	Organization development (OD) **Performance and change management**: to improve performance and make positive change, AI may be used to discover performance gaps and provide insights that are driven by data (Buck & Morrow, 2018). N.B.: Change management includes invention and innovation management. • Optimize performance management with AI algorithms that accurately identify performance gaps and provide data-driven insights.	• Improve employee understanding of performance evaluations by explaining how AI algorithms analyze performance data and offer insights. • Regularly evaluating AI algorithms minimizes biases and ensures that performance insights and recommendations are fair and unbiased for all employees. • Use AI ethically and responsibly for performance management. AI will help employees grow and develop, improving their quality of life.

continued on following page

Table 3. Continued

SRAI Considerations	AI-Enabled HRM Activities	Recommendations for HRM Professionals
	Employee engagement and retention: by using sentiment analysis, organizations can gain insights into the levels of employee engagement and satisfaction, as well as identify contributing factors to employee turnover and attrition (Speer et al., 2019).	• To assess employee engagement and experiences, use AI-powered sentiment analysis models that are accurate and reliable. • Ensure that sentiment analysis follows human rights, data protection, and labor laws to protect employees' rights and privacy throughout the process. • Assess sentiment analysis algorithms regularly to identify and remove biases, ensuring fair and unbiased employee sentiment analysis. • Employees should be able to provide feedback during sentiment analysis and opt out to protect their autonomy, and privacy. • Secure employee sentiment data with strict data privacy measures that comply with privacy legislation. • Explain rationale and process behind sentiment analysis to employees, to ensure they are aware of how their opinions are used.
	Succession planning: employing AI to identify high-potential employees for the purpose of succession planning and the implementation of leadership development programs (Tambe, Cappelli, & Yakubovich, 2019).	• Use AI algorithms and models that accurately identify high-potential employees for succession planning, using an array of data sources to make the best judgments. • Apply methods to identify and fix AI algorithm biases to ensure high-potential employees identification accuracy and dependability. • Keep employees informed about the purpose and procedures of AI-driven succession planning and high-potential employees and their identification. • Engage employees in succession planning by letting them volunteer and express their preferences and goals.

continued on following page

Table 3. Continued

SRAI Considerations	AI-Enabled HRM Activities	Recommendations for HRM Professionals
	Workforce planning: employing AI and competency analytics to predict upcoming workforce demands via analyzing data patterns, while empowering HR experts to strategize for talent requirements and address skill gaps (Wingard, 2019).	• HRM professionals must make informed talent planning decisions. AI algorithms and competency analytics tools that accurately predict workforce needs might achieve this. • Detecting and correcting AI algorithm and competency analytics biases is essential. Workforce projections will be accurate and reliable. • The use of AI and competency analytics in workforce planning should be explained clearly to employees and stakeholders to ensure transparency in decision-making.

Adapted from: Chang & Ke (2023).

It is important to keep in mind that the SRAI framework could undergo adjustments in the future, as the technology and its environment continue to develop (Chang & Ke, 2023).

In the next subsection we will explore core subjects of AI in the scope of social responsibility.

4.1 Core Subject of Social Responsibility in Artificial Intelligence Based on ISO 26000

ISO 26000 (2010) defines social responsibility as the "responsibility of an organization for the impacts of its decisions and activities on society and the environment, through transparent and ethical behavior that contributes to sustainable development, including health and the welfare of society; takes into account the expectations of stakeholders; is in compliance with applicable law and consistent with international norms of behavior; and is integrated throughout the organization and practiced in its relationships." More about social responsibility and its relation to HRM can be found in section 5.

According to ISO 26000 an organization should address the following core subjects to define the scope of its social responsibility, identify relevant issues, and establish priorities (Zhao, 2018):

- organizational governance,
- human rights,
- labor practices,
- the environment,

- fair operating practices,
- consumer issues, and
- community involvement and development.

1) Organizational Governance

Organizational governance is the most important factor in enabling AI to take responsibility for the consequences of its decisions and activities, as well as to integrate social responsibility throughout the organization and its relationships. Therefore, effective governance should integrate social responsibility principles into the decision making and implementation of AI. Leadership is also essential for effective governance of an organization. Not only does this hold true for the process of decision making, but also for the employee's motivation to engage in social responsibility practices and to incorporate social responsibility into the organizational culture (Zhao, 2018). With this in mind, organizations should consider three main aspects (Zhao, 2018):

- First, organizations should fully integrate SRAI concepts and strategies into their development strategy, organizational culture, and production management.
- Secondly, organizations should respect laws and regulations, stakeholders' interests, and international norms of behavior, as well as adhere to principles such as ethical behavior, transparency, and accountability.
- Thirdly, implementation of an all-dimensional social responsibility management is essential; this system should aim to foster the effective integration of CSR in achieving both social and economic benefits.

2) Human Rights

AI intelligence systems and robots increasingly resemble humans in both their internal mechanisms and external appearance (Zhao, 2021). Moreover, Hanson Robotics' robot Sofia was granted citizenship by Saudi Arabia in October 2017, marking the first time in history that a robot has been granted citizenship. While making the machines aware may be one of the key objectives of AI research, this raises significant concerns regarding AI (Zhao, 2018). This has also raised many questions regarding human rights (Zhao, 2021). Therefore, it is important to know that while AI can help us, we should not rely entirely on AI, and that we as humans are entirely responsible for our destiny (Zhao, 2018).

With AI entering human society, it is imperative for it to adhere to the legal, moral, and other norms and values of society. Researchers, policy makers, and product consumers must take the responsibility for AI, as it is unlikely to fall back, once it has been developed. Integration of social responsibility into AI technology is crucial; AI must be SRAI (Zhao, 2021).

3) Labor Practices

In the era of increasing modern technologies, including AI, it is crucial for organizations to have employees with certain capabilities. These include critical thinking, information literacy, self-awareness and self-regulations, creativity, problem solving, civic responsibility, social participation, and lifelong learning (Zhao, 2018). Therefore, to ensure that employees can adapt to the new employment and work paradigm brought about by AI, it is imperative to enhance the education and training of employees in the field of AI in the future (Zhao, 2021).

AI expresses and simulates the wisdom of humans. It exceeds the speed and accuracy of human work, and it helps people in solving a variety of issues, including those that occur in hazardous work situations and extreme environments (Zhao, 2018). Generally speaking, AI has many benefits on labor practices. In addition to taking over low-skill, low-creative work tasks, it is reducing recruitment and training costs, and it can aid organizations in the selection of suitable candidates, and in avoiding the risk of false resumes. In addition, it can foster a more honest and fairer competitive environment for job seekers (Zhao, 2021). While all this presents a big advantage, we should not neglect potential negative impacts AI has in work environment. For instance, it is a known fact that smart machines and robots will eventually eliminate numerous jobs (Zhao, 2018).

4) The Environment

AI has the potential to improve environmental protection by for example enabling the more effective control of air pollution. However, at the same time the advancement of AI has resulted in an increasing amount of electronic waste. Electronic waste emits radiation into the environment and worsens it; with modern technology developing quickly, the waste produced by replacing related products will be even worse, causing environmental degradation (Zhao, 2019).

E-waste must be recycled in a manner that is both environmentally friendly and rational. Establishing professional institutions for the disposal of electronic waste is therefore essential to provide employees with the necessary safety precautions to separate hazardous materials and precious metals from electronic waste in a secure environment and to dispose of toxic materials in an appropriate manner. Furthermore, educating stakeholders, advertising, and establishing e-waste collection sites on the streets etc. are all critical steps in raising awareness of the hazards associated with the private disposal of e-waste (Zhao, 2021).

5) Fair Operating Practices

The degree of automation in AI decision-making is increasing by the day. AI will replace humans in increasingly important decision-making roles, and the question is: How can fairness and justice be ensured? How can we guarantee that the algorithm is free of discrimination and injustice? Modern machines are incapable of consciously imposing prejudices or experiencing emotions. Nevertheless, they accurately capture the real prejudices that exist within the database and society. Ensuring transparency, accountability, participation, and accuracy in the development of AI systems is essential (Zhao, 2021).

6) Consumer Issues

AI provides consumers with many advantages. For example, the merchant may recommend the next item based on the consumer's shopping history and interests. However, the utilization of AI systems for automating personal data, performing intelligent analysis, and making decisions can impact consumers' personal rights and interests, often without their knowledge. The question that arises is how to achieve a balance between the commercial utilization of data and the rights and interests of consumers (Zhao, 2021)?

7) Community Involvement and Development

In the digital age, we frequently encounter a conflict between the preservation of public safety and the safeguarding of personal privacy, causing a moral and ethical dilemma. For tech giants with access to vast amounts of information, finding a balance between personal and public responsibility poses a significant challenge (Zhao, 2021).

It is important to note that the Internet is not a realm outside the jurisdiction of the law. Business innovation, as well as other actions within the network community, must adhere to the rule of law (Zhao, 2021).

After delving into SRAI, we will transition to discussing sustainability and sustainable development, because SRAI may be crucial for them.

5. SUSTAINABILITY AND SUSTAINABLE DEVELOPMENT

Since its introduction by the Brundtland Commission of the United Nations in the 1980s, the concept of sustainability has lately received renewed impetus and grown to be seen as "a hot topic" in the corporate world. Today, the success of an organization is not simply measured by its short-term financial gains; rather, it is evaluated based on its anticipated achievements for future generations. Although there are several definitions of sustainability and sustainable development, the concept of sustainability is often referred to as the "three pillars." These three pillars of sustainability commonly encompass economic, social, and environmental outcomes and developments (Kramar, 2022).

Sustainable development ensures the preservation of resources without causing harm to the well-being of future generations, robbing them of good environmental conditions and failing to meet their fundamental needs (Brundtland, 1987). Hence, organizations' social responsibility functions as a tool for achieving sustainable development (Slapnik, Hrast, & Mulej, 2018).

Nowadays, the concept of sustainability in the realms of business and politics is frequently associated with SDGs (Brundtland, 1987). These goals were introduced in the UN's document Transforming our World: The 2030 Agenda for Sustainable Development (United Nations, 2015). The 2030 Agenda, much like the concept of social responsibility, balances the economic, social, and environmental dimensions of sustainable development by interweaving them into 17 interlinked and indivisible Sustainable Development Goals (SDGs) and emphasizing that every person, organization and nation, both developed and developing, is responsible for achieving them, in order to help humankind to survive in the current global social-economic crisis. Within the 2030 Agenda, global leaders cautioned about the many negative impacts of climate change, the depletion of natural resources, and the adverse effects arising from a deteriorating environment. Therefore, the 2030 Agenda emphasizes incorporating the environment into all policies and adhering to planetary boundaries as essential measures for achieving a sustainable future for humanity (Hrast, 2018). VCEN must adopt this fact.

Therefore, the Agenda for Sustainable Development until 2030 includes 17 general and 169 concrete interconnected and inseparable goals for sustainable development (Hrast, 2018). These goals are:

- Goal 1. End poverty in all its forms everywhere
- Goal 2. End hunger, achieve food security and improved nutrition, and promote sustainable agriculture
- Goal 3. Ensure healthy lives and promote well-being for all at all ages

- Goal 4. Ensure inclusive and equitable quality education and promote lifelong learning opportunities for all
- Goal 5. Achieve gender equality and empower all women and girls
- Goal 6. Ensure availability and sustainable management of water and sanitation for all
- Goal 7. Ensure access to affordable, reliable, sustainable, and modern energy for all
- Goal 8. Promote sustained, inclusive and sustainable economic growth, full and productive employment and decent work for all
- Goal 9. Build resilient infrastructure, promote inclusive and sustainable industrialization and foster innovation
- Goal 10. Reduce inequality within and among countries
- Goal 11. Make cities and human settlements inclusive, safe, resilient and sustainable
- Goal 12. Ensure sustainable consumption and production patterns
- Goal 13. Take urgent action to combat climate change and its impacts
- Goal 14. Conserve and sustainably use the oceans, seas and marine resources for sustainable development
- Goal 15. Protect, restore and promote sustainable use of terrestrial ecosystems, sustainably manage forests, combat desertification, and halt and reverse land degradation and halt biodiversity loss
- Goal 16. Promote peaceful and inclusive societies for sustainable development, provide access to justice for all and build effective, accountable and inclusive institutions at all levels
- Goal 17. Strengthen the means of implementation and revitalize the global partnership for sustainable development

When asked »Why is sustainable development so relevant now?« Slapnik, Hrast, & Mulej (2018) state that the escalating population growth, rapid urbanization, and consequential urban pollution necessitate an increasing need to preserve the natural environment. Industrial development is causing a scarcity of resources and increased demand for innovative technologies. Besides that, uncontrolled production and consumption are emerging, and it is challenging to meet the needs of a growing population. Furthermore, emissions, global warming, and climate change are causing human and animal health problems; deforestation and natural disasters (such as floods and droughts) are also rising, and we are confronted with migration and adaptation issues.

HRM practitioners can find many of these SDGs relevant, including goals for economic growth and decent work, gender equality, good health, and well-being, reducing inequalities, and developing partnerships to achieve these goals (Kramar, 2022).

Concepts of sustainability and human development on a global scale are thoroughly intertwined with the SDGs. While the SDGs are connected to various fundamental human rights, the framework is different. The SDGs involve a more comprehensive focus on what is commonly known as the five P's: people, planet, prosperity, peace, and partnership (Brundtland, 1987).

Now that we have discussed sustainability and sustainable development, we will focus on the connection between sustainability and AI; it may be crucial for the survival of humankind.

5.1 Sustainability and AI

AI-driven technologies require regulatory oversight to continue their rapid development. Failure to do so could affect transparency, safety, and ethical norms (Vinuesa et al., 2020). AI has both potential positive and negative impacts on sustainable development. Upon reviewing the relevant evidence, Vinuesa et al. (2020) conclude that AI may act as an enabler on 134 targets (79%) across all SDGs. This is mostly due to technology that can help overcome certain limitations. Nevertheless, AI development may harm 59 targets - 35% of all SDGs. To provide an overview of AI's general influence, the authors divided SDGs into three main categories based on sustainable development's three pillars: society, economy, and environment.

AI and Societal Outcomes

AI-based technologies can benefit sixty-seven targets (82%) within the Society group. Furthermore, AI may serve as an enabler for all the targets by supporting the provision of health, water, food, and energy services to the population (for example, in SDG 1 on no poverty, SDG 4 on quality education, SDG 6 on clean water, SDG 7 on clean and affordable energy, and SDG 11 on sustainable cities). Additionally, AI can serve as a foundation for low-carbon systems by supporting the development of circular economies and smart cities that effectively use their resources (Vinuesa et al., 2020).

Figure 5 shows documented evidence of AI's positive or negative impact achieving each of the targets from SDGs 1, 2, 3, 4, 5, 6, 7, 11, and 16. AI could facilitate or inhibit objectives outlined in green or orange. The lack of highlighting signifies the absence of identified evidence, but this does not necessarily mean no relationship (Vinuesa et al., 2020).

Figure 5. Assessment of the impact of AI on the SDGs within the society group (Vinuesa et al., 2020)

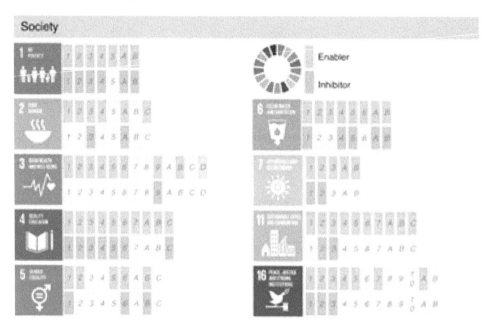

A rather small number of targets within the Society category, precisely 31 or 38%, are negatively impacted by AI. Nevertheless, consideration of them is crucial. Several of these relate to implementing AI-enabled technology improvements in countries with different cultural values and wealth (Vinuesa et al., 2020). While AI-enabled technology can accelerate progress toward the 2030 Agenda, it may also worsen inequalities that hinder the achievement of SDGs 1, 4, and 5. The duality can be seen in target 1.1 since AI may assist in identifying regions struggling with poverty and promote global actions through satellite imagery (Jean et al., 2016). In contrast, it may increase employment qualification for any job, which would increase inherent inequalities and hinder the achievement of this aim (Vinuesa et al., 2020; Nagano, 2018). Another significant limitation of AI-based developments is their traditional reliance on the demands and values of the nations (and organizations) where AI is developed. If AI and big data are used in regions with poor ethical scrutiny, transparency, and democratic control, it could encourage nationalism, hate towards minorities, and biased election outcomes (Helbing & Pournaras, 2015).

The term "big nudging" refers to using AI and big data to manipulate decisions by exploiting psychological weaknesses that threaten human rights, democracy, and social cohesiveness. Recently, AI has been used to develop citizen scores, which are used to control social behavior (Helbing et al., 2019). This score shows how AI misuse threatens human rights. One of its major issues is the inadequate disclosure of information to citizens that analyzed its potential impacts (Nagler et al., 2019). It is important to note that AI technology is not evenly distributed. For example, AI-enhanced agricultural equipment may not be affordable to small farmers, widening the gap between them and larger producers in developed countries. This imbalance affects Sustainable Development Goal 2, which aims to eliminate hunger (Wegren, 2018).

AI and Economic Outcomes

AI's technological benefits can also contribute positively to achieving several SDGs in the Economy groups. In light of this, AI has been shown to positively affect 42 out of 70% of the SDG targets. Figure 6 shows an assessment of documented evidence of AI's positive or negative impacts on achieving each of the targets from SDGs 8, 9, 10, 12, and 17 (Vinuesa et al., 2020).

Figure 6. Assessment of the impact of AI on the SDGs within the economy group (Vinuesa et al., 2020)

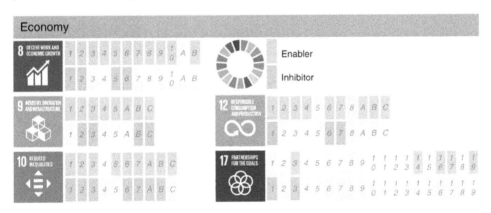

While the Economy group's relations are mainly positive, we can't neglect trade-offs. If future markets heavily rely on data analysis and these resources are not equally accessible in low- and middle-income nations, the SDG Economy group could widen the economic gap. This newly introduced inequality would significantly affect SDGs 8, which concerns decent work and economic growth, SDG 9, which concerns industry, innovation, and infrastructure, and SDG 10, which concerns the reduction of inequalities (Bissio, 2018; Brynjolfsson & McAfee, 2014). Besides that, AI can worsen inequalities within nations. Namely, AI can disproportionately reward the educated by replacing old jobs with more skill-intensive ones: US grad-

Socially Responsible Application of Artificial Intelligence

uate degree holders have seen their income increase by 25% since the mid-1970s, whereas the average high school dropout has seen their pay drop by 30% (Brynjolfsson & McAfee, 2014).

AI and Environmental Outcomes

The final set of SDGs, specifically those related to the environment, is analyzed in Figure 7.

This group identified 25 AI-enabled targets (93%). The ability of AI to analyze large, interconnected databases and develop collective environmental preservation strategies is beneficial. When examining SDG 13, a goal that focuses on climate action, AI advances could be beneficial in understanding climate change and predicting its potential consequences. Furthermore, AI will promote low-carbon energy systems with high renewable energy and energy efficiency, which is essential to addressing climate change (International Energy Agency, 2017; Vinuesa et al., 2016; World Economic Forum, 2018).

Figure 7. Assessment of the impact of AI on the SDGs within the environmental group (Vinuesa et al., 2020)

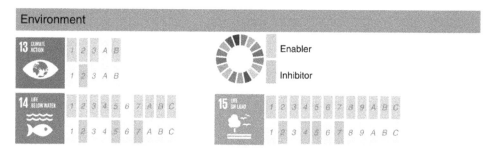

CSR and sustainable development are interconnected concepts that have three essential characteristics: responsibility, interdependence, and a holistic approach (ISO, 2010). Therefore, the next section focuses on Corporate social responsibility (CSR).

6. CORPORATE SOCIAL RESPONSIBILITY

CSR refers to the organization's commitment to improving the community's well-being through ethical business practices and the contribution of corporate resources. The term "community well-being" covers environmental issues and human conditions (McWilliams et al., 2006). Moreover, CSR refers to the organization's voluntary involvement in endeavors that contribute to achieving social good. Therefore, it may be argued that CSR is a voluntary dedication of an organization to promote the welfare of its employees and society as a whole while exercising discretion in conducting business (Bučiūnienė & Kazlauskaitė, 2012). Furthermore, World Business Council for Sustainable Development offered one of the most widely used definitions of CSR (WBCSD, 1999): CSR is the voluntary commitment of an organization to behave in a way that promotes economic development and increases the well-being of its employees, the local community, and society in general. CSR expresses VCEN and impacts skills and use of them. It is a non-technological innovation process, which can be supported by AI as SRAI.

Not all organizations recognize the significance of CSR. Various factors, including industry, geographic location, and government regulation, can impact a company's motivation to do the right thing. The International Organization for Standardization introduced the ISO 26000 standard to help guide organizations in a more socially conscious direction in order to address this issue (Robichaud, 2024). There was no official definition of CSR until ISO 26000 defined it as organization's responsibility to society and the environment through honest and ethical behavior that fosters sustainable development, including the health and well-being of society, takes into account stakeholder expectations, complies with applicable law, and follows international standards of behavior ((ISO, 2010; Zore, 2018).

CSR can be defined in both a narrow and broad sense. The broader definition covers firms' environmental and social responsibility, while the narrower definition entails responsibility towards stakeholders—customers, business partners, interest groups, colleagues, shareholders, etc. (Zore, 2018).

According to the definitions provided, CSR entails not only adhering to legal obligations but also investing "more" in human capital, the environment, and stakeholder relations; this goes beyond simple compliance. Taking social responsibilities beyond the bare minimum, for example, working conditions, training, and management-employee relations can significantly impact productivity. Namely, it opens up a way towards effectively managing change and balancing social development with enhanced competitiveness. The latter includes several strategies highly related to HRM, such as (European Comission, 2001):

- enhancing employability,
- reducing skill gaps,
- providing access to learning opportunities,
- promoting specialized programs that help unemployed individuals in filling skill gaps,
- prioritizing lifelong learning by focusing on the synergy between it and adaptability and through flexible working time management and job rotation,

- promoting all aspects of equal opportunities, and
- supporting the work-life balance, specifically by establishing a new benchmark for improved childcare provisions.

CSR should not be viewed as an obligation, expense, or some sort of charity work. Instead, it is a source of opportunity, innovation, and competitive advantage. Besides that, it is important to note that CSR does not diminish a company's financial performance; instead, it has the potential to enhance it. The concept entails the (r)evolution of the VCEN (values, culture, ethics, and norms) and the path towards a new socio-economic structure. This is why asserting it is challenging; it necessitates breadth and long-term standards of behavior (Zore, 2018).

Today, when acting socially responsibly seems to be the only viable option for the survival of everyone who live on this planet, research on SR has never been more relevant (Zore, 2018). Besides that, privilege issues, social justice concerns, and business operations' environmental and societal impacts are more important than ever. Consequently, CSR presently refers to a significantly broader set of issues than during its early stages (Verbin, 2020).

Responsibility towards society, i.e., towards people and the environment, or CSR, reflects the essence that, as humanity (should) emerge from the current global socio-economic crisis, it is about (1) interdependence and (2) a holistic approach rather than independence, dependence, and one-sidedness (ISO, 2010; Mulej et al., 2016b). Besides that, CSR is crucial (Hrast, 2018):

- In the interest of business – CSR offers advantages to companies in various areas, including risk management, innovation capacity, human resource management, customer relations, savings, and access to capital;
- In the interest of the economy – CSR enhances the sustainability and innovation of companies, thereby contributing to the development of more sustainable economies;
- In the interest of society – CSR provides a foundation for a transition to a sustainable economic system and a set of values upon which a more cohesive community can be developed.

Modern businesses remain concerned about CSR's lack of apparent benefits. Besides, one may think allocating company resources toward non-commercial activities could negatively impact shareholder value. One might further argue that social responsibility should belong to the government and the individuals rather than being placed on businesses (WBCSD, 1999). In his theory of stakeholders, Freeman (2004) further developed the concept of CSR. He proposed that, for an organization to be considered socially responsible, it should consider the issues and needs of all its stakeholders, including consumers, employees, suppliers, investors, and the community. Namely, these stakeholders, in turn, influence the organization's performance outcomes. Greenwood (2007), on the other hand, criticized Freeman's viewpoint, claiming that stakeholder inclusion does not inherently imply that an organization will act responsibly. His perspective opposes the previous claim and calls it a myth. In light of this, Hrast (2018) states that companies should establish a process that integrates social, environmental, ethical, human rights, and consumer issues into their operations and core strategy in close cooperation with stakeholders to fulfill their social responsibility consistently. The process should establish an array of shared values for owners/shareholders, stakeholders, and society as a whole while also identifying, preventing, and mitigating potential adverse effects.

Despite the growing awareness among organizations about their social responsibility, many have not yet implemented management methods that would align with that. In order to effectively incorporate it into their daily operations, organizations must provide training and retraining to their employees and managers, involving their whole supply chain. This is crucial to gaining essential skills and competency. CSR is the organization's responsibility, although stakeholders like employees, consumers, and investors can influence them to embrace socially responsible practices. This influence can be exerted in areas such as working conditions, the environment, or human rights, either in the organization's own interest or on behalf of other stakeholders (European Comission, 2001).

In regard to social responsibility, ISO 26000 addresses seven fundamental principles[3] (ISO, 2010; Mulej et al., 2016a):

1. accountability - (formal) accountability/responsibility for influence;
2. transparency - data transparency;
3. ethical behavior - integrity, fairness, and equality; concern for the environment, animals, and people; and a commitment to assessing the effects of decisions and actions on the interests of stakeholders;
4. respect for stakeholder interests - not only owners and managers;
5. respect for the rule of law - but not monopolistic, biased;
6. respect for international norms of behavior - but not monopolistic, biased;
7. respect for human rights - unfortunately, this has to be taken into account with globally valid documents.

They support two principles derived from system theory, as shown in Figure 8 (ISO, 2010):

1. Interdependence
2. Holistic approach.

Figure 8. Seven core subjects of social responsibility (ISO, 2010)

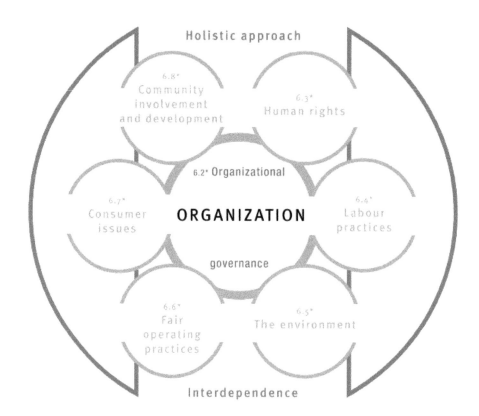

CSR is a valuable tool for gaining competitive advantage. Furthermore, the application of social responsibility has been found to have several benefits. When it comes to communication with stakeholders, publishing CSR-related information continues to rise in importance. In addition, numerous studies have proven that investing in CSR and sharing SR information leads to better financial indicators, as well as an increase in market share and value (Damnjanović, 2018). Social responsibility is of utmost importance from an economic standpoint, as it reduces the expenses associated with strikes and poor work, replacing bad business partners and markets, restoring natural conditions, and enabling the long-term survival of humanity (Mulej & Mihec, 2020).

In addition to its positive impact on financial performance, nonfinancial benefits include (Zlatanović, Mulej, & Ženko, 2022; Zore, 2018; Hrast & Mulej, 2009):

- strengthening the company's image and reputation,
- improvements in brand image,

- better consumer relationships,
- competitive advantage,
- stronger relationships with stakeholders,
- reduced employee turnover,
- improved relations with companies and communities,
- better control and risk management,
- improved reputation,
- due to innovation and differentiation in the market, the company opens up new business opportunities,
- increased efficiency in daily operations,
- increased employee motivation,
- increased sales,
- consumer loyalty,
- greater ability to attract capital, partners, customers, etc.,
- higher level of customer satisfaction and loyalty,
- more loyal, innovative, and satisfied employees, and
- positive consumer and investor relations.

Socially responsible behavior has the potential to provide substantial advantages by fostering positive attitudes towards the organization and competitive advantages. On the contrary, the observed socially irresponsible and unethical business practices could lead to a decline in reputation, increased expenses, and reduced value of the company's shares (Maon et al., 2008).

CRS brings many benefits to society in various areas (Mulej, 2018):

- In *management*, CSR promotes decision-making based on a better understanding of society's expectations and the opportunities that come with them. It also improves the organization's risk management, competitiveness (including its ability to obtain financial resources), and status as a preferred partner. It makes business more reliable and fair by promoting fair competition, responsible political engagement, and an absence of corruption. Additionally, these practices improve the organization's relationships with its stakeholders, exposing it to new expectations and a diverse range of stakeholders.
- In *recruitment*, CSR contributes to increased employee morale, affiliation, integration, participation, and innovation of employees. There is a reduction in employee turnover, absenteeism, presentism and employee health and safety are enhanced. Additionally, it positively affects a company's ability to recruit, motivate, and retain employees.
- Socially responsible companies detect *an increasing number of innovative ideas* from various stakeholders (employees, customers, suppliers, etc.) with

the aim of improving their business. Furthermore, more innovation is generated, and learning of good practices from other companies is adopted to enhance the work of one's own business.

- In the field of *finance*, SR has the potential to reduce expenses through increased productivity and resource efficiency, reduce energy and water consumption, reduce waste, and acquire valuable by-products. Furthermore, gaining new business with suppliers and customers, increasing customer purchases, and receiving suggestions for enhancing cooperation all (indirectly) contribute to a higher profit at the end of the year.
- CSR not only *improves relations with customers and suppliers* but also enhances the organization's reputation and promotes greater public confidence in the organization and its products and services.
- In the field of *communication*, the organization's visibility, and reputation among stakeholders (including the media, general public, consumers, suppliers, and business partners) are growing, resulting in increased media coverage about the organization. Regularly reporting on the non-financial aspects of the organization's performance improves the organization's reputation and enhances its relationship with stakeholders.

A company that effectively incorporates the concept of SR into its integrated business strategy builds a reputation for trustworthiness, honesty, and integrity while also generating valuable resources that enable and generate new aspects of business performance, which are reflected in financial outcomes; this approach averts numerous costs (Zore, 2018).

Social responsibility is essential for (business) excellence and individuals striving to achieve success rather than failure (Mulej et al., 2016b). Nevertheless, given the crucial role of human resources in the success of organizations, the following subsection examines the connection between CSR and HRM.

6.1 The Relationship Between Corporate Social Responsibility and Human Resource Management

HRM is essential to CSR as it aligns the organization's economic and social goals with performance (Bučiūnienė & Kazlauskaitė, 2012). Moreover, HRM is crucial in CSR since employee collaboration and reciprocity are necessary for policy and practice implementation. To foster value/VCEN attunement, firms must hire employees with specific moral values, establish performance evaluation methods that promote employee social responsibility, and provide employee training and development (Orlitzky & Swanson, 2006).

Organizations can be found everywhere, including government, business, and civil society—furthermore, every household functions as a small organization. Organizations consist of individuals characterized by their values, beliefs, interests, and other characteristics. Thus, social responsibility begins with each person's responsibility (Slapnik, Hrast, & Mulej, 2018). As said, social responsibility is realized (or not) by people, so their sufficient and necessary, i.e. requisite (personal and/or) team holism is vital (RPH). As the individual is also an employee and coworker, organizational conditions such as management, HRM, organizational culture, motivational culture, internal communication, ethics, and organizational behavior significantly impact an individual's RPH in these roles. It is important to acknowledge that RPH is essential in effectively managing stress, achieving job satisfaction, fostering creative collaboration, enhancing work motivation, and promoting psychological well-being (PWB). Indirectly, all of this enhances the performance of organizations, which must take steps to provide opportunities for employees to implement RPH-enhancing techniques (Šarotar Žižek & Mulej, 2016).

The concept of social responsibility in ISO 26000 indirectly exposes the state of well-being (ISO, 2010). Among the seven ISO 26000 principles, RPH has a particularly strong connection to the principles of respecting stakeholders, upholding human rights, and adhering to ethical conduct (Šarotar Žižek & Mulej, 2016). Nevertheless, all principles are expected to support endeavors for the well-being of people as the main goal of all human activities. Therefore, ISO 26000 establishes a connection between economics and psychology, especially the positive psychology (Šarotar Žižek S., 2015).

Many companies now realize that improved job quality improves environmental performance, which can be achieved via clean technologies. Clean technology is frequently connected to more technologically advanced and rewarding employee jobs. Therefore, the adoption of clean technology can, at the same time, increase profitability, job satisfaction, and environmental performance (European Comission, 2001).

Employees are the primary stakeholders of organizations. The implementation of CSR requires top-level managers' commitment, innovative thinking, the development of new skills, and closer involvement of employees and their representatives in a two-way dialogue that can promote permanent feedback and adjustment. Social dialogue with employees' representatives, one of the main ways to build a relationship between a company and its employees, is crucial to promoting socially responsible behaviors. Moreover, employee representatives must be actively consulted in developing policies, plans, and initiatives because CSR concerns affect almost every aspect of business operations. Also, social dialogue should contain issues and tools to enhance organizations' social and environmental performance. These include management and employee awareness-raising, training programs, social and environmental performance advisory campaigns, and strategic management systems integrating economic, social, and environmental factors (European Comission, 2001).

As the internal aspect of CSR refers to actions inside the organization, where managers and employees have the main role (e.g., health and safety, managing change, investments in human capital, etc.), a major challenge for modern businesses is to attract and retain skilled employees. In this context, useful measures might include (European Comission, 2001):

- life-long learning,
- increased workforce diversity,
- concerns over employability and job stability,
- improved information through the entire organization,
- profit sharing and share ownership schemes,
- a more harmonious balance between work, family, and leisure,
- empowerment of employees, and
- equal pay and more opportunities for women in the workforce.

Today's dynamic labor economy, where employee commitment is low and absenteeism is high, requires CSR policies and practices that focus on employees, such as corporate investment and engagement in CSR-related HRM practices. Research shows that engagement in CSR-related HRM approaches positively impacts employee commitment. This may reduce absenteeism and turnover since employees appreciate the organization's dedication. Furthermore, positive word-of-mouth would help improve the employer's brand. In light of labor market developments and changing worker attitudes regarding their job and workplace, CSR-related HRM strategies are vital for organizational effectiveness (Bučiūnienė & Kazlauskaitė, 2012).

At a time when talent shortages are an issue in several industries and more job seekers are asking about firms' employment policies, many tools can improve HRM best practices information and transparency. Additionally, firms must use responsible hiring practices to meet the European Employment Strategy goals of reducing unemployment, expanding employment, and fighting social exclusion. Facilitating the recruitment of ethnic minorities, older workers, women, the long-term unemployed, and the disadvantaged can promote nondiscrimination (European Comission, 2001).

Based on the aforementioned findings, it can be concluded that organizations that prioritize CSR are likely to have more advanced HRM practices, which may be referred to as CSR-related HRM practices. Such organizations should (Bučiūnienė & Kazlauskaitė, 2012):

- prioritize responsible recruitment, training, and career management practices,
- implement flexible working time and job rotation,
- improve communication within the company,
- define training needs more effectively,
- introduce profit sharing and share ownership schemes,
- involve and empower employees,
- prioritize employee health and well-being,
- provide better childcare facilities to support work-life balance, and
- demonstrate concern for job security.

7. CONCLUSION

AI is a type of modern technology that, when used correctly, can help employers, HR professionals, and employees. It can take over dull, routine, and long-lasting tasks that require precise analysis of vast amounts of data and information. We could say that implementing AI systems into various HRM functions might significantly save an organization's time, effort, and money. Furthermore, AI supports organizations in remaining aware of global, technological, and market trends and promptly adapting to them while fostering the organization's overall growth and development. Besides that, applying AI to HRM functions can enhance employee performance, productivity, retention, commitment, and talent development, all while minimizing employee turnover and fluctuation.

The connection between HRM and CRS is vital for developing an organizational culture based on ethics, social and environmental responsibility, and responsible behavior toward employees and other important stakeholders of the organization. This connection supports the implementation of sustainable approaches that emphasize the well-being of employees and the community while also being in line with the organization's fundamental values/VCEN.

Despite many benefits, AI is still not implemented into as many organizations' HRM functions as possible. Besides high costs, one of the main reasons lies in the lack of trust in modern technologies, which is often due to a lack of knowledge. Nevertheless, to keep up with the competition and successfully follow and adapt to trends in the ever-changing environment, organizations must implement AI into their HRM functions sooner rather than later. However, besides solely implementing and then using AI, organizations must do it in a socially responsible manner. Doing so, both sustainable and CRS organization efforts would benefit from it. Therefore, it is vital that when implementing AI into their HRM functions, businesses follow principles of sustainable development and social responsibility. This will not only drive organizations towards better overall success, employee job satisfaction, etc. but will also lead towards greater sustainable development and progress and generate a picture of a great employer.

AI has great potential in the field of HRM, especially in recruitment, training, and development of employees, talent acquisition, and employee performance and management. However, using AI brings many challenges, including improper implementation, which can lead to chaos and overall disaster, lack of trust and knowledge in modern technologies, and the belief that AI will simply replace many HRM functions, thus causing job losses. Besides that, a common concern is associated with potential biases in previous data, which AI users apply to assess and make predictions. This issue has already caused many unintentional discriminatory problems in the workplace. In conclusion, AI should become more SRAI.

Socially responsible artificial intelligence (SRAI) is a novel concept, essential for HRM. It entails CRS, ESG, and SDGs to ensure responsible and ethical implementation and use of AI in HRM practices in any organization. It encompasses five key aspects, namely economic, legal, ethical, philanthropic, and environmental. In addition to these, the guiding principles of fairness, inclusiveness, transparency, reliability, accountability, reliability, privacy, and security, are also essential. Each of these plays a crucial role in achieving sustainability and social responsibility. With the focus on responsibility and ethics within HR practices, socially responsible AI provides guidelines for implementing and utilizing AI in a way that benefits both individuals and society as a whole.

SRAI is a dynamic concept that is still evolving and adapting to new opportunities as well as challenges in the field of HRM. Therefore, it would be useful for future research to investigate and provide examples and practical advice for organizations wishing to implement SRAI in their HRM practices. It would also be very useful to develop an assessment method that would enable organizations to assess their performance in implementing SRAI initiatives and principles in HRM practices.

REFERENCES

Agrawal, A., Gans, J., & Goldfarb, A. (2018). *Prediction Machines: The Simple Economics of Artificial Intelligence*. Harvard Business Press.

Bankins, S. (2021). The ethical use of artificial intelligence in human resource management: A decision-making framework. *Ethics and Information Technology*, 23(4), 841–845. 10.1007/s10676-021-09619-6

Bhatia, T. (2018). *Artificial intelligence in HR*. HR Strategy and Planning Excellence Essentials.

Birdi, K., Clegg, C., Patterson, M., Robinson, A., Stride, C. B., Wall, T. D., & Wood, S. J. (2008). The impact of human resource and operational management practices on company productivity: A longitudinal study. *Personnel Psychology*, 61(3), 467–501. 10.1111/j.1744-6570.2008.00136.x

Bissio, R. (2018). Vector of hope, source of fear. *Spotlight Sustain. Dev.*, 77-86.

Bowen, D., & Ostroff, C. (2004). Understanding HRM–firm performance linkages: The role of the "strength of the HRM system". *Academy of Management Review*, 22(2), 203–221.

Boxall, P. (2013). Mutuality in the management of human resources: Assessing the quality of alignment in employment relationships. *Human Resource Management Journal*, 1(23), 3–17. 10.1111/1748-8583.12015

Boxall, P. (2014). The future of employment relations from the perspective of human resource management. *The Journal of Industrial Relations*, 56(4), 578–593. 10.1177/0022185614527980

Boxall, P., & Purcell, J. (2011). *Strategy and Human Resource Management*. Palgrave Macmillan.

Brundtland, G. H. (1987). *Report of the World Commission on environment and development: our common future*. United Nations.

Brynjolfsson, E., & McAfee, A. (2014). *The Second Machine Age: Work, Progress, and Prosperity in a Time of Brilliant Technologies*. WW Norton & Company.

Bučiūnienė, I., & Kazlauskaitė, R. (2012). The linkage between HRM, CSR and performance outcomes. *Baltic Journal of Management*.

Buck, B., & Morrow, J. (2018). AI, performance management and engagement: Keeping your best their best. *Strategic HR Review*, 17(5), 261–262. 10.1108/SHR-10-2018-145

Bujold, A., Roberge-Maltais, I., Parent-Rocheleau, X., Boasen, J., Sénécal, S., & Léger, P. M. (2023). Responsible artificial intelligence in human resources management: A review of the empirical literature. *AI and Ethics*, 1–16. 10.1007/s43681-023-00325-1

Carroll, A. B. (1991). The pyramid of corporate social responsibility: Toward the moral management of organizational stakeholders. *Business Horizons*, 34(4), 39–48. 10.1016/0007-6813(91)90005-G

Chang, Y. L., & Ke, J. (2023). Socially Responsible Artificial Intelligence Empowered People Analytics: A Novel Framework Towards Sustainability. *Human Resource Development Review*.

Cheng, L., & Liu, H. (2023). *Socially Responsible AI: Theories and Practices*. World Scientific. 10.1142/13150

Cheng, L., Varshney, K. R., & Liu, H. (2021). Socially Responsible AI Algorithms: Issues, Purposes, and Challenges. *Journal of Artificial Intelligence Research*, 71, 1137–1181. 10.1613/jair.1.12814

Chowdhury, S., Dey, P., Joel-Edgar, S., Bhattacharya, S., Rodriguez-Espindola, O., Abadie, A., & Truong, L. (2023). Unlocking the value of artificial intelligence in human resource management through AI capability framework. *Human Resource Management Review*, 33(1), 100899. 10.1016/j.hrmr.2022.100899

Cognizant. (2024). *Human-centered artificial intelligence*. Retrieved from Cognizant: https://www.cognizant.com/us/en/glossary/human-centered-ai

Colbert, A., George, G., & Yee, N. (2016). The digital workforce and the workplace of the future. *Academy of Management Journal, 59*(3), 731-739.

Dalenberg, D. J. (2018). Preventing discrimination in the automated targeting of job advertisements. *Computer Law & Security Report*, 34(3), 615–627. 10.1016/j.clsr.2017.11.009

Damnjanović, M. (2018). Relation between corporate social responsibility reporting and financial indicators of Serbian companies. *the Proceedings of the 5th International Scientific Conference on Contemporary Issues in Economics, Business and Management (EBM 2018)*, 473-480.

De Sousa, W. G., de Melo, E. P., Bermejo, P. D., Fairas, R. S., & Gomes, A. O. (2019). How and where is artificial intelligence in the public sector going? *A literature review and research agenda. Government Information Quarterly*, 36(4), 101392. 10.1016/j.giq.2019.07.004

Dwivedi, Y. K., Hughes, L., Ismagilova, E., Aarts, G., Coombs, C., Crick, T., & Galanos, V. (2021). Artificial intelligence (AI): Multidisciplinary perspectives on emerging challenges, opportunities, and agenda for research, practice and policy. *International Journal of Information Management*, 57, 57. 10.1016/j.ijinfomgt.2019.08.002

Eckhaus, E. (2018). Measurement of organizational happiness. *Advances in Human Factors, Business Management and Leadership*, 266-278.

Eubanks, B. (2022). *Artificial Intelligence for HR: Use AI to Support and Develop a Successful Workforce* (2nd ed.). Kogan Page Publishers.

European Comission. (2001). *GREEN PAPER: Promoting a European framework for Corporate Social Responsibility*. Commission of the European Communities.

Feige, I. (2019). *What is AI Safety? Towards a better understanding.* Retrieved from https://faculty.ai/blog/what-is-ai-safety/

Ghatak, R. (2022). *People Analytics: Data to Decisions*. Springer. 10.1007/978-981-19-3873-3

Ghosh, S., Majumder, S., & Peng, S. L. (2023). Aritificial Intelligence Techniques in Human Resource Management. In *An empirical study on adoption of Aritificial Intelligence in Human Resource Management* (pp. 29-85). CRC Press.

Go, E., & Sundar, S. (2019). Humanizing chatbots: The effects of visual, identity and conversational cues on humanness perceptions. *Computers in Human Behavior*, 97, 304–316. 10.1016/j.chb.2019.01.020

Graßmann, C., & Schermuly, C. C. (2021). Coaching with artificial intelligence: Concepts and capabilities. *Human Resource Development Review*, 20(1), 106–126. 10.1177/1534484320982891

Greenwood, M. (2007). Stakeholder engagement: Beyond the myth of corporate responsibility. *Journal of Business Ethics*, 74(4), 315–327. 10.1007/s10551-007-9509-y

Guest, D. E. (1997). Human resource management and performance: A review and research agenda. *International Journal of Human Resource Management*, 8(3), 263–276. 10.1080/095851997341630

Guest, D. E., Michie, J., Sheehan, M., & Conway, N. (2000). *Employee Relations, HRM and Business Performance: An analysis of the 1998 Workplace Employee Relations Survey*. CIPD.

Guo, P., Xiao, K., Ye, Z., Zhu, H., & Zhu, W. (2022). Intelligent career planning via stochastic subsampling reinforcement learning. *Scientific Reports*, 12(1).35585154

Helbing, D., Frey, B. S., Gigerenzer, G., Hafen, E., Hagner, M., Hofstetter, Y., . . . Zwitter, A. (2019). Will democracy survive big data and artificial intelligence? *Towards digital enlightenment: Essays on the dark and light sides of the digital revolution*, 73-98.

Helbing, D., & Pournaras, E. (2015). Society: Build digital democracy. *Nature*, 527(7576), 33–34. 10.1038/527033a26536943

Hrast, A. (2018). Usmeritve Evropske Unije na področju družbene odgovornosti in trajnostnega razvoja. In *Uvod v politično ekonomijo družbeno odgovorne družbe* (pp. 285-294). Maribor: Kulturni center, zavod za umetniško produkcijo in založništvo.

Hrast, A., & Mulej, M. (2009). *Družbena odgovornost in izzivi časa 2009: Delo -odnosi do zaposlenih in različnih starostnih generacij. Zbornik prispevkov*. IRDO.

International Energy Agency. (2017). *Digitalisation and Energy*. Retrieved from International Energy Agency: https://www.iea.org/reports/digitalisation-and-energy

ISO. (2010). *ISO 26000, Social Responsibility*. Retrieved from ISO: https://www.iso.org/iso/discovering_iso_26000.pdf

Jarrahi, M. H., Askay, D., Eshraaghi, A., & Smith, P. (2023). Artificial intelligence and knowledge management: A partnership between human and AI. *Business Horizons*, 66(1), 87–99. 10.1016/j.bushor.2022.03.002

Jean, N., Burke, M., Xie, M., Davis, W. M., Lobell, D. B., & Ermon, S. (2016). Combining satellite imagery and machine learning to predict poverty. *Science*, 353(6301), 790–794. 10.1126/science.aaf789427540167

Jobin, A., Ienca, M., & Vayena, E. (2019). The global landscape of AI ethics guidelines. *Nature Machine Intelligence*, 1(9), 389–399. 10.1038/s42256-019-0088-2

Jones, D., Molitor, D., & Reif, J. (2019). What do workplace wellness programs do? Evidence from the Illinois workplace wellness study. *The Quarterly Journal of Economics*, 134(4), 1747–1791. 10.1093/qje/qjz02331564754

Köchling, A., Riazy, S., Wehner, M. C., & Simbeck, K. (2021). Highly accurate, but still discriminatory: A fairness evaluation of algorithmic video analysis in the recruitment context. *Business & Information Systems Engineering*, 63(1), 39–54. 10.1007/s12599-020-00673-w

Kramar, R. (2022). Sustainable human resource management: Six defining characteristics. *Asia Pacific Journal of Human Resources*, 60(1), 146–170. 10.1111/1744-7941.12321

Kuo, T. C., & Smith, S. (2018). A systematic review of technologies involving eco-innovation for enterprises moving towards sustainability. *Journal of Cleaner Production*, 192, 207–220. 10.1016/j.jclepro.2018.04.212

Langer, M., König, C. J., & Hemsing, V. (2020). Is anybody listening? The impact of automatically evaluated job interviews on impression management and applicant reactions. *Journal of Managerial Psychology*, 35(4), 271–284. 10.1108/JMP-03-2019-0156

Leong, C. (2018). Technology & recruiting 101: How it works and where it's going. *Strategic HR Review*, 17(1), 50–52. 10.1108/SHR-12-2017-0083

Li, F. (2020). The digital transformation of business models in the creative industries: A holistic framework and emerging trends. *Technovation*, 92-93, 92. 10.1016/j.technovation.2017.12.004

Luo, X., Tong, S., Fang, Z., & Qu, Z. (2019). Frontiers: Machines vs. humans: The impact of artificial intelligence chatbot disclosure on customer purchases. *Marketing Science*, •••, 937–947. 10.1287/mksc.2019.1192

Majumder, S., Ghosh, S., & Polkowski, Z. (2023). A brief introduction to Human Resource Management. In *Artificial Intelligence Techniques in Human Resource Management* (pp. 1-25). CRC Press. 10.1201/9781003328346-1

Makridakis, S. (2017). The forthcoming Artificial Intelligence (AI) revolution: Its impact on society and firms. *Futures*, 90, 46–60. 10.1016/j.futures.2017.03.006

Maon, F., Lindgreen, A., & Swaen, V. (2008). Thinking of the organization as a system: The role of managerial perceptions in developing a corporate social responsibility strategic agenda. *Systems Research and Behavioral Science*, 25(3), 413–426. 10.1002/sres.900

Matsa, P., & Gullamajji, K. (2019). To Study Impact of Artificial Intelligence on Human Resource Management. *International Research Journal of Engineering and Technology*, 6(8), 1229–1238.

McCarthy, J., Minsky, M. L., Rochester, N., & Shannon, C. E. (1955). A proposal for the Dartmouth summer research project on artificial intelligence. *AI Magazine*, 27(4).

McWilliams, A., Siegel, D. S., & Wright, P. M. (2006). Corporate social responsibility: Strategic implications. *Journal of Management Studies*, 43(1), 1–18. 10.1111/j.1467-6486.2006.00580.x

Mehrabi, N., Morstatter, F., Saxena, N., Lerman, K., & Galstyan, A. (2021). A survey on bias and fairness in machine learning. *ACM Computing Surveys*, 54(6), 1–35. 10.1145/3457607

Mikaelf, P., Framnes, V. A., Danielsen, F., Krogstie, J., & Olsen, D. (2017). *Big data analytics capability: antecedents and business value.*

Miller, T. (2019). Explanation in artificial intelligence: Insights from the social sciences. *Artificial Intelligence*, 267, 1–38. 10.1016/j.artint.2018.07.007

Mithas, S., Tafti, A., & Mitchell, W. (2013). How a firm's competitive environment and digital strategic posture influence digital business strategy. *MIS Quarterly*, 511-536.

Mulej, M. (2018). Politična ekonomija in družbena odgovornost. In *Uvod v politično ekonomijo družbeno odgovorne družbe* (pp. 197-284). Maribor: Kulturni center, zavod za umetniško produkcijo in založništvo.

Mulej, M., Hrast, A., & Mihec, M. (2024a). *Inovativna trajnostna družbeno odgovorna družba, Knj. 2: Nekaj informacijskih spodbud za obstoj človeštva.* Ljubljana: IRDO-Inštitut za razvoj družbene odgovornosti.

Mulej, M., Hrast, A., & Šarotar Žižek, S. (2024b). *Bases of the innovative sustainable socially responsible society (ISSR Society). Volume 2: some information management viewpoints of ISSR SOCIETY.* Ljubljana: IRDO - Inštitut za razvoj družbene odgovornosti.

Mulej, M., Hrast, A., & Štrukelj, T. (2024c). *Inovativna trajnostna družbeno odgovorna družba, Knj. 1: Nekaj ekonomskih spodbud zoper propad človeštva.* Ljubljana: IRDO – Inštitut za razvoj družbene odgovornosti.

Mulej, M., Hrast, A., & Štrukelj, T. (2024d). *Inovativna trajnostna družbeno odgovorna družba, Knj. 3: Skrb za starejše.* Ljubljana: IRDO - Inštitut za razvoj družbene odgovornosti.

Mulej, M., Hrast, A., Štrukelj, T., Likar, B., & Šarotar Žižek, S. (2024e). *Bases of an innovative sustainable socially responsible (ISSR) society. Volume 1: Social responsibility - a non-technical innovation process toward ISS SOCIETY.* Ljubljana: IRDO - Inštitut za razvoj družbene odgovornosti.

Mulej, M., Merhar, V., & Žakelj, V. (2016a). Prihodnost - preživetje brez ekosocializma in družbene odgovornosti? In *Nehajte sovražiti svoje otroke in vnuke. Knj. 1, Družbenoekonomski okvir in osebne lastnosti družbeno odgovornih* (pp. 11-23). Maribor: Kulturni center, zavod za umetniško produkcijo in založništvo.

Mulej, M., & Mihec, M. (2020). Social Responsibility as a Precondition of Innovation in Higher Education. In *Handbook of Research on Enhancing Innovation in Higher Education Institutions* (pp. 49-74). IGI Global. 10.4018/978-1-7998-2708-5.ch003

Mulej, M., Ženko, Z., Hrast, A., & Merhar, V. (2016b). Prihodnost ekonomije v obdobju izobilja in krize izobilja brez zdravih ambicij. In *Nehajte sovražiti svoje otroke in vnuke. Knj. 1, Družbenoekonomski okvir in osebne lastnosti družbeno odgovornih* (pp. 35-85). Maribor: Kulturni center, zavod za umetniško produkcijo in založništvo.

Nagano, A. (2018). Economic Growth and Automation Risks in Developing Countries Due to the Transition Toward Digital Modernity. *Proceedings of the 11th International Conference on Theory and Practice of Electronic Governance*, 42-50. 10.1145/3209415.3209442

Nagler, J., van den Hoven, J., & Helbing, D. (2019). Towards Digital Enlightenment. *Springer International Publishing*, 41-46.

Naim, M. F. (2023). Reinventing workplace learning and development: Envisaging the role of AI. In *The adoption and Effect of artificial intelligence on human resources management* (pp. 215–227). Part A. 10.1108/978-1-80382-027-920231011

Noe, R., Hollenbeck, J., Gerhart, B., & Wright, P. (2007). *Fundamentals of Human Resource Management*.

Oracle & Future Workplace. (2019). From Fear to Enthusiasm. *Artificial Intelligence Is Winning More Hearts and Minds in the Workplace*, 1-18.

Orlitzky, M., Schmidt, F. L., & Rynes, S. L. (2003). Corporate Social and Financial Performance: A Meta-Analysis. *Organization Studies*, 24(3), 403–441. 10.1177/0170840603024003910

Orlitzky, M., & Swanson, D. L. (2006). Socially responsible human resource management. In *J. R. Deckop*. Human Resource Management Ethics.

Patterson, M. G., West, M. A., Lawthom, R., & Nickell, S. (1997). *Impact of People Management Practice*. IPD.

Pawar, Y. (2019). *Impact of Artificial Intelligence in Performance Management*.

Penrose, E. (1959). *The Theory of the Growth of the Firm*. Blackwell.

Purcell, J., Kinnie, K., Hutchinson, R., Rayton, B., & Swart, J. (2003). *Understanding the People and Performance Link: Unlocking the black box*. CIPD.

Ransbotham, S., Khodabandeh, S., Kiron, D., Candelon, F., Chu, M., & LaFountain, B. (2020). Expanding AI's impact with organizational learning. *MITSloan Managemen review*.

Rao, P., & Teegen, H. (2009). *Human Resource Management*. Academic Press.

Raub, M. (2018). Bots, bias and big data: Artificial intelligence, algorithmic bias and disparate impact liability in hiring practices. *Arkansas Law Review*, 71, 529–570.

Ravipolu, A. (2017). Role of Artificial Intelligence in Recruitment. *International Journal of Engineering Technology, Management and Applied Sciences*, 115-117.

Robert, L. P., Pierce, C., Marquis, L., Kim, S., & Alahmad, R. (2020). Designing fair AI for managing employees in organizations: A review, critique, and design agenda. *Human-Computer Interaction*, 35(5-6), 545–575. 10.1080/07370024.2020.1735391

Robichaud, F. (2024). *ISO 26000: 7 Core subjects of Corporate Social Responsibility*. Retrieved from Borealis: https://www.boreal-is.com/blog/iso-26000-social-responsibility/

Rudin, C. (2019). Stop explaining black box machine learning models for high stakes decisions and use interpretable models instead. *Nature Machine Intelligence*, 1(5), 206–215. 10.1038/s42256-019-0048-x35603010

Sætra, H. S. (2021). A framework for evaluating and disclosing the ESG related impacts of AI with the SDGs. *Sustainability (Basel)*, 13(15), 8503. 10.3390/su13158503

Šarotar, Ž. S., Zolak Poljašević, B., Mulej, M., Šket, R., Kovše, K., Preskar, M., . . . Hrast, A. (2023a). *Aktualne teme managementa človeških virov. Knj. 1, Izzivi sodobnega časa, poslovanje prihodnosti, trajnostni management človeških virov, blagovna znamka delodajalca*. Ljubljana: IRDO - Inštitut za razvoj družbene odgovornosti.

Šarotar, Ž. S., Zolak Poljašević, B., Mulej, M., Treven, S., Šket, R., & Preskar, M. (2023b). *Aktualne teme managementa človeških virov. Knj. 3, Agilnost in zavzetost zaposlenih, čuječnost vodij in zaposlenih, organizacijska energija, delovna sreča, psihično dobro počutje zaposlenih*. Ljubljana: IRDO - Inštitut za razvoj družbene odgovornosti.

Šarotar, Ž. S., Zolak Poljašević, B., Treven, S., Šket, R., Preskar, M., & Senekovič, P. (2023c). *Aktualne teme managementa človeških virov. Knj. 2, Oblike sodelovanja med zaposlenimi, ključni kazalniki uspešnosti, osebni in osebnostni razvoj človeka*. Ljubljana: IRDO - Inštitut za razvoj družbene odgovornosti.

Šarotar Žižek, S. (2015). Well-Being as the Basic Aim of Social Responsibility. In *Social responsibility—Range of perspectives per topics and countries (Social responsibility beyond neoliberalism and charity, Vol. 4)* (pp. 49-76). London, Uk: Bentham Books.

Šarotar Žižek, S., & Milfelner, B. (2014a). *Vpliv menedžmenta človeških virov na uspešnost*. IRDO.

Šarotar Žižek, S., & Mulej, M. (2016). Zadostna in potrebna (osebna) celovitost posameznika (ZIPOC). In *Nehajte sovražiti svoje otroke in vnuke. Knj. 1, Družbenoekonomski okvir in osebne lastnosti družbeno odgovornih* (pp. 86-140). Maribor: Kulturni center, zavod za umetniško produkcijo in založništvo.

Šarotar Žižek, S., Mulej, M., & Treven, S. (2014b). *Zagotavljanje zadostne in potrebne osebne celovitosti človeka*. IRDO.

Šarotar Žižek, S., Treven, S., & Mulej, M. (2014c). *Psihično dobro počutje zaposlenih*. IRDO.

Schermerhorn, J. R., Hunt, J. G., & Osborn, R. N. (2012). *Comportement humain et organisation*. Pearson - Village Mondial.

Scholz, T. M. (2019). Big data and human resource management. In *Big data: Promise, application and pitfalls* (pp. 69-89). 10.4337/9781788112352.00008

Singh, R., Vatsa, M., & Ratha, N. (2021). Trustworthy ai. *Proceedings of the 3rd ACM India Joint International Conference on Data Science & Management of Data (8th ACM IKDD CODS & 26th COMAD)*, 449-453.

Slapnik, T., Hrast, A., & Mulej, M. (2018). Nacionalna strategija družbene odgovornosti v Sloveniji - idejni osnutek. In *Uvod v politično ekonomijo družbeno odgovorne družbe* (pp. 295-328). Maribor: Kulturni center, zavod za umetniško produkcijo in založništvo.

Speer, A. B., Dutta, S., Chen, M., & Trussel, G. (2019). Here to stay or go? Connecting turnover research to applied attrition modeling. *Industrial and Organizational Psychology: Perspectives on Science and Practice*, 12(3), 277–301. 10.1017/iop.2019.22

Stahl, B. C., Schroeder, D., & Rodrigues, R. (2023). AI for good and the SDGs. In *Ethics of artificial intelligence: Case studies and Options for addressing ethical challenges* (pp. 95-106). Springer International Publishing. 10.1007/978-3-031-17040-9_8

Tambe, P., Cappelli, P., & Yakubovich, V. (2019). Artificial intelligence in human resources management: Challenges and a path forward. *California Management Review*, 61(4), 15–42. 10.1177/0008125619867910

Thiebes, S., Lins, S., & Sunyaev, A. (2021). Trustworthy artificial intelligence. *Electronic Markets*, 31(2), 447–464. 10.1007/s12525-020-00441-4

Thompson, M. (2002). *High Performance Work Organization in UK Aerospace*. The Society of British Companies.

Tursunbayeva, A., Di Lauro, S., & Pagliari, C. (2018). People analytics—A scoping review of conceptual boundaries and value propositions. *International Journal of Information Management*, 43(1), 224–247. 10.1016/j.ijinfomgt.2018.08.002

Ulrich, D. (1997). *Human resource champions: The next agenda for adding value and delivering results*. Harvard Business Press.

United Nations. (2015). *Transforming Our World: The 2030 Agenda for Sustainable Development*. Division for Sustainable Development Goals.

Upadhyay, A. K., & Khandelwal, K. (2018). Applying artificial intelligence: Implications for recruitment. *Strategic HR Review*, 17(5), 255–258. 10.1108/SHR-07-2018-0051

van den Broek, E., Sergeeva, A., & Huysman, M. (2021). When the machine meets the expert: An ethnography of developing AI for hiring. *Management Information Systems Quarterly*, 45(3), 1557–1580. 10.25300/MISQ/2021/16559

van der Broek, E., Sergeeva, A., & Huysman, M. (2019). Hiring Algorithms: An Ethnography of Fairness in Practice. *ICIS 2019 Proceedings, 6*.

Verbin, I. (2020). *Corporate Responsibility in the Digital Age: A Practitioner's Roadmap for Corporate Responsibility in the Digital Age*. Routledge. 10.4324/9781003054795

Vinuesa, R., Azizpour, H., Leite, I., Balaam, M., Dignum, V., Domisch, S., . . . Fuso Nerini, F. (2020). The role of artificial intelligence in achieving the Sustainable Development Goals. Academic Press.

Vinuesa, R., Luna, M., & Cachafeiro, H. (2016). Simulations and experiments of heat loss from a parabolic trough absorber tube over a range of pressures and gas compositions in the vacuum chamber. *Journal of Renewable and Sustainable Energy*, 2(8). Advance online publication. 10.1063/1.4944975

Walker, J., Pekmezovic, A., & Walker, G. (2019). *Sustainable Development Goals: Harnessing Business to Achieve the SDGs through Finance, Technology and Law Reform*. John Wiley & Sons. 10.1002/9781119541851

WBCSD. (1999). *Corporate social responsibility: Meeting Changing Expectations, World Business Council for Sustainable Development*. World Business Council for Sustainable Development.

Wegren, S. K. (2018). The "left behind": Smallholders in contemporary Russian agriculture. *Journal of Agrarian Change*, 18(4), 913–925. 10.1111/joac.12279

West, M. A., Borrill, C. S., Dawson, C., Scully, J., Carter, M., Anclay, S., & Waring, J. (2002). The link between the management of employees and patient mortality in acute hospitals. *International Journal of Human Resource Management*, 13(8), 1299–1310. 10.1080/09585190210156521

Wingard, D. (2019). Data-driven automated decision-making in assessing employee performance and productivity: Designing and implementing workforce metrics and analytics social sciences, sociology, management and complex organizations. *Psychosociological Issues in Human Resource Management*, 7(2), 13–18. 10.22381/PIHRM7220192

World Economic Forum (WEF). (2018). Harnessing Artificial Intelligence for the Earth. *Fourth Industrial Revolution for the Earth Series Harnessing Artificial Intelligence for the Earth.*

Zhao, W. (2021). Artificial Intelligence and ISO 26000 (Guidance on Social Responsibility). In A. a.-I. Directions. IntechOpen.

Zhao, W. W. (2018). Improving social responsibility of artificial intelligence by using ISO 26000. *Iop conference series: Materials science and engineering*, 428(1).

Zhao, W. W. (2019). Research on social responsibility of artificial intelligence based on ISO 26000. *Recent Developments in Mechatronics and Intelligent Robotics:Proceedings of International Conference on Mechatronics and Intelligent Robotics (ICMIR2018)*, 130-137.

Zlatanović, D., Mulej, M., & Ženko, Z. (2022). Corporate Social Responsibility Considered With Two Systems Theories: A Case from Serbia. *Naše gospodarstvo*, 68(3), 10-17.

Zore, M. (2018). Družbena odgovornost podjetij: soavtorska monografija. In *Uvod v politično ekonomijo družbeno odgovorne družbe* (pp. 153-192). Maribor: Kulturni center, zavod za umetniško produkcijo in založništvo.

ADDITIONAL READING

Chang, Y. L., & Ke, J. (2023). Socially Responsible Artificial Intelligence Empowered People Analytics: A Novel Framework Towards Sustainability. *Human Resource Development Review*.

Cheng, L., & Liu, H. (2023). *Socially Responsible AI: Theories and Practices*. World Scientific. 10.1142/13150

Cheng, L., Varshney, K. R., & Liu, H. (2021). Socially Responsible AI Algorithms: Issues, Purposes, and Challenges. *Journal of Artificial Intelligence Research*, 71, 1137–1181. 10.1613/jair.1.12814

Mulej, M., Hrast, A., & Šarotar Žižek, S. (2024b). *Bases of the innovative sustainable socially responsible society (ISSR Society). Volume 2: some information management viewpoints of ISSR SOCIETY*. Ljubljana: IRDO - Inštitut za razvoj družbene odgovornosti.

Mulej, M., Hrast, A., Štrukelj, T., Likar, B., & Šarotar Žižek, S. (2024e). *Bases of an innovative sustainable socially responsible (ISSR) society. Volume 1: Social responsibility - a non-technical innovation process toward ISRS SOCIETY*. Ljubljana: IRDO - Inštitut za razvoj družbene odgovornosti

Šarotar Žižek, S. (2015). Well-Being as the Basic Aim of Social Responsibility. In M. Mulej, & R. G. Dyck, *Social responsibility—Range of perspectives per topics and countries (Social responsibility beyond neoliberalism and charity, Vol. 4)* (pp. 49-76). London, UK: Bentham Books.

KEY TERMS AND DEFINITIONS

Artificial Intelligence (AI): Artificial intelligence is a type of modern technology that enables machines to perform SOME tasks or behave in a way that would typically require human knowledge and intelligence.

Corporate Social Responsibility (CSR): Corporate social responsibility is a voluntary commitment of organization to behave in an environmentally and socially responsible way, and thus promote the sustainable development. It focuses on improving the long term environmental and social well-being with actions that go over and beyond the legal requirements.

Human Resource Management (HRM): Human resource management is the process of managing the organization's most valuable asset e. g. its employees, encompassing various function including recruitment and selection, training and development, performance management, talent acquisition, pay and reward system etc.

Innovative Sustainable Socially Responsible Society (ISSRS): Innovative sustainable socially responsible society is a society that with the aim of promoting the well-being of all people, prioritizes social responsibility, sustainability and innovation, via facing the economic, social, and environmental issues with a requisitely holistic approach to overcome the neoliberalist one-sided monopolies endangering humankind by ruining the free market and democracy.

Social Responsibility (SR): Social responsibility means that organizations, individuals, and institutions behave in a way that benefits other people, organizations, and wider society, and thus ensure their actions have a positive impact on the environment, society, and economy.

Socially Responsible Artificial Intelligence (SRAI): Socially responsible artificial intelligence a novel and still evolving concept that by encompassing SDGs, ESG, and CSR, and thus aiming for ethical and socially responsible implementation of AI, benefits both individuals, organizations, and society.

Transparency: Transparency means being open and clear about how humans, their business and other organizations, including their use of ai systems operate, and what data they are using, to help people to understand it, as well as to ensure fairness and build trust among stakeholders and employees.

ENDNOTES

[1] In other texts we are suggesting the transition from the corporate SR, limited to organizations or even to businesses, to 'innovative sustainable socially responsible society (ISSRS)'. There we include practices indicating that such a process is going on. (Mulej et al., 2019; Mulej et al., 2024, a, b, c d, e; Šarotar Žižek et al, 2014, a, b, c; Hrast et al, 2006-2023, whose IRDO conference proceedings can be found online; www.irdo.si). The point is in the alternative: either ISSRS or the third world war ending humankind.

[2] There are many other aspects of SRAI considerations and recommendations for HRM activities, such as sociological and psychological aspects, which we do not highlight here due to the limitations.

[3] We mentioned them earlier, but now the context and comment are different due to another selected viewpoint.

Chapter 5
Legal Aspects of Digital Ethics in the Age of Artificial Intelligence

Hemendra Singh
https://orcid.org/0000-0003-1160-8284
Jindal Global Law School, O.P. Jindal Global University, Sonipat, India

ABSTRACT

This chapter explores the legal dimensions of digital ethics amidst the proliferation of artificial intelligence technologies. It explores the evolution of legal frameworks governing digital technologies and their adaptation to address emerging ethical dilemmas in artificial intelligence. Focusing on the European Union's Artificial Intelligence Act and OECD's Principles on Artificial Intelligence, it assesses their implications for global AI governance and effectiveness in addressing ethical concerns. The intersection of data protection laws and artificial intelligence ethics is analyzed, emphasizing their role in safeguarding human rights. The chapter also examines legal challenges and solutions in ensuring the ethical use of AI, particularly regarding liability and accountability issues. By identifying emerging challenges and advocating for collaborative governance approaches, it outlines pathways to enhance legal frameworks in addressing evolving ethical concerns.

INTRODUCTION

Legal frameworks are essential in addressing the myriad ethical dilemmas that arise with the development and deployment of artificial intelligence (AI) and other digital technologies. As these technologies increasingly permeate various aspects of society, they bring about complex ethical challenges related to privacy, bias, accountability, and transparency. Robust legal frameworks provide the necessary structure to ensure that these technologies are developed and used in ways that are ethical, fair and respectful of human rights.

The problems of ethics in the application of AI are particularly complex and multifaceted (Bankins & Formosa, 2023). Recently, criminals utilized AI-based voice technology to impersonate a chief executive's voice and successfully demand a fraudulent transfer of $243,000 (Stupp, 2019). AI voice impersonation for fraud is just one concern. Another significant issue is the rise of deepfake technology, which uses machine learning to superimpose and synthesize existing images and videos onto new ones. This technology can be used to create pornographic content with superimposed human faces or fabricate videos of political leaders to incite violence and panic. Additionally, deepfakes may be employed during election cycles to influence and bias the voters (Libby, 2019).

One of the primary ethical concerns in AI and digital technologies is privacy and data protection. AI systems often rely on large amounts of data, which can include personal and sensitive information. Without proper legal safeguards, this data can be misused, leading to privacy violations and potential harm to individuals. Regulations like the General Data Protection Regulation (GDPR) in the European Union establish strict guidelines for data collection, processing, and storage, ensuring that individuals have control over their personal information (Tanaka, 2019). These legal requirements mandate that organizations implement appropriate security measures and obtain explicit consent from individuals before processing their data, thereby protecting privacy and mitigating the risk of data breaches. The GDPR also talks about concepts such as Data Protection by Design and by Default, assigning organizations accountability to integrate privacy safeguards into their procedures (Bygrave, 2017).

Bias and fairness in AI are also critical ethical issues that legal frameworks help to address. AI systems can inadvertently perpetuate and even exacerbate existing biases if they are trained on biased data sets (Nassar & Kamal, 2021). This can lead to unfair treatment and discrimination in areas such as hiring, lending, and law enforcement. Legal frameworks can mandate the implementation of fairness assessments and bias mitigation techniques in AI development. For instance, the proposed EU Artificial Intelligence Act requires that high-risk AI systems undergo rigorous testing for biases and are designed to ensure non-discrimination and fairness. Such regulations compel organizations to take proactive steps in identifying and correcting biases, thereby promoting equitable outcomes.

Accountability and transparency are fundamental to ethical AI deployment. When AI systems make decisions that significantly impact individuals' lives, it is crucial to have mechanisms in place to ensure accountability (Pan, 2024). Legal frameworks can establish clear lines of responsibility, making it possible to hold developers, operators, and users of AI systems accountable for their actions. For example, the EU AI Act includes provisions for transparency, requiring that AI systems provide clear information about their capabilities, limitations, and the logic behind their decisions (Almada & Petit, 2022). This transparency enables affected individuals and regulatory bodies to understand how decisions are made and to challenge them, if necessary.

Moreover, legal frameworks play a vital role in fostering public trust in AI and digital technologies. Trust is a cornerstone of the successful adoption and integration of new technologies. When individuals know that robust legal protections are in place to safeguard their rights and interests, they are more likely to trust and accept these technologies (Siau & Wang, 2018). Laws and regulations that emphasize ethical considerations signal to the public that ethical issues are being taken seriously and that there are mechanisms to prevent and address potential harms, if any.

In addition to protecting individuals, legal frameworks also provide guidance and clarity for organizations developing and deploying Artificial Intelligence and digital technologies. They set clear standards and expectations, which can help companies innovate responsibly and avoid costly legal pitfalls. By adhering to established legal requirements, organizations can ensure that their technological advancements align with societal values and ethical principles.

Legal frameworks are indispensable in addressing the ethical dilemmas associated with AI and digital technologies. They protect privacy, promote fairness, ensure accountability and transparency, and foster public trust. As Artificial Intelligence continues to evolve and integrate into various aspects of society, the development and enforcement of comprehensive legal frameworks will be crucial in ensuring that these technologies are used ethically and responsibly. Understanding and adhering to these legal frameworks is essential for navigating the complex ethical landscape of the digital age.

Methodology and Research Question

The author has adopted doctrinal approach to formulate this book chapter. The author will analyse the problem with the help of law books, existing statutes, proposed bills, conventions and decided cases pertaining to the issue. The author has selected bibliographic databases to identify secondary sources ranging from text books, bar review, scholarly articles and legal encyclopaedias. The book chapter aims to answer the following research question: How do current legal frameworks address the ethical dilemmas posed by the development and deployment of AI and digital technologies?

EVOLUTION OF LAWS AND REGULATIONS IN RESPONSE TO AI

The governance of digital technologies involves a complex interplay of legal frameworks that aim to protect individual rights, ensure security, and promote ethical standards. These frameworks operate at various levels - international, regional, and national and cover a broad spectrum of issues from data protection to cybersecurity. At the international level, several treaties and agreements set standards for digital technology governance. These frameworks facilitate cross-border cooperation and establish common principles to address global challenges in the digital age.

Digital technologies have become integral to modern life, influencing various sectors such as communication, healthcare, finance, and governance. As these technologies proliferate and evolve, legal frameworks have been developed to address the unique challenges they pose. In the initial stages of the digital revolution, existing legal frameworks were adapted to cover emerging digital issues. For example, the Electronic Communications Privacy Act (ECPA) of 1986 extended restrictions on government wiretaps of telephone calls to include electronic data transmission. Similarly, the Computer Fraud and Abuse Act (CFAA) of 1986, aimed at combating hacking, made it illegal to access a computer without authorization, reflecting early concerns about computer security and data breaches. It criminalizes accessing computer systems without authorization or exceeding authorized access and addresses issues such as data theft, fraud, and damage to computer systems.

As digital technologies advanced and became more pervasive, specific regulations were introduced to address the challenges. The General Data Protection Regulation, implemented in the European Union in 2018, is one of the most comprehensive data protection regulations globally. It sets stringent requirements for data privacy and gives individuals significant control over their personal data. In the United States, the Digital Millennium Copyright Act (DMCA) of 1998 addresses copyright issues in the digital realm, particularly focusing on digital rights management and the responsibilities of internet service providers in handling copyright infringement.

The GDPR is a landmark regulation that has significantly influenced global data protection standards. It grants individuals rights to access their data, correct inaccuracies, and request deletion (the "right to be forgotten"). Organizations are required to implement appropriate technical and organizational measures to ensure data security and compliance with GDPR, with substantial fines for non-compliance (Mendoza, 2019). The DMCA, on the other hand, addresses digital copyright issues, prohibiting the circumvention of digital rights management (DRM) technologies and limiting the liability of internet service providers for user-generated content, provided they comply with specific requirements, such as removing infringing content upon notification. Both the ECPA and CFAA in the US and the GDPR in the EU address privacy and data protection concerns but with different emphases. The ECPA and CFAA focus on regulating access to electronic communications and preventing unauthorized access to computer systems, whereas the GDPR focuses on protecting personal data and ensuring individuals' rights regarding their data.

More recently, the new European Union Artificial Intelligence Act aims to establish a comprehensive regulatory framework for artificial intelligence within the European Union. This Act categorizes AI systems based on their risk levels, with stricter regulations for higher-risk applications, mandates transparency in AI operations and establishes accountability mechanisms for AI developers and users, including compliance obligations and penalties for non-compliance.

At the global level, these EU regulations set benchmarks and influence regulatory developments in other regions. Many countries are considering or have already adopted data protection laws inspired by the GDPR to bolster their own digital governance frameworks. The AI Act's emphasis on risk-based regulation and accountability mechanisms for AI developers and users is also informing international discussions on AI ethics and governance. Organizations worldwide are increasingly mindful of complying with EU standards to maintain market access and uphold reputational integrity, underscoring the GDPR and AI Act's global impact on digital regulation. Thus, while rooted in the EU's regulatory framework, these legislations are shaping global norms and practices in digital ethics, data protection, and AI governance.

The OECD Guidelines on AI provide a global framework for the responsible development and deployment of AI technologies. These guidelines emphasize human-centred values, ensuring that AI systems respect human rights, democracy, and the rule of law. They also stress the importance of robustness and safety, requiring AI systems to be technically robust and safe throughout their lifecycle, with mechanisms to ensure accountability. Additionally, the guidelines advocate for transparency and explainability, ensuring that AI operations are transparent and that stakeholders have access to information that allows for understanding and scrutiny of AI systems.

The development of specific legal frameworks to govern digital technologies has been crucial in addressing the unique challenges posed by the digital age. Regulations such as GDPR, DMCA, CFAA, the new EU AI Act, and OECD guidelines on AI represent significant steps towards ensuring that digital technologies are used responsibly and ethically. As we move forward, understanding these foundational regulations will be essential for navigating the complex landscape of digital ethics and AI governance.

EU'S ARTIFICIAL INTELLIGENCE ACT AND OECD'S AI PRINCIPLES

The development and deployment of artificial intelligence technologies present numerous ethical, legal, and societal challenges. To address these, comprehensive regulatory frameworks and guidelines are essential. Two significant efforts in this direction are the EU's Artificial Intelligence Act and the OECD's AI Principles. This section explores the key provisions, objectives, and implications of these regulatory approaches.

The EU's Artificial Intelligence Act

The European Union's Artificial Intelligence Act (AI Act) is a groundbreaking legislation aimed at creating a comprehensive regulatory framework for artificial intelligence within the European Union. The Act is designed to address the ethical, legal, and societal challenges posed by AI technologies while fostering innovation and protecting fundamental rights. With AI technologies rapidly advancing and becoming integral to various sectors, this legislation marks a significant step towards responsible AI governance.

The AI Act adopts a risk-based approach, categorizing AI systems into four risk levels: unacceptable risk, high risk, limited risk, and minimal risk. This tiered structure allows for proportionate regulatory measures based on the potential impact of AI systems (Kop, 2021). AI systems deemed to pose unacceptable risks are prohibited outrightly. This category includes AI applications that manipulate human behavior, exploit vulnerabilities, or employ subliminal techniques that could result in psychological or physical harm. By banning such high-risk applications, the EU aims to protect individuals from potentially dangerous or exploitative AI technologies.

High-risk AI systems, such as those used in critical infrastructure, healthcare, education, employment, and law enforcement, are subject to stringent requirements. These systems must undergo rigorous testing, documentation, transparency, and human oversight to ensure safety and compliance. For instance, an AI system used in healthcare for diagnostic purposes must be thoroughly tested to ensure its accuracy and reliability. Similarly, AI used in law enforcement for surveillance or predictive policing must be transparent and subject to human oversight to prevent abuses and ensure accountability.

Limited-risk AI systems, like chatbots and virtual assistants, are subject to transparency obligations. Users must be informed when they are interacting with an AI system, which enhances awareness and informed consent. This transparency helps build trust between users and AI developers by ensuring that users understand they are engaging with a machine rather than a human. Minimal-risk AI systems, such as AI-enabled video games or spam filters, are largely exempt from regulatory requirements due to their lower potential for harm. However, even these systems are encouraged to adhere to best practices to ensure ethical deployment.

Transparency is a cornerstone of the AI Act, especially for high-risk AI systems. Developers and operators must provide clear information about the capabilities, limitations, and intended uses of their AI systems. This includes detailed documentation and disclosure of the system's logic and decision-making processes. These transparency requirements aim to enhance user trust and facilitate accountability. Transparency ensures that AI-generated decisions are not shrouded in opacity, empowering individuals to make informed choices (Musch, Borrelli, & Kerrigan, 2023). For instance, if an AI system makes a decision that significantly impacts an individual, the person affected has the right to understand how that decision was made and challenge it, if necessary.

The AI Act also establishes robust accountability mechanisms to ensure responsible AI development and deployment. Key measures include implementing comprehensive risk management systems, ongoing monitoring and evaluation of AI performance and impact, and ensuring effective human oversight for high-risk AI systems. Compliance and enforcement are critical components, with national supervisory authorities tasked with monitoring adherence and imposing substantial penalties for non-compliance, including fines of up to 7% of a company's global annual revenue. This strong enforcement mechanism underscores the EU's commitment to upholding high standards in AI governance.

While imposing regulatory requirements to ensure safety and ethical use, the AI Act also aims to foster innovation and competitiveness within the EU. It encourages the development of AI technologies through support for research and innovation initiatives. For example, the Act includes provisions for funding AI research and development projects, facilitating collaboration between academia, industry, and government bodies. This balanced approach promotes technological advancement while safeguarding fundamental rights.

The Act is designed to achieve several key objectives: protecting fundamental rights by ensuring AI systems respect privacy, non-discrimination, and human dignity; enhancing trust and accountability through transparency and clear lines of responsibility; promoting innovation by providing a clear regulatory framework; and ensuring the safety and reliability of AI systems to minimize potential harm. These objectives reflect the EU's broader commitment to ethical AI governance and the protection of individual rights.

The AI Act has significant implications for various stakeholders, including AI developers, operators, users, and regulatory authorities. Developers and operators will need to make substantial investments in risk management, transparency, and human oversight to comply with the stringent requirements for high-risk AI systems.

This may involve adopting new technologies and practices to ensure compliance, which could drive innovation and improve AI system performance. For users, the Act aims to protect their rights and build trust in AI systems by ensuring transparency, accountability, and safety, providing mechanisms to challenge and seek redress for harmful AI decisions. Users will benefit from greater transparency and accountability, leading to increased confidence in AI technologies.

Regulatory authorities are empowered to monitor and enforce compliance with AI regulations, requiring significant resources and expertise to effectively oversee the diverse and rapidly evolving AI landscape. These authorities will play a crucial role in ensuring that AI systems adhere to the standards set out in the AI Act, providing oversight and enforcement to maintain high ethical standards. The establishment of a European Artificial Intelligence Board, composed of representatives from each member state's supervisory authority, will facilitate cooperation and coordination among regulatory bodies, ensuring consistent application of the AI Act across the EU (Khajuria, 2024).

The EU's Artificial Intelligence Act represents a pioneering effort to regulate AI technologies comprehensively. By adopting a risk-based approach, mandating transparency, and ensuring accountability, the Act aims to protect fundamental rights, enhance public trust, and promote innovation. As AI continues to evolve, the AI Act will play a crucial role in shaping the ethical and legal landscape of AI governance within the European Union and potentially influence global standards. The Act's emphasis on ethical considerations, combined with its support for innovation, positions the EU as a leader in responsible AI governance, setting a benchmark for other regions to follow.

The OECD's Principles on Artificial Intelligence

The Organisation for Economic Co-operation and Development (OECD) has developed comprehensive guidelines known as the OECD Principles on AI, which were adopted in May 2019 and later updated in 2024. These recommendations provide a global framework for the responsible development, deployment, and governance of artificial intelligence technologies. Endorsed by OECD member countries and several non-member countries, the principles aim to promote trustworthy AI that is aligned with human rights, democratic values, and inclusivity.

Key Principles of the OECD on AI

Human-Centered Values and Fairness

The OECD emphasizes that AI systems should be designed and operated in a way that respects human rights and democratic values. This includes ensuring that AI technologies do not perpetuate discrimination or bias. Fairness is a core principle, advocating for AI systems that promote inclusivity and diversity. For example, AI used in hiring practices should be scrutinized to prevent bias against certain demographic groups, ensuring that all candidates are evaluated fairly.

Robustness, Security, and Safety

The principles highlight the importance of technical robustness and security for AI systems throughout their lifecycle. This involves implementing safety measures to prevent malfunctions, ensuring resilience against cyber threats, and conducting regular assessments to identify and mitigate risks. AI systems should be designed to operate safely under normal conditions and to fail gracefully when encountering unexpected situations. For instance, autonomous vehicles must undergo rigorous testing to ensure they can handle diverse and unpredictable road conditions safely.

Transparency and Explainability

Transparency and explainability are crucial for building trust in AI systems. The OECD recommends that AI systems provide clear and understandable information about their capabilities, limitations, and decision-making processes (Agarwal, 2023). This transparency allows stakeholders, including users and those affected by AI decisions, to understand how decisions are made and to challenge or appeal those decisions if necessary. For example, an AI system used in credit scoring should explain the factors that influence credit decisions, enabling individuals to understand and potentially contest their scores.

Accountability

Accountability mechanisms are essential to ensure that those who design, develop, deploy, or operate AI systems can be held responsible for their actions. The OECD advocates for clear governance frameworks that delineate responsibilities and ensure oversight throughout the AI system's lifecycle. Developers and operators must implement processes to monitor and audit AI systems, ensuring they adhere to ethical standards and legal requirements.

Collaboration and Stakeholder Engagement

The OECD underscores the importance of collaborative approaches to AI governance. This involves engaging a broad range of stakeholders, including governments, industry, academia, civil society, and the public to address the diverse challenges posed by AI. Collaborative governance can facilitate the sharing of best practices, harmonize standards, and ensure that AI development aligns with societal values. For instance, multi-stakeholder forums can help shape policies that balance innovation with ethical considerations, ensuring that AI benefits all segments of society.

Implications of the OECD Principles

The OECD Principles have significant implications for the global AI landscape. By providing a common framework, they aim to harmonize AI governance across different countries, promoting international cooperation and reducing regulatory fragmentation. This harmonization is crucial for global companies that operate across multiple jurisdictions, as it simplifies compliance and fosters a level playing field.

For governments, the recommendations offer a blueprint for developing national AI strategies and policies. By aligning with OECD principles, countries can ensure that their AI policies promote innovation while safeguarding ethical standards and human rights. By emphasizing transparency, ethical considerations, human rights frameworks, multi-stakeholder collaboration, risk assessment, and international cooperation, the OECD helps ensure that innovations benefit society while minimizing potential harm.

This alignment can also enhance international collaboration, enabling countries to address cross-border AI challenges more effectively. For the private sector, the OECD Principles provide guidelines for responsible AI development and deployment. Companies can use these principles to design AI systems that are ethical, transparent, and accountable, thereby building trust with consumers and stakeholders (Camilleri, 2023). Adhering to OECD principles can also mitigate legal and reputational risks, as companies demonstrate their commitment to ethical AI practices.

Academia and research institutions play a critical role in advancing AI technologies and understanding their societal impacts. The OECD recommendations encourage interdisciplinary research that integrates technical, ethical, and social perspectives, fostering innovations that are both cutting-edge and ethically sound. Civil society organizations and the public are also essential stakeholders in AI governance. The OECD emphasizes the importance of public engagement and awareness, ensuring that the benefits and risks of AI are widely understood. By involving civil society, the recommendations help ensure that AI development is transparent and aligned with public interests.

The OECD's Principles on AI provide a robust framework for the ethical and responsible development of AI technologies. By emphasizing human-centered values, robustness, transparency, accountability, and collaboration, the principles aim to ensure that AI serves the public good and upholds democratic principles. As countries and organizations around the world adopt these principles, they contribute to the development of trustworthy AI systems that can drive innovation while protecting fundamental rights and societal values. The OECD's leadership in promoting international cooperation and best practices in AI governance sets a global standard that can inspire and guide efforts to harness the benefits of AI responsibly.

DATA PROTECTION AND PRIVACY LAWS

The General Data Protection Regulation (GDPR) is a cornerstone of data protection legislation within the European Union, and its implications extend globally, significantly influencing digital ethics. Enforced since May 2018, GDPR establishes rigorous standards for data privacy and protection, emphasizing individual rights and corporate accountability. Its comprehensive framework addresses the collection, processing, storage, and transfer of personal data, aiming to safeguard privacy in an increasingly digital world.

GDPR enshrines several rights for individuals, known as data subjects, aimed at giving them control over their personal data. These rights include the right to access, which allows individuals to request access to their personal data held by organizations, ensuring transparency and accountability in data processing activities. The right to rectification lets individuals request corrections to inaccurate or incomplete personal data, ensuring data accuracy. The right to erasure, or the right to be forgotten, enables individuals to request the deletion of their personal data under certain conditions, enhancing their privacy. The right to restrict processing allows individuals to limit how their data is processed in specific situations, providing more control over their data. The right to data portability enables individuals to receive their personal data in a structured, commonly used, and machine-readable format, facilitating the transfer of data between service providers. Lastly, the right to object allows individuals to challenge the processing of their personal data for specific purposes, such as direct marketing.

GDPR outlines fundamental principles for data processing, ensuring that organizations handle personal data responsibly and ethically. These principles include lawfulness, fairness, and transparency, mandating that data be processed in a lawful, fair, and transparent manner (Dumas, 2019). Purpose limitation requires that data be collected for specified, explicit, and legitimate purposes and not further processed in a manner incompatible with those purposes. Data minimization mandates that data collection be limited to what is necessary for the intended purposes. The principle of accuracy requires data to be accurate and kept up to date. Storage limitation stipulates that data should be kept in a form that permits identification of individuals only as long as necessary for processing purposes. Lastly, integrity and confidentiality demand that data be processed securely, protecting against unauthorized or unlawful processing and accidental loss, destruction, or damage.

GDPR imposes strict accountability obligations on organizations, requiring them to demonstrate compliance with its principles. Key measures include conducting Data Protection Impact Assessments (DPIAs) for high-risk data processing activities to identify and mitigate privacy risks (Blume, 2018). Organizations engaged in large-scale data processing must appoint a Data Protection Officer (DPO) to oversee compliance with GDPR (Smouter-Umans, 2017). Additionally, GDPR mandates that organizations report data breaches to supervisory authorities within 72 hours and notify affected individuals without undue delay, ensuring prompt and transparent handling of data breaches.

The impact of GDPR on digital ethics is profound, significantly enhancing privacy protection and fostering greater trust between individuals and organizations. By granting individuals control over their personal data, GDPR empowers them to make informed decisions about how their data is used, promoting respect for individuals' autonomy and privacy. The stringent requirements for data protection and accountability have led organizations to prioritize ethical data practices. Companies are now more vigilant about data security, accuracy, and transparency reflecting a broader commitment to ethical standards. This shift towards corporate responsibility aligns with the ethical principle of accountability ensuring that organizations are answerable for their data processing activities.

GDPR's emphasis on data minimization and purpose limitation directly impacts AI and machine learning practices. AI systems often require large datasets for training and improving algorithms. GDPR mandates that data collection be limited to what is necessary, challenging organizations to balance data needs with privacy requirements (Singh, 2024). This has led to the development of more privacy-preserving techniques, such as federated learning and differential privacy, aligning AI development with ethical standards of privacy and data protection.

Although GDPR is an EU regulation, its impact is global. Companies worldwide, especially those dealing with European customers, must comply with GDPR standards (Wang, 2020). This has set a global benchmark for data protection, influencing legislation in other jurisdictions, such as the California Consumer Privacy Act (CCPA) in the United States (Harris, 2020). The international influence of GDPR promotes a more standardized approach to data protection, fostering global digital ethics. Despite its positive impact, GDPR has faced criticisms and challenges. Small and medium-sized enterprises (SMEs) often struggle with the complexity and cost of compliance. Additionally, there are concerns about the regulation's impact on innovation, particularly in data-driven sectors. Balancing robust data protection with the need for innovation remains an ongoing ethical challenge.

GDPR has profoundly impacted digital ethics by enhancing privacy protection, promoting accountability, and influencing global data protection standards. Its comprehensive framework ensures that personal data is handled ethically, respecting individuals' rights and fostering trust. As technology continues to evolve, GDPR's principles will remain crucial in guiding ethical data practices, ensuring that the benefits of digital innovation are realized without compromising fundamental rights and freedoms.

Intersection of Data Protection Laws and Ethical Considerations in Digital Technologies

The Cambridge Analytica scandal is a landmark case that underscores the critical intersection of data protection laws and ethical considerations in digital technologies. In 2018, it was revealed that Cambridge Analytica had harvested personal data from millions of Facebook users without their consent. This data was used to create detailed voter profiles and influence political campaigns, including the 2016 U.S. presidential election and the Brexit referendum. The scandal highlighted several GDPR violations, such as the failure to obtain explicit user consent and the lack of transparency regarding data usage. Ethically, the unauthorized use of personal data for political manipulation raised significant concerns about privacy and the integrity of democratic processes. The fallout from the scandal led to increased awareness and stricter enforcement of data protection laws, emphasizing the need for robust frameworks to safeguard individual rights and ensure ethical data practices.

Another pertinent example is Google's Street View project, which faced controversy in 2010 when it was discovered that Google vehicles collected personal data from unsecured Wi-Fi networks during the mapping process. This unauthorized data collection violated privacy laws and the GDPR principle of data minimization. Additionally, Google failed to inform individuals about the data collection, breaching transparency requirements. Ethically, this incident underscored the importance of obtaining informed consent and ensuring data security. In response, Google faced fines and committed to improving its data collection practices, highlighting the critical need for transparency and responsible data management in digital technologies.

These case studies illustrate the vital role of data protection laws like GDPR in addressing ethical considerations in digital technologies. They highlight the importance of transparency, informed consent, accountability, and respect for privacy in data practices. As digital technologies continue to evolve, robust data protection frameworks and ethical guidelines will remain essential to safeguard individual rights and promote trust in digital ecosystems. There are probably more risks with respect to the AI than the Cambridge Analytica. This requires greater emphasis on overall education in the area of AI and ICT at all levels of international polity and more transparency of the ICT companies, platforms, and networks globally.

ETHICAL USE OF AI: LEGAL CHALLENGES AND SOLUTIONS

In the rapidly evolving landscape of artificial intelligence, the ethical use of AI has emerged as a paramount concern across diverse sectors. As AI technologies permeate industries ranging from healthcare to finance to criminal justice, they bring with them a multitude of ethical considerations that must be carefully navigated. The integration of AI into decision-making processes raises profound questions about fairness, transparency, and accountability, prompting a critical examination of existing legal frameworks. In this section, we embark on a comprehensive exploration of the legal challenges and solutions surrounding the ethical use of AI. By delving into specific sectors and analyzing the intersection of AI with legal regulations, this section aims to elucidate the complexities inherent in AI decision-making and offer insights into how legal frameworks can be adapted to ensure responsible and ethical AI deployment.

Liability and Accountability Issues in AI Decision-Making Processes

As artificial intelligence becomes increasingly integrated into various decision-making processes, it raises critical questions regarding liability and accountability. These issues are particularly complex due to the autonomous nature of AI systems, the potential for unforeseen outcomes, and the difficulty in tracing decisions back to specific human actions or intents. Many AI systems, especially those based on machine learning and deep learning, operate through intricate algorithms that are not easily interpretable by humans. This opacity, often referred to as the "black box" problem, makes it difficult to pinpoint the exact cause of an error or harmful decision, complicating efforts to assign responsibility (Huang, 2023). For instance, if an AI system in healthcare misdiagnoses a patient, it can be challenging to determine whether the fault lies with the data used to train the algorithm, the design of the algorithm itself or its implementation in the clinical setting.

Legal frameworks are gradually evolving to address these challenges. The European Union's AI Act and the General Data Protection Regulation are at the forefront of these efforts, aiming to provide guidelines for transparency, accountability and risk management in AI applications. The AI Act, for example, classifies AI systems based on their risk levels and imposes stricter requirements on high-risk AI systems, including robust documentation, transparency obligations and human oversight mechanisms. These regulations aim to ensure that AI systems are designed and deployed responsibly, reducing the likelihood of harmful outcomes and providing clearer pathways for accountability. The United States, meanwhile, is considering similar regulatory approaches, focusing on the ethical implications of AI and the need for transparent and fair AI systems.

Determining liability in AI decision-making often involves multiple parties, including AI developers, deployers, and users. In the event of an adverse outcome, such as a financial loss due to an AI-driven trading algorithm, liability could potentially be attributed to the developers who created the algorithm, the financial institution that deployed it or even the end-users who relied on its recommendations. Legal principles such as product liability, negligence, and breach of duty can come into play, but traditional notions of liability may need adaptation to account for the unique characteristics of AI. This complexity is further illustrated by the advent of autonomous vehicles, which has prompted significant debate over who should be held responsible in the event of an accident. Manufacturers, software developers and vehicle operators all play a role, and liability can depend on factors such as the level of autonomy of the vehicle and the circumstances of the incident.

Case studies from various sectors highlight the complexities of assigning liability. In 2018, an Uber autonomous vehicle was involved in a fatal accident, leading to extensive investigations and discussions on liability. The incident underscored the need for clear regulatory guidelines and accountability frameworks (Nersessian & Mancha, 2020). Similarly, in healthcare, a misdiagnosis by an AI system leading to patient harm prompts legal proceedings to determine liability, involving AI developers, healthcare providers, and insurers. Such cases emphasize the necessity for rigorous testing, clear documentation and human oversight in AI decision-making processes. In the financial sector, an AI algorithm discriminating against a minority group in lending decisions can trigger regulatory scrutiny and legal action under anti-discrimination laws. These examples show that assigning liability is not straightforward and requires a nuanced understanding of how AI systems operate and interact with human decisions.

To mitigate liability risks and ensure accountability in AI decision-making, several best practices have emerged. These include developing AI systems with explainable algorithms that allow stakeholders to understand how decisions are made, ensuring that AI systems undergo rigorous testing and validation before deployment, particularly in high-stakes sectors like healthcare and finance, and maintaining comprehensive documentation of the AI system's development, training data, and decision-making criteria. Human oversight and intervention mechanisms are also crucial, particularly for high-risk applications where automated decisions can have significant consequences. Promoting ethical AI development practices that prioritize fairness, non-discrimination, and respect for user rights is essential. These practices involve incorporating ethical considerations into the design and deployment phases and ensuring that AI systems are aligned with societal values and norms.

In a nutshell, addressing liability and accountability in AI decision-making processes requires a multifaceted approach that combines robust legal frameworks, best practices for transparency and oversight, and ongoing dialogue among regulators, developers, and users. As AI technology continues to evolve, so must our approaches to ensuring that its deployment is both responsible and accountable. The future of AI holds great promise but realizing its potential while safeguarding against its risks demands a concerted effort to develop and implement effective regulatory and ethical guidelines. This will not only help in managing the legal complexities of AI but also foster public trust in these transformative technologies.

Legal Frameworks Addressing the Ethical Use of AI in Various Sectors

Artificial intelligence has become integral to various sectors, raising substantial ethical and legal concerns. The healthcare sector, for instance, is governed by stringent regulations such as the Health Insurance Portability and Accountability Act (HIPAA) in the U.S. and the Medical Device Regulation (MDR) in the EU. These laws oversee the use of AI in medical devices and patient data management (Hsieh, 2014). Key ethical issues include ensuring patient privacy, obtaining informed consent for AI-driven diagnostics, and maintaining transparency in AI decision-making processes (Wilkes, 2014). Legal challenges involve balancing innovation with patient safety and privacy, addressing biases in AI algorithms that may lead to discriminatory practices, and defining liability for AI-driven medical errors. Solutions lie in rigorous testing and validation of AI tools, clear guidelines for AI use in patient care, and fostering interdisciplinary collaboration to align AI development with ethical standards.

In the financial services sector, regulations like the General Data Protection Regulation in the EU and the Fair Credit Reporting Act (FCRA) in the U.S. govern the use of AI in credit scoring, fraud detection, and trading. Ethical concerns revolve around ensuring fairness and transparency in AI-driven financial decisions, preventing discriminatory practices in lending and credit scoring, and protecting consumer data. Legal challenges include defining accountability for AI-driven financial decisions, managing the risk of algorithmic bias, and ensuring compliance with diverse international regulations. Addressing these issues requires developing transparent AI models, conducting regular bias audits and establishing international regulatory standards for AI in finance.

The criminal justice sector also sees significant AI application, with constitutional protections against discrimination and unjust punishment, along with data protection laws like GDPR, guiding AI use. Ethical concerns include ensuring AI-driven law enforcement tools do not perpetuate biases, maintaining fairness in sentencing algorithms, and protecting individual rights. Legal challenges encompass addressing potential biases in AI systems used for predictive policing, ensuring transparency and accountability in AI-driven judicial decisions, and safeguarding against privacy violations. Effective solutions involve stringent oversight mechanisms, promoting transparency in AI algorithm development, and ensuring robust legal recourse for individuals affected by AI decisions.

Liability and accountability in AI decision-making are crucial areas of concern. Legal frameworks, such as the EU's Product Liability Directive and emerging AI-specific regulations, aim to clarify accountability. Ethical concerns here include determining who is liable for AI errors or harms, ensuring AI developers and deployers are accountable, and providing redress mechanisms for affected individuals. Legal challenge is to establish clear lines of responsibility in complex AI systems, addressing the "black box" nature of AI decision-making and ensuring timely and effective legal remedies. A possible solution could be to include crafting comprehensive liability laws covering various AI use cases, encouraging the development of explainable AI and fostering interdisciplinary collaboration to address legal and technical complexities.

Case studies highlight the legal challenges and best practices in different sectors. In healthcare, a misdiagnosis by an AI system leading to patient harm prompts legal proceedings to determine liability, involving AI developers, healthcare providers, and insurers. Best practices will include robust validation processes, transparency with patients about AI use and human oversight in critical decisions. In financial services, an AI algorithm discriminating against a minority group in lending decisions can trigger regulatory scrutiny and legal action under anti-discrimination laws. Regular bias audits, transparent AI decision-making processes, and inclusive data sets for training AI models are crucial best practices. In predictive policing, an AI system disproportionately targeting minority communities may face legal challenges based on civil rights violations and lack of transparency. Developing AI tools with community input, ensuring transparency in AI deployment, and conducting independent audits to detect and mitigate biases are essential measures.

The ethical use of AI across various sectors presents significant legal challenges, including defining liability, ensuring transparency and protecting individual rights. Legal frameworks must evolve to address these complexities, fostering an environment where AI can be developed and deployed responsibly. Collaboration between regulators, industry stakeholders, and civil society is crucial to create effective and adaptable legal frameworks that promote ethical AI use while safeguarding human rights and social justice.

FINAL REFLECTIONS: SHAPING A SUSTAINABLE AND EQUITABLE DIGITAL FUTURE

To effectively address the evolving ethical concerns associated with AI and digital technologies, enhancing existing legal frameworks is imperative. The following recommendations aim to provide a comprehensive approach to updating and refining these frameworks to ensure they are robust, adaptable, and capable of addressing the complex ethical issues that arise in the digital age:

Develop Adaptive Regulatory Mechanisms

Current regulatory frameworks often struggle to keep pace with the rapid evolution of AI technologies. Therefore, it is essential to create adaptive regulatory mechanisms that can quickly respond to new developments. This could include establishing specialized regulatory bodies or task forces dedicated to monitoring AI advancements and updating regulations as needed. These bodies should have the authority to issue interim guidelines and adapt existing regulations in response to emerging technologies and practices.

Foster International Cooperation

Given the global nature of AI development and deployment, international cooperation is crucial. Harmonizing AI regulations across jurisdictions can help ensure consistent ethical standards and reduce regulatory fragmentation. International organizations, such as the OECD and the United Nations, should facilitate dialogue and collaboration among nations to develop common ethical principles and regulatory standards for AI. Bilateral and multilateral agreements can also play a role in aligning national regulations with global best practices.

Enhance Transparency and Accountability

Legal frameworks should mandate greater transparency and accountability in AI systems. This includes requirements for explainability, where AI developers must provide clear documentation and explanations of how their systems make decisions. Transparency can be further enhanced by implementing audit trails that track AI decision-making processes and data usage. Establishing accountability frameworks that clearly delineate responsibilities among developers, deployers, and users of AI systems is also essential. These frameworks should include provisions for liability in cases of harm or ethical violations.

Address Bias and Fairness

To combat bias and ensure fairness in AI systems, legal frameworks should include stringent requirements for bias detection, mitigation, and reporting. Regular audits of AI systems for bias should be mandated, and developers should be required to use diverse and representative data sets. Additionally, frameworks should promote the development of standardized metrics for assessing fairness in AI decisions. Creating an independent oversight body to review and address complaints related to AI bias and discrimination can also help enforce these standards.

Strengthen Data Protection and Privacy

Robust data protection and privacy laws are crucial for safeguarding individuals' rights in the digital age. Legal frameworks should enforce strict compliance with data protection regulations, such as the GDPR, and ensure that AI systems adhere to principles of data minimization, purpose limitation, and user consent. Enhancing cybersecurity measures to protect AI systems from data breaches and malicious attacks is also essential. Regulators should work to close any gaps in existing data protection laws to address new challenges posed by AI technologies.

Promote Ethical AI Development and Use

Legal frameworks should encourage the integration of ethical considerations throughout the AI development lifecycle. This includes establishing ethical guidelines for AI research and development, promoting ethical training for AI developers, and incentivizing the creation of AI systems that prioritize human well-being. Governments can support ethical AI initiatives through funding and partnerships with academia and industry. Additionally, legal frameworks should require companies to conduct ethical impact assessments for AI applications and publish the results for public scrutiny.

Implement a Multistakeholder Approach

Regulating AI ethics effectively requires input from a diverse range of stakeholders, including government agencies, industry leaders, academic experts, civil society organizations, and the public. Legal frameworks should facilitate multistakeholder engagement in the regulatory process, ensuring that different perspectives and interests are considered. Public consultations, advisory committees, and collaborative platforms can help gather input and foster consensus on ethical standards and regulations. An example is the collaboration between the UK government and tech companies in developing the AI Code of Conduct to promote transparency, accountability, and fairness in AI systems

Facilitate Continuous Learning and Adaptation

AI ethics and regulatory landscapes are continuously evolving. Legal frameworks should support ongoing education and training for regulators, policymakers, and other stakeholders to keep them informed about the latest developments in AI and digital technologies. Establishing centers of excellence or think tanks focused on AI ethics can provide valuable insights and recommendations for regulatory updates. Encouraging research on the societal impacts of AI and funding studies on ethical issues can also inform future regulatory decisions.

In conclusion, shaping a sustainable and equitable digital future requires a proactive, flexible, and inclusive approach to legal regulation. By prioritizing ethical considerations, stakeholders can ensure that AI and digital technologies contribute positively to society. Policymakers, industry leaders, and civil society must work together to create a regulatory environment that supports responsible innovation, protects human rights, and promotes social responsibility. Further research and policy development are essential to advance human rights, social responsibility, and digital ethics. This ongoing effort will help address new ethical challenges as they arise and ensure that the benefits of AI and digital technologies are realized in a manner that is fair, transparent, and aligned with societal values. By committing to these principles, we can shape a future where technology serves humanity and fosters a more just and equitable world.

DEFINITIONS

1. Artificial Intelligence - It refers to the simulation of human intelligence processes by machines, typically through the use of algorithms and data analysis, enabling them to perform tasks that traditionally require human cognition.
2. Digital Ethics - It refers to the moral principles and guidelines that govern the responsible and ethical use of digital technologies, including considerations of privacy, data protection, transparency, fairness and accountability.

3. General Data Protection Regulation - EU regulation which sets out guidelines for responsible data management, enhances individuals' privacy rights, and imposes penalties for failure to comply with its provisions.
4. Organisation for Economic Co-operation and Development (OECD) - It is an international organization comprising 38 member countries, dedicated to promoting policies that foster economic growth, prosperity, and sustainable development worldwide.
5. Multi-stakeholder approach - It involves including various groups, such as governments, businesses, civil society, and individuals, in decision-making processes to tackle complex issues collaboratively.
6. Data protection – It refers to the practices and legal measures implemented to safeguard personal information from unauthorized access, misuse, or disclosure, ensuring individuals' privacy and the security of their data.
7. Data Protection Impact Assessments (DPIAs) – These are comprehensive evaluations conducted to assess and mitigate potential risks to individuals' privacy and data protection rights arising from data processing activities.

REFERENCES

Agarwal, L. (2023). Defining Organizational AI Governance and Ethics. *SSRN*. https://ssrn.com/abstract=4553185

Almada, M., & Petit, N. (2022). The EU AI Act: Between product safety and fundamental rights. SSRN Electronic Journal. https://doi.org/10.2139/ssrn.4308072

Bankins, S., & Formosa, P. (2023). The ethical implications of artificial intelligence (AI) for meaningful work. *Journal of Business Ethics*, 185(4), 1–16. 10.1007/s10551-023-05339-7

Blume, J. (2018). Contextual extraterritoriality analysis of the DPIA and DPO provisions in the GDPR. *Georgetown Journal of International Law*, 49(4), 1425–1460.

Bygrave, L. A. (2017). Data protection by design and by default: Deciphering the EU's legislative requirements. *Oslo Law Review*, 4(2), 105–122. 10.18261/issn.2387-3299-2017-02-03

Camilleri, M. (2023). Artificial Intelligence Governance: Ethical Considerations and Implications for Social Responsibility. *Expert Systems: International Journal of Knowledge Engineering and Neural Networks*. Advance online publication. 10.1111/exsy.13406

Dumas, J. (2019). General Data Protection Regulation (GDPR): Prioritizing resources. *Seattle University Law Review*, 42(3), 1115–1128.

Harris, R. (2020). Forging path towards meaningful digital privacy: Data monetization and the CCPA. *Loyola of Los Angeles Law Review*, 54(1), 197–234.

Hsieh, R. (2014). Improving HIPAA enforcement and protecting patient privacy in digital healthcare environment. Loyola University Chicago Law Journal, 46(1), 175-224.

Huang, C., Zhang, Z., Mao, B., & Yao, X. (2023). An overview of artificial intelligence ethics. *IEEE Transactions on Artificial Intelligence*, 4(4), 799–819. 10.1109/TAI.2022.3194503

Khajuria, A. (2024). The EU AI Act: A landmark in AI regulation. Observer Research Foundation. https://www.orfonline.org/research/the-eu-ai-act-a-landmark-in-ai-regulation

Kop, M. (2021). EU Artificial Intelligence Act: The European approach to AI. Stanford - Vienna Transatlantic Technology Law Forum, Transatlantic Antitrust and IPR Developments, Stanford University, Issue No. 2/2021. https://law.stanford.edu/publications/eu-artificial-intelligence-act-the-european-approach-to-ai/

Libby, K. (2019). This Bill Hader Deepfake video is Amazing. It's also Terrifying for our Future. Popular Mechanics. Retrieved from https://www.popularmechanics.com/technology/security/a28691128/deepfaketechnology/

Mendoza, S. (2019). GDPR Compliance - It Takes a Village. *Seattle University Law Review*, 42(3), 1155–1162.

Musch, S., Borrelli, M., & Kerrigan, C. (2023). The EU AI Act as global artificial intelligence regulation. *SSRN*. https://doi.org/10.2139/ssrn.4549261

Nassar, A., & Kamal, M. (2021). Ethical dilemmas in AI-powered decision-making: A deep dive into big data-driven ethical considerations. *International Journal of Responsible Artificial Intelligence*, 11(8), 1–11.

Nersessian, D., & Mancha, R. (2020). From Automation to Autonomy: Legal and Ethical Responsibility Gaps in Artificial Intelligence Innovation. *SSRN*, 27(55). Advance online publication. 10.2139/ssrn.3789582

Pan, K. (2024). Ethics in the Age of AI: Research of the Intersection of Technology, Morality, and Society. *Lecture Notes in Education Psychology and Public Media*, 40(1), 259–262. 10.54254/2753-7048/40/20240816

Siau, K., & Wang, W. (2018). Building Trust in Artificial Intelligence, Machine Learning, and Robotics. *Cutter Business Technology Journal*, 31(2), 47–53.

Singh, H. (2024). Navigating the Legal Landscape of Information Governance. In Helge, K., & Rookey, C. (Eds.), *Creating and Sustaining an Information Governance Program* (pp. 217–235). IGI Global. 10.4018/979-8-3693-0472-3.ch012

Smouter-Umans, K. L. (2017). Research, GDPR, and the DPO how GDPR changes the game for those conducting research and the data supervisors. International Journal for the Data Protection Officer *Privacy Officer and Privacy Counsel*, 1(1), 26.

Stupp, C. (2019). Fraudsters Used AI to Mimic CEO's Voice in Unusual Cybercrime Case. The Wall Street Journal. Retrieved from https://www.wsj.com/articles/fraudsters-use-ai-to-mimic-ceos-voice-in-unusualcybercrime-case-11567157402

Tanaka, H. (2019). Impact of the GDPR on Japanese companies. *Business Law International*, 20(2), 137–146.

Wang, F. Y. (2020). Cooperative Data Privacy: The Japanese model of data privacy and the EU-Japan GDPR adequacy agreement. *Harvard Journal of Law & Technology*, 33(2), 661–692.

Wilkes, J. (2014). The creation of HIPAA culture: Prioritizing privacy paranoia over patient care. *Brigham Young University Law Review*, 2014(5), 1213–1250.

Chapter 6
Sovereignty Over Personal Data:
Legal Frameworks and the Quest for Privacy in the Digital Age

Andreja Primec
https://orcid.org/0000-0002-5615-7299
University of Maribor, Slovenia

Gal Pastirk
https://orcid.org/0009-0001-7877-639X
University of Maribor, Slovenia

Igor Perko
https://orcid.org/0000-0002-6101-1091
University of Maribor, Slovenia

ABSTRACT

The contribution focuses on an individual's right to control their personal data and the right to privacy, emphasizing the legal aspects and developments in the context of modern technological and digital challenges. These rights become especially critical in the context of swift advancements in technology and digitalization, enabling the accumulation, examination, distribution, and sharing of extensive data volumes in previously unattainable manners. There has been a surge in sensors embedded within smart devices, gathering data from various aspects of our professional and personal lives. Furthermore, artificial intelligence (AI) has emerged, capable of analysing and interpreting data in unthinkable ways. These shifting dynamics present a challenge to reconsider how data is regulated. It's not merely about updating the rules themselves; it involves reimagining the process of creating regulatory frameworks.

DOI: 10.4018/979-8-3693-3334-1.ch006

INTRODUCTION

Artificial intelligence overwhelms us in our private and business lives at every turn. The data that fuels its operation and further development is becoming increasingly valuable. Big companies managing vast amounts of data are therefore more successful in developing new tools and devices, making their products more efficient and reliable. In this battle for data, private companies, which started to acquire data with the advent of the Internet and have since seen their collections "skyrocket", are generally the winners. Personal data, which in its most general sense means "any information relating to an identified or identifiable individual" (Article 4(1) of the GDPR), occupies an essential place in private databases. Understanding and protecting this data is becoming increasingly important in the digital age, where privacy and personal data protection remain at the forefront of ethical and legal debates.

The right of individuals to control their data is a fundamental concept in modern data protection and privacy. This right is recognised and guaranteed in many international and national legal frameworks. One of the most well-known and influential examples is the European Union's General Data Protection Regulation (GDPR), cited above, which came into force in May 2018. The GDPR has set new privacy and data protection standards for individuals within the European Union (EU) and the European Economic Area (EEA). It sets out the rights of individuals about their personal data and the obligations of organisations that process that data. Of particular note is its extraterritorial reach, as it also applies to controllers or processors outside the EU who process individuals' data in the EU or monitor their behaviour within the EU (Ryngaert & Taylor, 2020). This means that the provisions of the GDPR must be respected by any company that processes European citizens' data (Martins Gonçalves, Mira da Silva & Rupino da Cunha, 2023) to strengthen the protection of fundamental rights and create fair competition in the EU market (Gömann, 2017).

The GDPR is considered the most modern data protection legislation in the world. The California Consumer Privacy Act of 2018 (CCPA) in the US (State of California, Department of Justice, 2018), the Personal Information Protection and Electronic Documents Act (PIPEDA) in Canada (Government of Canada, 2019) and the Lei Geral de Proteção de Dados Pessoais (LGPD) in Brazil (Presidência da República, Secretaria-Geral Subchefia para Assuntos Jurídicos, 2018), which, like the GDPR, lay down rules for the protection of citizens' privacy and personal data, and the obligations of companies that process the data, also play an essential role. However, there are also differences between them. In the EU and Canada, there is centralised control over the use of personal data in the private sector. At the same time, the US has minimal regulation, focusing on protecting freedom from government intrusion (Levin & Nicholson, 2023). As the authors of this paper are from an EU member state, they have limited their focus to the normative framework of the GDPR.

In light of the development of technology, which allows the accumulation of ever more significant amounts of data day by day, the Internet of things (IoT) and increasingly sophisticated artificial intelligence tools that will enable profiling and the automated use of big data, it is necessary to consider whether the right of the individual to control their data is genuinely guaranteed by the consent of the data subject, which the GDPR "offers" for this purpose? Obtaining consent for every processing may not be reasonable or feasible, while limiting consent by purpose may inhibit the further development of otherwise necessary and effective technologies. The implementation of the GDPR poses challenges for both data subjects and data controllers. In this context, data processors are also to be noted. While data controllers are natural or legal persons who, alone or jointly with others, determine the purposes and means of the processing, processors are persons who process personal data on behalf of the controller (Article 4(7) and (8) GDPR). Consequently, the data controller is responsible for establishing appropriate safeguards and ensuring that data subjects can exercise their rights. The data processor, in turn, is responsible for carrying out the personal data processing task set by the data controller (Peyrone & Wichadakul, 2023).

The main research question of the chapter is therefore whether the right of individuals to control their own data is adequately regulated in the changed context of the digital age and virtual reality. To this end we looked at how the existing regulatory framework, led by the GDPR, ensures individuals' personal data protection. In the first part of the paper (the theoretical background), we traced the development of the right to protect personal data and placed it in the context of the broader right to privacy. The exercise of the right to data control is based on the consent of the data holder, which can be problematic nowadays from several angles.

Furter, in the research for this paper, we examined the unresolved and complex issues of the GDPR, particularly concerning improving the data economy and the issue of data sharing. In this context, the new EU legislative package presents attractive solutions that build on and complement the GDPR. The paper thus introduces a new regulation called the Data Act, which is closely related to the IoT. The relation of the Data Act to the existing legislation is also the subject of the research. At the same time, the issue of non-personal data and mixed data sets is presented, closely intertwined with the problems above and the emergence of new technologies.

In the context of the research question, the technologies used in data collection and analysis cannot be ignored. Technologies used in collecting and analysing data used by the data collectors create a state of informational asymmetry. Therefore, we have analysed the options for balancing it in the section titled 'Mechanisms for the Management and Exchange of Personal Data'.

METHODOLOGY

In this paper, we have used classical methods of legal research. We employed historical and teleological methods to trace the development of the right to the protection of personal data and to situate it within the broader right to privacy. Simultaneously, we used comparative legal methods to examine the interconnectedness, as well as the diversity, of the two rights (Strban, 2022). The systematic method (Perenič, 1998) confirmed the right to control personal data as a complex right guaranteed by several individual rights in the GDPR.

We demonstrated that the GDPR does not directly provide a right to personal data protection. Instead, a thorough understanding of its key provisions guarantees the right to data control, which in turn ensures the exercise of the right to the protection of personal data, guaranteed by the highest national instruments: national constitutions, either as a stand-alone right or as an integral part of the right to privacy.

The research begins by analysing and systematically interpreting several EU legal and political documents (Pavčnik, 2015). We also used descriptive and historical methods to present the evolution of EU data protection and its relevance for development (Novak, 2021). Furthermore, the research employs a comparative method to analyse some features of the GDPR and the Data Act. We examined fundamental differences between personal and non-personal data in law using analytical and comparative methods. Mechanisms of personal data governance and sharing are examined, including existing conceptual and technological designs that could be integrated into data governance mechanisms of the personal data digital ecosystem. More specifically, it examines a conceptual proposal for personal data-sharing.

THEORETICAL BACKGROUND

Right to Control Personal Data

The key rights guaranteed by the GDPR to individuals (referred to as "data subject") include the right to information (Articles 13 and 14), the right of access by the data subject (article 15), right to rectification (article 16), right to erasure ('right to be forgotten') (article 17), right to restriction of processing (article 18), right to data portability (article 20) and s right to object (article 21). These rights give individuals autonomy and power over decisions relating to their data, including deciding how it is collected, used and shared. Consequently, they can be grouped under a common denominator: the "right to data self-determination" or, more generally, the "right to control personal data". As can be seen from the description of the individual rights (the GDPR articles just quoted), the right to control over data belongs to individuals if the data are processed based on their explicit consent to the processing, as well as if the processing is carried out on any other legitimate basis set out in Article 6 of the GDPR.

The interaction of individual rights, in particular data portability, the right to be forgotten and the right to information, presents the potential for a more comprehensive EU regulation on personal data, which could eventually lead to a better balance between the rights of data subjects and the economic needs of the big data economy (Uršić, 2018). With these rights, the GDPR makes an essential contribution to ensuring the protection of individuals over their personal data. Still, specific limitations exist, such as difficulties in managing consent and ensuring effective control over individuals' data. As Verhulst (2022, 3-4) points out, these shortcomings lead to persons who could benefit from access to data having limited access (data asymmetries), further to an imbalance of information between two or more parties such as individuals, large companies, researchers (information asymmetries) and to an unequal balance of power between the two parties, resulting in one party having a more remarkable ability to influence the decisions of the other (agency asymmetries). Organisations with access to large databases use these data in ways that may not be transparent or fully understood by the individuals whose data are collected and analysed. As a result, trust in technology is eroded, leading to feelings of powerlessness among individuals. Verholst (2022, 6) sees the solution in a multifaceted approach involving processes, people and organisations, policies, and products and technologies for the responsible implementation of digital self-determination as a new premise in personal data protection.

The right to control personal data, as guaranteed by modern legal frameworks in personal data protection, is strongly linked to the right to personal data protection. The former enables the exercise of the latter right. Historically, the right to personal data protection has developed as a 'sub-right' of a fundamental human right: the right to privacy or another aspect of privacy (Kovalenko, 2022), as we explain below.

The Right for Data Privacy

The origins of the right to privacy can be traced back to ancient Mesopotamia, India and Egypt (Sharari & Faqir, 2014). However, the seminal work that stimulated its legal development is the article "The Right to Privacy" by Samuel D. The authors argue for recognising a right to privacy as a response to technological advances and increased media attention. They conceptualise the right to privacy as a fundamental right for the dignity and autonomy of the individual (Warren & Brandeis, 1890). It gradually evolved from a bourgeois right to a universal human right, influenced by changing social norms, political ideologies and the rise of liberal democracies (Richardson, 2017). After the Second World War, privacy became an international human right, even before it was established in national legislation (Diggelmann & Cleis, 2014). The authors note that national legal frameworks only protected particular aspects of privacy, particularly the inviolability of the home and correspondence. The concept of privacy as public policy emerged in the 1960s and 1970s, coinciding with the rise of data protection legislation in Europe and the need for safeguards against large-scale government data processing (Bennet, 2018, 239).

The OECD initiated personal data protection in 1980 with the Guidelines on the Protection of Privacy and Transborder Flows of Personal Data (O'Leary et al., 1995). The OECD based personal data protection on eight principles (OECD, 2002, 14-16), which have been adopted in the national laws of several countries. Data controllers are required to obtain the consent of data subjects for the collection of personal data. More specifically, the collection of personal data and any such data should be obtained by lawful and fair means and, where appropriate, with the knowledge or consent of the data subject (Paragraph 7). The European Convention on Human Rights in Protocol 108 introduced a similar rule a year later. As Kovalenko (2022) points out, the ECHR does not guarantee adequate personal data protection due to the blurred distinction between the right to protection of personal data and the right to privacy.

The right to privacy is one of the most essential personal rights, but is also a powerful barrier and guarantee for everyone's personal sphere. The right to the protection of personal data includes the protection of identifiable information, while the right to privacy covers the protection of intimate aspects of an individual's life (Pintiliuc, 2018).

Important as it is, the right to privacy cannot and should not be unlimited. Usually, an individual's rights are limited by another individual's rights. Legal protection of personal data as a form of privacy is no exception to this, and it is therefore also possible to limit the right to data protection, for example, in the case of a conflict between one person's right to data protection and another person's right to intellectual property (Milosavljević, 2023).

Another example of a restriction on informational privacy is the right of access to public information. The information of a selected candidate on their eligibility for a senior position in a state-owned company (diploma, programme of work, certificates of work experience, certificate of no criminal record, etc.) is of a public nature, as the public has the right to know who will occupy such an essential position in the state, even at the expense of reducing their privacy (Information Commissioner, 2022, 31-32).

Recent events like the Cambridge Analytica scandal have focused global attention on privacy issues, leading companies like Facebook to adopt the EU's General Data Protection Regulation (GDPR) standards.

The European Charter of Fundamental Rights defines the right to privacy and data protection as two separate rights. Still, the substantive distinction between the two rights is not very clear. The confusing and interchangeable use of the terms 'privacy' and 'data protection' can be found in the case law of the Court of Justice of the EU (Milaj-Weishaar, 2020, p. 6). Although the EU Charter recognises data protection and the right to privacy (or respect for private life), they generally have very different constitutional histories. The right to privacy is often seen as a traditional right, while data protection is seen as a new right. A comprehensive analysis of the rights in the EU Member States' constitutions has shown that this distinction is inappropriate. Constitutional rights to data protection emerged around the same time and were often linked to a general right to privacy, but they are still found in only about half of EU countries (Erdos, 2021, 22-23).

Explicit consent as a freely given statement based on the informed consent of the individual (as nowadays present in Article 6(1)(a) and Article 7 of the GDPR) has been succeeded by Directive 95/46/EC of the European Parliament and of the Council on the protection of individuals in 1995. Directive 2002/58/EC of the European Parliament and of the Council on Privacy and Electronic Communications in 2002, among the acts cited above (Oh et al., The GDPR, which is based on and repealed the European Directive 95/46/EC on the protection of individuals concerning the processing of personal data, has built significantly on the existing legal framework for the protection of personal data, which is based on the right to control personal data. Recent events, such as the Cambridge Analytica scandal, have focused global attention on privacy issues, leading companies such as Facebook to adopt the standards set out in the EU General Data Protection Regulation (GDPR) (Bennet, 2018, 239).

Consent as a Legal Basis for Data Processing

The concept of the right to control personal data is most clearly expressed in the condition of consent for the lawfulness of data processing (Article 6(1)(a) GDPR) and in the individual rights of data subjects (Article 12 et seq. of the GDPR) (Thouvenin & Tamò-Larrieux, 2021).

Informed consent is the cornerstone of legislative efforts to ensure the right to data control. However, practice shows certain shortcomings. Legal requirements for consent lead to 'desensitisation of consent', undermining privacy protection and trust in data processing. Explicit consent further reduces the effectiveness of the consent mechanism (Schermer, Custersn & Hof, 2014). Complex data processing (mainly in algorithmic systems), manipulative design (such as dark patterns) and cognitive limitations of individuals (such as bounded rationality) make it difficult or impossible for data subjects, both adults and children, to understand how data is processed, what the potential risks are, and how to balance these risks with the benefits of data processing (Custers, 2022). Besides, data sharing may result in a potential loss of control over personal data, as data are across boundaries between software services, which could be addressed using blockchain technology(Peyrone & Wichadakul, 2023). Data mining approaches are typically binary, operating on an opt-in or opt-out basis (de Man et al., 2023), meaning they do not provide individuals with a choice. Data processing based on initial consent is often subject to change, while reuse issues are unclear, although they are key to unlocking the potential of data (Francis & Francis, 2017).

As an alternative, theoretical studies offer many new solutions and concepts. Data ownership can be understood as a form of ownership and as a form of control. The first concept would counter the EU's efforts to promote the free flow of data. Moreover, the position of individuals would worsen since once data ownership is transferred to companies, the latter would be excluded from using it. The GDPR establishes the second concept, although it raises the doubts we have outlined above. Whether data control balances economic and individual interests remains open (Thouvenin & Tamò-Larrieux, 2021).

In contrast to data ownership, the Data Commons model operates based on sharing data resources in a community, where data and information resources are shared between organisations and individuals under certain conditions. The Data Commons model supports open science and collaboration and encourages innovation. Besides, it includes ensuring consent for data use, controlling personal data and ensuring data security (Wong, Henderson & Ball, 2022). From a technological perspective, personal information management systems (PIMS) are mentioned in addition to blockchain technology. PIMS allow individuals to become 'keepers' of their own personal data. In addition to giving individuals more control over their data, it encourages the development of new human-centred business models and protects unlawful tracking and profiling techniques (European Data Protection Supervisor, 2021). As Verhulst (2023) notes, all the shortcomings of both existing and hypothetical methods of intervention call for a new approach to address the asymmetries of our age. He proposes a new conceptual solution based on the principle of digital self-determination. As a working definition of this principle, digital self-determination can be understood as "the principle of respecting, embedding, and enforcing people's and peoples' agency, rights, interests, preferences, and expectations throughout the digital data life cycle in a mutually beneficial manner for all parties involved (Verhulst, 2023, e14-5).

The legal framework of personal data governance should be closely connected to the executive part of implementing governance mechanisms. In executing personal data governance, we enter the digital world, which differs from the physical in the speed of data flow, the high capacity of focused organisations to manage data, and the limited capacity of individuals to participate actively in the data-sharing process. These particular circumstances should be considered when creating the regulative framework and designing the mechanisms of personal data governance and sharing (Perko, 2022).

Technological Framework

Some ground technologies that enable personal data analysis and usage are Machine Learning, which can be applied to analyse data in various forms, including pictures, sound and videos (Mitchell, 1997); Large Language Models, which can understand large data volume sets and provide means to personal insights (Brown et al. 2020); and Generative AI, which can actively produce new data, based on existing frameworks and a set of instructions Hacker, P., et al. (2023). Thus, artificial intelligence technologies are becoming capable of collecting, analysing, synthesising and acting upon personal data.

To mitigate the potentially negative effects of AI-based personal data use and to govern the digital data landscape a set of similar technologies can be used

Intelligent Agents (IA), which can act autonomously based on the underlying framework and instructions (Wooldridge, 1999). To mitigate the limited capacity of a single IA, they can work together in Multi-agent systems. Aligned with the agency theory, they are capable of representing their contractor interests, mostly in the digital environment within a certain domain (Russell and Norvig, 2016). By employing IA capacities, entities from the real world with limited digital capacities can be represented (Calegari et al. 2021).

Smart Contracts are self-executing contracts with the terms of the agreement between parties being directly written into lines of code (Szabo, 1997). These contracts automatically enforce and execute the terms of the agreement (Kolvart et al. 2016) and can be the product of an automated negotiation process. Using smart contracts, the relations between private data ecosystems can be clarified.

Underlying technologies used to democratise document storage, creation, and usage are Distributed Ledger and Blockchain (El Ioini and Pahl, 2018), while the framework for digitally protecting authors rights and digital assets ownership be found in Non-fungible Tokens (Arlowski, 2021).

RESEARCH

Enhancing Data Protection Beyond the Scope of the GDPR

In addition to its positive contribution to the protection of human rights, the GDPR has also strengthened the European Union's internal market. The GDPR was, therefore, born out of an awareness of the increasing importance of data for developing the digital society and the consequent economic power it brings. This is especially evident in the competitive advantages it brings and the boost it gives to innovation.

It is important to emphasise that all data processors, including SMEs, must comply with GDPR requirements, not just large corporations. Implementing GDPR presents significant challenges for SMEs, particularly in having proper systems, staff, and other resources to do so effectively (Klinger, Wiesmaier, & Hinzmann, 2023). Various tools have been developed to assist SMEs in GDPR compliance to provide guidance and support in understanding and implementing the regulation effectively, ultimately saving costs and improving data management strategies (Cambronero, 2022).

Despite normalising the data economy through the GDPR, several unresolved issues remain. These include the improvement of the data economy and the issue of data sharing. (König, 2022). This is noticeable in Article 20 of the GDPR, which grants data subjects a new right to data portability between different data controllers. Even if the idea was to facilitate data sharing and interoperability, the new portability right has proven problematic on many levels, especially regarding data reuse (Van Ooijen & Vrabec, 2019).

Furthermore, the rapid advancement of technology has led to new challenges that the GDPR does not adequately address. In this context, the recently adopted EU legislative package under the European data strategy addresses areas that have not yet been sufficiently normalised, and relates to regulating different areas of data governance and meeting the challenges of the digital age. The following section presents the recently adopted EU legislative acts that address the issue of data protection in a digital and technologically evolving society and build significantly on the GDPR.

Navigating Recent EU Legislation: Implications for Personal Data

In recent years, the EU regulatory framework has focused on large multinationals based outside the EU, mainly in the US. The most significant technological development race is between Europe, China and the US. The latter two countries rapidly develop innovations while expanding their influence in the global data market. Large technology companies in the US control and dominate the data space. In the United States, the majority of data infrastructure is under the control of the private sector.

In contrast, in China, government control and firm control over data in technology companies is prevalent, often resulting in a lack of protection of individuals' rights (European Commission, 2020). The challenge for the EU is to find its way, where the key is to follow democratic standards that maintain a high level of privacy, security, safety and ethics. It necessitates establishing a balance between the flow of data and its widespread use (Graef & Gellert, 2021). Despite its accumulated knowledge and sophisticated technology, the EU lags behind the US and China in the digital race. A significant concern for the EU is its overdependence on US cloud providers. As much as 70% of the EU market depends on US cloud providers. Policies, including the Data Act, have been implemented to reduce this dependency and promote greater EU autonomy in this area (European Commission, 2020).

For a considerable period, the European Commission has emphasised its ambition to lead the EU into a digital future. While the GDPR represents a significant achievement, other legislative initiatives to transition Europe into digitalisation and supranational connectivity have either stalled or failed to reach their high ambitions (Craig & de Bùrca, 2020). Nevertheless, despite this, more precise guidelines are emerging within the European framework, and regulation in this area is strengthening and becoming increasingly complex.

As part of the Second Digital Agenda for Europe, the European Union presented key strategic documents from the European Commission, including the European data strategy (European Commission, 2020), which sets out the Commission's vision for creating a European data economy over the next 5-10 years. One of the key ideas is the use of data in different sectors of the economy, which is crucial in creating this data economy (Graef & Gellert, 2021). The document presents data as a valuable resource that can be harnessed for the benefit of European citizens (Craig & de Bùrca, 2020). Furthermore, the creation of the data strategy was motivated by the need to establish the foundations for EU competitiveness in the coming decades. This was to be achieved by creating a European data space and enabling European businesses, in particular, to build on the scale of the single market. In addition, the possibility of external, international data flows was to be tied to European values (König, 2022).

Research has demonstrated that the usage and sharing of data is inadequate due to several factors, including a lack of data availability, the absence of administrative structures and technical infrastructure, or the absence of appropriate tools to enable consumers to exercise their rights based on data sharing (Graef & Gellert, 2021). However, alternative methods of cooperating on data sharing could confer a significant competitive advantage to the EU, leading to higher economic growth.

In alignment with its overarching objectives, the EU has introduced a suite of regulations that facilitate access and reuse of information (Mylly, 2024).

Following the European Commission's Digital Strategy, two significant pieces of legislation have been enacted: the Digital Services Act (DSA) and the Digital Markets Act (DMA). These Acts have a comparable objective to the GDPR in data protection, establishing uniform regulations to supersede existing frameworks. They aim to create a level playing field and remove barriers that prevent EU companies from competing with companies outside the EU. These two pieces of legislation regulate the provision of digital services within the EU, with a particular focus on those services that involve processing personal data. They have significant implications for consumer protection (König, 2022).

Although the current political discourse revolves around the DSA, the DMA and the Artificial Intelligence Act (AI Act), there is an additional need to consider two additional legislative proposals that have the potential to have a similarly profound effect on the digital landscape: the Data Governance Act and the Data Act (Consideration, 2021).

The concept of facilitating data sharing was already considered when the GDPR was created. However, the new right to data portability has proven problematic in several ways, particularly regarding data reuse (Van Oojien & Vrabec, 2019). To address this issue, the Data Governance Act introduces the concept of data altruism, which will facilitate the further sharing of personal data. The primary objective of the DGA is to promote the exchange of data across sectors and EU countries, thereby increasing the availability of data (European Commission, 2024).

On the other hand, the Data Act will regulate the area relating to effective rights of access to and use of data. One of the key elements of the EU's data strategy, the Data Act represents a significant step forward in creating a single market for data in the EU while at the same time stimulating Europe's data economy by fostering innovation and promoting new data-driven business models.

Personal and Non-Personal Data: Legal Implications in the EU

A clear distinction between personal and non-personal data is a fundamental sine qua non of European data protection law. The concept of non-personal data can be considered implicitly recognised in the GDPR, as it pertains to personal data rather than non-personal data (Graef & Gellert, 2021). Consequently, non-personal data is explicitly excluded from the scope of the GDPR. The central focus of data protection law has traditionally been directed towards personal data.

The relationship between personal and non-personal data came to the forefront of regulation in 2018 (Graef & Gellert, 2021). At that time, legislation adopted by the EU expanded its scope to include direct data to achieve economic and social benefits unimpeded by data protection constraints. The Free Flow of Non-Personal Data (NPDR) regulation represents an initial step in shaping European law on non-personal data (Craig & de Bùrca, 2021).

The concept of non-personal data encompasses information initially considered personal data but has been anonymised to the extent that it can no longer be linked to a specific individual.

While pseudonymisation is defined in the GDPR, anonymisation is only mentioned (Recital 26 of the GDPR, 2016), as anonymised data can no longer be attributed to an individual and consequently ceases to be personal data. Pseudonymised data protects data by adhering to the principle of data minimisation (i.e., using only the data necessary to achieve lawful purposes). However, it qualifies as personal data because it remains possible to attribute individuals objectively. The capacity to re-identify individuals from anonymised data has been significantly enhanced due to technological advancement (Craig & de Bùrca, 2021).

Non-personal data may also include information that cannot be linked to or does not relate to a specific or identifiable person. As an illustration, data relating to weather forecasts could be regarded as non-personal data (Your Europe, 2024). The European Commission has determined that data can transform in nature, whereby non-personal data can potentially become personal data. The concept of personal data is not only extensive but also highly contextual.

It is possible for data that initially appears to be non-personal to become personal data due to the information derived from it (Graef & Gellert, 2021). In addition, concerns can be raised that draw their strength from rapidly advancing technologies, which enable increasingly sophisticated analysis and linking of data, potentially leading to the identification of individuals (Purtova, 2018).

In some situations, data may be both personal and non-personal. In such cases, "mixed datasets" describe a dataset that includes individual and non-personal data. It is important to note that there is a distinction between non-personal data, which falls under the NPDR, and personal data, which falls under the GDPR.

Furthermore, the lack of definition of this concept in EU legislation also presents a challenge. The European data strategy also fails to address the issue of mixed datasets. The Strategy mentions Mixed datasets only once, limiting itself to the NPDR Regulation (Graef & Gellert, 2021).

In contrast to the GDPR, the NPDR lacks a legal basis for fundamental rights. Unlike the GDPR, the NPDR focuses on the ability of Member States to require data localisation rather than on protecting individuals' rights about processing their personal data (Craig & de Bùrca, 2021). Consequently, data localisation requirements imposed by Member States about the geographical location for storing or processing non-personal data are prohibited. There are exceptions to this rule if justified on the grounds of public security and considering the principle of proportionality (Article

4, NPDR, 2018). It is also important to note the long-term impact of this NPDR legislation on the EU legal system with non-personal data as a separate category.

The European Data Act: Legal Perspectives and Implementation Challenges of Internet of Things

The Internet of Things (IoT) refers to devices that combine different technologies, including sensors and software. The design of these products enables data to be recorded, received and exchanged over the Internet. With the advent of the IoT, the Internet has extended its reach to physical objects, fundamentally changing numerous areas. The IoT has significantly impacted the development of businesses, industry, and infrastructure while influencing consumer habits (Sinclair, 2017). However, such technological advances raise several significant data protection issues. Nevertheless, a defining characteristic of the IoT is the vast quantity of data that such technologies can collect, both personal and non-personal (Bolognini & Balboni, 2019). The increased use of IoT devices and the generation of new data has led to new legal regulations in various areas.

A Quick Overview of the Data Act

The benefits of data governance have been widely recognised by the European Union (EU), which has therefore supported the creation of a data ecosystem. The EU considers data sharing to be one of the critical elements that will contribute to the growth of the European economy (Mylly, 2024).

The immense power of data has prompted the emergence of new questions regarding ownership of data, namely, who is entitled to control, utilise and create value from it. While a data protection legislative framework has been established for personal data, the issue of non-personal data for which legal rights often do not yet exist has remained almost untouched. The discourse has evolved from the initial proposal of exclusive data ownership to introducing enhanced rights to access and share data to advance innovation and competitiveness (Eckardt & Kerber, 2024). In addition, European legislators have identified a problem with data management in IoT devices. This problem arises because manufacturers of IoT devices often gain exclusive de facto control over the data generated by users' IoT devices through their technical design of these devices. This hinders the development of the data economy and innovations that would otherwise be possible.

The recently adopted Data Act introduces new rights for users of IoT devices to access, use and share IoT data. Therefore, the Data Act is essential in evolving new bundles of rights concerning non-personal IoT data. The DA's primary tool to address this problem is the introduction of new legal rights for users of IoT devices (Eckardt & Kerber, 2024).

A fundamental new right in the Data Act is the right of users to access, share and use data generated by the use of products or related services (Articles 4 and 5, Data Act, 2023).

The user of the product or related services may promptly access, use, and share the data with third parties of their choice. The Data Act obligates the data holder to make the data available to users and third parties designated by users in certain circumstances. Furthermore, the Data Act ensures that data holders make data available to data recipients in the Union on fair, reasonable and non-discriminatory terms and transparently (Articles 4, 5 and 8, Data Act, 2023). It is important to note that the Data Act should not be interpreted to give data holders new rights to use data generated by the use of a related product or service. Consequently, the data holder is not granted a legal basis to collect or create personal data; instead, they are obliged to make personal data available to users or third parties of the user's choice upon the user's request (Recitals 5, 7 and 25, Data Act, 2023).

Furthermore, the Data Act stipulates that data holders may only utilise IoT data based on a contractual commitment with the user (Section 4 (13, 14), Data Act, 2023). This represents a significant departure from the previous situation, wherein data holders were permitted to utilise non-personal data under their effective control without any contract or explicit consent from the user. In essence, the new framework established by the Data Act bears a resemblance to the GDPR, where data controllers are required to obtain the consent of data subjects following Article 6(1)(a) of the GDPR before processing, utilising, or sharing the data in question (Eckardt & Kerber, 2024).

They Balanced Personal and Non-Personal Data: Intersection of GDPR and the Data Act

The frequent use of connected products and services in everyday life entails numerous situations where a natural person generates a substantial amount of personal data. In instances where a set of personal and non-personal data is inextricably linked, the rules of the GDPR apply. The data subject may be the user or another natural person. Personal data may only be requested by the controller or the data subject. The Data Act does not affect the right of a user to access personal data in certain circumstances under the GDPR. The new legislative framework broadens the scope of data to which a user who is a natural person is entitled to access.

Consequently, a user can now access all data, whether personal or non-personal, that is generated by the use of a connected product. In the second situation, where the user is not a data subject but a business, including a sole trader, but not where a joint household uses the connected product, the user is considered the controller. If a controller requests personal data generated by a linked product, it must have a legal basis for processing the data per Article 6(1) of the GDPR. This may be demonstrated by demonstrating that the data subject has consented to the processing of their data or that the processing is necessary for the performance of a contract to which the data subject is a party (Recital 34, Data Act, 2023).

The issue of distinguishing between personal and non-personal data is more complex than the rule that when data is unknown, the GDPR rules should be followed. Treating personal data as non-personal data would be in breach of the GDPR. To avoid this dilemma, a company could label all data as personal. Even without explicit requirements in the Data Act, it is arguably the most appropriate course of action for companies to consider which data is personal and which is not. It may be advisable to document the rationale behind the decision-making process for these challenging cases in the most complex cases (Fritz & Dannhausen, 2024).

Mechanisms of Personal Data Governance and Sharing

Personal data variety analysed by IT organisations is expanding to understand people better. They can combine data from social media (Lin and Wang, 2023), stored documents, personal communication, pictures and videos, cookies, smart device recordings, and publicly accessible intelligent devices, to name the most obvious ones (Henriksson, 2018).

Quite often, personal data are analysed without the data producer's knowledge or understanding of the purpose of the analysis. The policies informing people about the data collecting processes, such as cookie banners, have little or no impact on their understanding of the data processes (Kretschmer et al., 2021; Laine, 2021). Most people do not read the terms of use or policies, understand them, or react to their contents. On the other hand, the control if data collectors comply with the clauses written in the same document is rarely, if not executed.

Invoking AI in analysing personal data dramatically increased the IT organisation's capacity to analyse personal data use (Zhang et al., 2022; Kataev et al., 2022). Huge, detailed micro datasets are gathered to fuel the AI-based analysis process, resulting in a much better understanding of the behaviour patterns of individuals in certain situations. These analyses go well beyond academic research or a single website traffic analysis – large information-focused organisations are applying population-scaled behavioural studies with no or limited ethical considerations involved (Perko, 2021). The speed and the complexity of intelligent data processing reach well beyond any personal data ecosystem participants but large IT organisations, resulting in a strong data governance asymmetry.

We claim that the regulative frameworks should be coordinated with the executive part, clearly keeping in mind and utilising the capacity of digital technology, especially AI and the limited capacity of individuals to deal with them. Since it is nearly impossible to reduce the capacity of IT organisations, it is imperative to fundamentally augment the data governance capacity of other stakeholders to data governance asymmetry.

A short example of a personal data governance open question is a video of people crossing the red light: https://youtu.be/PzKsHg0-eI4, recorded with a smart device, which can be easily streamed in the cloud and analysed using AI technologies.

Figure 1. Crossing the red light

The video was recorded as a part of a research project, with persons on the video agreeing to share the recording in the academic environment. The video opens several questions: who should have permission to analyse the video, who should take ownership, and how should governance mechanisms be organised to be aligned with the regulatory frameworks in the digital environment? It is intended to open questions. It should make the reader understand that every photo, audio, EKG, blood pressure or eye fixation is personal data and can be analysed using AI algorithms. The new paradigm should make us revisit all questions regarding personal data, and it should be kept in our minds when partial proposals are being made to address this complex issue.

There are several technical proposals on how to democratise personal data, including using blockchains, smart contracts, Non-fungible tokens (NFT), and AI technology (Miyachi and Mackey, 2021, Wang et al., 2023, Zou et al., 2021, Jobin et al., 2019) in several personal data domains. Most of these proposals are proposed independently of the regulative frameworks. They can, therefore, be used as building blocks but lack conceptual coherence to address the issue of augmenting capacity to govern personal data.

In the EU landscape, efforts to create a conceptually correct and feasible personal data governance are framed in multiple documents, including the Artificial Intelligence Act (European Commission 2023), which defines AI-related risks and can be used to propose prevention and mitigation mechanisms. Aligned with this document, several proposals are being conceptualised.

Figure 2. Dynamic data negotiations in data-sharing process (Perko, 2022)

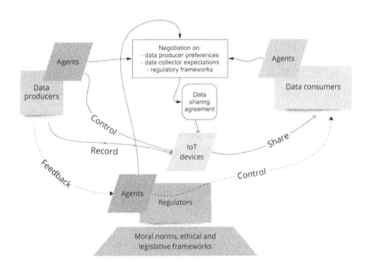

For example, we can point out an AI-supported data-sharing concept, where the relationship between personal data stakeholders is being arranged in the negotiation process between the data producers, the data collectors and the regulators (Perko, 2022). The intelligent agents address the asymmetry in the stakeholders' capacity to govern data, representing their standpoints.

As depicted in Figure 2, producers, consumers, and regulators, represented by intelligent data agents, negotiate data-sharing rules based on data producer preferences and the interests of data consumers, following frameworks introduced by the regulators. On the negotiation level, the rights and responsibilities of all participants are defined. Data structure and contents are agreed upon, as well as the rules of how data is to be employed by data consumers and the responsibilities of data consumers toward data producers – sharing the research results, providing reimbursement, or others. Several tools enabling successful data governance are implemented if data-sharing agreements are negotiated on an instance level before actual data sharing. By applying this and similar mechanisms, the data-related frameworks can shift from ex-post data-related accident mitigation to ex-ante establishing governing rulesets embedded within the data governance execution, aligned with data governance regulatory frameworks.

DISCUSSION

Given the significant amount of EU legislation concerning digitalisation, ensuring that all relevant stakeholders are adequately addressed and that rights holders are made aware of the new legal possibilities is crucial. While the AI Act has recently been a prominent topic of discussion in the EU, some argue that the Data Act may have at least as much potential, if not more (Niebuhr, 2024).

Research has indicated that data protection is undergoing a significant transformation, with the emergence of novel approaches that have not been previously addressed. This evolution is particularly evident in the context of data generated through the utilisation of IoT technology. Moreover, the research has highlighted the necessity to distinguish between personal and non-personal data produced by IoT technology. Identifying the data sets generated by a specific IoT device will be crucial for stakeholders who must comply with the GDPR and the Data Act. In addition, it will be essential to remain apprised of any new case law that may resolve outstanding issues.

The paradigm shift where the user plays a central role in data sharing, as exemplified by the IoT, represents a new challenge for all stakeholders. This could enable the user to exercise their rights based on data sharing. However, many unresolved issues could severely limit the user's data rights. One of these concerns is security and trade secrets that the data holder is not obliged to share with the user under certain conditions (Myll, 2024). The law must delineate which data can and cannot be accessed at this juncture. The overarching principle of the Data Act is that users should be granted access to data. Consequently, denial of access should be exceptional, even in instances of trade secrets or security concerns (Niebuhr, 2024).

In conclusion, the primary focus of future discussions will be on the practical implementation of user rights and the provision of adequate legal protection for users. Otherwise, the objectives of the Data Act, particularly innovation and users' legal capacity, will not be achieved (Kerber, 2024; Perko, 2021).

CONCLUSION

The GDPR is one of the European Union's most successful legislative initiatives and has served as a model for numerous countries worldwide. It effectively preserves the right to self-determination by ensuring secure data use and personal data protection. This regulation enables individuals to manage and control their personal information, enhancing citizens' rights to informational self-determination. However, the consent-based framework and targeted restriction of processing that underpin the GDPR have proven ineffective in safeguarding individuals' privacy rights in the context of new technologies such as big data, the IoT, and machine learning. These technologies rely on processing vast amounts of data, which may harm individuals' interests when it comes to data concerning them. However, obtaining consent for data processing is not feasible or efficient, and the purpose limitation principle does not fit these technologies' dynamic and evolving nature. Therefore, to achieve a balanced and effective data protection framework, it is necessary to balance the risks to data protection and privacy and the benefits of new technologies. It is important to note that data protection rules do not exist in isolation; they must be placed in the context of the broader protection of privacy and other fundamental rights. These include the right to liberty and security, the freedom to conduct a business, the right to choose one's profession and to work, the freedom of expression, and the freedom of the arts and sciences. It is important to note that data protection rules are not isolated; they must be systematically placed within the broader protection of privacy and other fundamental rights. These include the right to liberty and security, the freedom to conduct a trade or business, the right to choose a profession and a job, and the freedom of artistic and scientific expression. Information self-determination is fundamental to human dignity but must be balanced against other fundamental rights and values.

However, to continue to protect human rights, democracy and the reduced concentration of capital in the hands of individual corporations, as promoted by the GDPR, this legislation urgently needs successors. Among them, the Data Act is worthy of note, as it addresses crucial challenges to data protection and human rights. It aims to ensure a fair distribution of the data value between the players in the data economy and to promote access to and use of data. This legislation represents the first instance of a legal framework that guarantees users of IoT devices access to their data. Before this, the producers largely controlled the data generated by such devices. The primary mechanism to address this challenge is the introduction of new rights for users of IoT devices to access and share this data. This will strengthen the data space and competition within the EU, thereby empowering the innovation economy through data sharing and sharing. In addition, the paper addresses the relationship between personal and non-personal data, which has been somewhat neglected by legislation. These issues will be crucial for the successful implementation of the Data Act.

REFERENCES

Arlowski, M. (2021). Personal Data as a New Form of Intellectual Property. *Journal of the Patent and Trademark Office Society*, 102, 649.

Bennett, C. J. (2018). The European General Data Protection Regulation: An Instrument for the Globalization of Privacy Standards? *Information Polity*, 23(2), 239–246. 10.3233/IP-180002

Bolognini, L., & Balboni, P. (2019). *IoT and Cloud Computing: Specific Security and Data Protection Issues*. In the Internet of Things Security and Data Protection., 10.1007/978-3-030-04984-3_4

Brown, T., Mann, B., Ryder, N., Subbiah, M., Kaplan, J. D., Dhariwal, P., Neelakantan, A., Shyam, P., Sastry, G., & Askell, A. (2020). Language models are few-shot learners. *Advances in Neural Information Processing Systems*, 33, 1877–1901. 10.48550/arXiv.2005.14165

Calegari, R., Ciatto, G., Mascardi, V., & Omicini, A. (2021). Logic-based technologies for multi-agent systems: A systematic literature review. *Autonomous Agents and Multi-Agent Systems*, 35(1), 1. 10.1007/s10458-020-09478-3

Cambronero, M.-E., Martínez, M. I., de la Vara, J. L., Cebrian, D., & Valero, V. (2022). GDPRValidator: A tool to enable companies using cloud services to be GDPR compliant. *PeerJ*, 8, e1171. 10.7717/peerj-cs.117136532816

Considerati. (2021). European Data Act & Data Governance Act: the kids with gamechanging potential. https://www.considerati.com/publications/eu-data-act-data-governance-act.html

Craig, P., & de Bùrca, G. (2020). *EU Law. Text, Cases, and Materials*. Oxford University Press, 10.1093/he/9780198856641.001.0001

Craig, P., & de Bùrca, G. (2021). *The Evolution of EU Law*. Oxford University Press. 10.1093/oso/9780192846556.001.0001

Custers, B., Fosch-Villaronga, E., van der Hof, S., Schermer, B. W., Sears, A. M., & Tamò-Larrieux, A. (2022). The Role of Consent in an Algorithmic Society – Its Evolution, Scope, Failings and Re-conceptualisation. In Kostas, E., Leenes, R., & Kamara, I. (Eds.), *Research Handbook on EU Data Protection* (pp. 455–473). Edward Elgar Publishing. 10.4337/9781800371682.00027

De Man, Y., Wieland-Jorna, Y., Torensma, B., de Wit, K., Francke, A. L., Oosterveld-Vlug, M. G., & Verheij, R. A. (2023). Opt-In and Opt-Out Consent Procedures for the Reuse of Routinely Recorded Health Data in Scientific Research and Their Consequences for Consent Rate and Consent Bias: Systematic Review. *Journal of Medical Internet Research*, 25, e42131. 10.2196/4213136853745

Diggelmann, O., & Cleis, M. N. (2014). How the Right to Privacy Became a Human Right. *Human Rights Law Review*, 14(3), 441–458. 10.1093/hrlr/ngu014

Directive 2002/58/EC of the European Parliament and of the Council of 12 July 2002 concerning the processing of personal data and the protection of privacy in the electronic communications sector (Directive on privacy and electronic communications),OJ L 201, 31.7.2002, 37–47.

Directive 95/46/EC of the European Parliament and of the Council of 24 October 1995 on the protection of individuals with regard to the processing of personal data and on the free movement of such data, OJ L 281, 23.11.1995, 31–50.

Eckardt, M., & Kerber, W. (2024). Property Rights Theory, Bundles of Rights on IoT Data, and the EU Data Act. 10.1007/s10657-023-09791-8

El Ioini, N., & Pahl, C. (2018). A review of distributed ledger technologies. On the Move to Meaningful Internet Systems. OTM 2018 Conferences: Confederated International Conferences: CoopIS, C&TC, and ODBASE 2018, Valletta, Malta, October 22-26, 2018, Proceedings, Part II, Springer. 10.1007/978-3-030-02671-4_16

Erdos, D. (2021). Comparing Constitutional Privacy and Data Protection Rights within the EU University of Cambridge, Faculty of Law Research Paper No. 21/2021, Available at *SSRN*: https://ssrn.com/abstract=3843653 or http://dx.doi.org/10.2139/ssrn.3843653

European Commission. (2020a). Communication From the Commission to the European Parliament, the Council, the European Economic and Social Committee and the Committee of the Regions. Shaping Europe's digital future, COM(2020) 67 final, 19. 2. 2020.

European Commission. (2020b). Communication From the Commission to the European Parliament, The Council, The European Economic and Social Committee and the Committee of the Regions a European Strategy for Data, COM/2020/66 final, 19. 2. 2020.

European Commission. (2023). Artificial Intelligence act. Available: https://eur-lex.europa.eu/legal-content/EN/TXT/?uri=CELEX%3A32023L2225&qid=1714133317618

European Commission. (2024). European Data Governance Act. https://digital-strategy.ec.europa.eu/en/policies/data-governance-act

European Data Protection Supervisor. (2021). TechDispatch #3/2020 - Personal Information Management Systems, https://www.edps.europa.eu/data-protection/our-work/publications/techdispatch/techdispatch-32020-personal-information_en

Francis, L. P., & Francis, J. G. (2017). Data Reuse and the Problem of Group Identity. Studies in Law, Politics, and Society, 73. University of Utah College of Law Research Paper No. 311. 10.2139/ssrn.3371134

Fritz, G., & Dannhausen, E. (2024). When GDPR and Data Act clash: what Businesses need to know. https://www.lexology.com/library/detail.aspx?g=6b3f477e-ab49-415d-bfb4-1d27e8b8db53

Gömann, M. (2017). The new territorial scope of EU data protection law: Deconstructing a revolutionary achievement. *Common Market Law Review*, 54(2), 567–590. 10.54648/COLA2017035

Government of Canada. (2019). Personal Information Protection and Electronic Documents Act. https://laws-lois.justice.gc.ca/eng/acts/p-8.6/

Graef, I., & Gellert, R. (2021). The European Commission's proposed Data Governance Act: some initial reflections on the increasingly complex EU regulatory puzzle of stimulating data sharing. TILEC Discussion Paper No. DP2021-006. 10.2139/ssrn.3814721

Graef, I., Gellert, R., & Husovec, M. (2018). Towards a Holistic Regulatory Approach for the European Data Economy: Why the Illusive Notion of Non-Personal Data is Counterproductive to Data Innovation. TILEC Discussion Paper No. 2018-029. 10.2139/ssrn.3256189

Hacker, P., Engel, A., & Mauer, M. (2023). Regulating ChatGPT and other large generative AI models. *Proceedings of the 2023 ACM Conference on Fairness, Accountability, and Transparency*. 10.1145/3593013.3594067

Henriksson, E. A. (2018). Data protection challenges for virtual reality applications. *Interactive Entertainment Law Review*, 1(1), 57–61. 10.4337/ielr.2018.01.05

Informacijski pooblaščenec. (2022). Letno poročilo informacijskega pooblaščenca. https://www.ip-rs.si/fileadmin/user_upload/Pdf/porocila/LP2022.pdf

Jobin, A., Ienca, M., & Vayena, E. (2019). The global landscape of AI ethics guidelines. *Nature Machine Intelligence*, 1(9), 389–399. 10.1038/s42256-019-0088-2

Kataev, M., Bulysheva, L., & Mosiaev, A. (2022). Application of artificial intelligence methods in the school educational process. *Systems Research and Behavioral Science*, 39(3), 531–541. 10.1002/sres.2873

Klinger, E., Wiesmaier, A., & Hinzmann, A. (2023). A review of existing GDPR solutions for citizens and SMEs. arXiv.org. https://doi.org//arXiv.2302.0358110.48550

Kolvart, M., Poola, M. & Rull, A. (2016). "Smart contracts." The Future of Law and eTechnologies, 133-147. 10.1007/978-3-319-26896-5

König, P. D. (2022). Fortress Europe 4.0? An analysis of EU data governance through the lens of the resource regime concept. *European Policy Analysis*, 8(4), 484–504. 10.1002/epa2.1160

Kovalenko, Y. (2022). The Right to Privacy and Protection of Personal Data: Emerging Trends and Implications for Development in Jurisprudence of European Court of Human Rights. *Masaryk University Journal of Law and Technology*, 16(1), 37–57. 10.5817/MUJLT2022-1-2

Kretschmer, M., Pennekamp, J., & Wehrle, K. (2021). Cookie Banners and Privacy Policies: Measuring the Impact of the GDPR on the Web. *ACM Transactions on the Web*, 15(4), 20. Advance online publication. 10.1145/3466722

Laine, J. (2021). There is no decision: design of cookie consent banner and its effect on user consent (Publication Number URN:NBN:fi:tuni-202111208544) Tampere University). https://trepo.tuni.fi/handle/10024/135598

Levin, A., & Nicholson, M. J. (2023). Privacy Law in the United States, the EU and Canada: The Allure of the Middle Ground. Toronto Metropolitan University. 10.32920/22227775.v1

Lin, X. L., & Wang, X. Q. (2023). Following too much on Facebook brand page: A concept of brand overload and its validation. *International Journal of Information Management*, 73, 19. 10.1016/j.ijinfomgt.2023.102682

Martins Gonçalves, R., Mira da Silva, M., & Rupino da Cunha, P. (2023). Implementing GDPR-Compliant Surveys Using Blockchain. *Future Internet*, 15(4), 143. 10.3390/fi15040143

Milaj-Weishaar, J. (2020). Safeguarding Privacy by Regulating the Processing of Personal Data – An EU Illusion? *European Journal of Law and Technology*, 11(2).

Milosavljević, N. (2023). The limitation of the data protection right for the purpose of intellectual property rights protection. In Ćalić, R. (Ed.), *Uporednopravni izazovi u savremenom pravu, In memoriam dr* (pp. 587–617). Stefan Andonović. 10.56461/ZR_23.SA.UPISP_NM

Mitchell, T. M. (1997). *Machine Learning*. McGraw-Hill.

Miyachi, K., & Mackey, T. K. (2021). HOCBS: A privacy-preserving blockchain framework for healthcare data leveraging an on-chain and off-chain system design. *Information Processing & Management*, 58(3), 24. 10.1016/j.ipm.2021.102535

Myll, U.M. (2024). Trade Secrets and the Data Act. IIC - International Review of Intellectual Property and Competition Law, 55, 368-393. 10.1007/s40319-024-01432-0

Novak, M. (2021). Pravna argumentacija v teoriji in praksi. Ljubljana: Uradni list RS.

O'Leary, D. E., Bonorris, S., Klosgen, W., Khaw, Y.-T., Lee, H.-Y., & Ziarko, W. (1995). Some privacy issues in knowledge discovery: The OECD personal privacy guidelines. *IEEE Expert*, 10(2), 48–59. 10.1109/64.395352

OECD. (2002). *OECD Guidelines on the Protection of Privacy and Transborder Flows of Personal Data*. OECD Publishing. 10.1787/9789264196391-

Oh, J., Hong, J., Lee, C., Lee, J. J., Woo, S. S., & Lee, K. (2021). Will EU's GDPR act as an effective enforcer to gain consent? *IEEE Access : Practical Innovations, Open Solutions*, 9, 79477–79490. 10.1109/ACCESS.2021.3083897

Olesen-Bagneux, O. (Host). (2024). EU Data Act Special: Decoding the Future Impact of the Act on Technology and Society [Audio podcast]. https://podcast.zeenea.com/eu-data-act-special-w-matthias-niebuhr-decoding-the-future-impact-of-the-act-on-technology-and-society

Pavčnik, M. (2015). *Teorija prava: prispevek k razumevanju prava*. GV Založba.

Perenič, A. (1998). Razlaga pravnih aktov. In Uvod v pravoznanstvo (pp. 191-217). Uradni list Republike Slovenije.

Perko, I. (2021). Hybrid reality development-can social responsibility concepts provide guidance? *Kybernetes*, 50(3), 676–693. 10.1108/K-01-2020-0061

Perko, I. (2022). Data sharing concepts: A viable system model diagnosis. *Kybernetes*, 52(9), 2976–2991. 10.1108/K-04-2022-0575

Peyrone, N., & Wichadakul, D. (2023). A formal model for blockchain-based consent management in data sharing. *Journal of Logical and Algebraic Methods in Programming*, 134, 100886. 10.1016/j.jlamp.2023.100886

Pintiliuc, I.-G. (2018). Protection of Personal Data. Logos Universality Mentality Education Novelty. *Law*, 6(1), 37–40. 10.18662/lumenlaw/05

Presidência da República. Secretaria-Geral Subchefia para Assuntos Jurídicos. (2018). Lei Geral de Proteção de Dados Pessoais. https://www.planalto.gov.br/ccivil_03/_ato2015-2018/2018/lei/l13709.htm

Purtova, N. (2018). The law of everything. Broad concept of personal data and future of EU data protection law. *Law, Innovation and Technology*, 10(1), 40–81. 10.1080/17579961.2018.1452176

Regulation (EU) 2016/679 of the European Parliament and of the Council of 27 April 2016 on the protection of natural persons with regard to the processing of personal data and on the free movement of such data, and repealing Directive 95/46/EC (General Data Protection Regulation), OJ L 119, 4. 5. 2016, 1-88.

Regulation (EU) 2018/1807 of the European Parliament and of the Council of 14 November 2018 on a framework for the free flow of non-personal data in the European Union, OJ L 303, 28. 11. 2018, 59-68.

Regulation (EU) 2023/2854 of the European Parliament and of the Council on harmonised rules on fair access to and use of data and amending Regulation (EU) 2017/2394 and Directive (EU) 2020/1828 (Data Act), OJ L, 2023/2854, 22. 12. 2023.

Richardson, M. (2017). The Right to Privacy. In *The Right to Privacy: Origins and Influence of a Nineteenth-Century Idea*. Cambridge University Press. 10.1017/9781108303972

Russell, S. J., & Norvig, P. (2016). *Artificial intelligence: a modern approach*. Pearson.

Ryngaert, C., & Taylor, M. (2020). The GDPR as Global Data Protection Regulation? *AJIL Unbound*, 114, 5–9. 10.1017/aju.2019.80

Schermer, B. W., Custers, B., & van der Hof, S. (2014). The crisis of consent: How stronger legal protection may lead to weaker consent in data protection. *Ethics and Information Technology*. Advance online publication. 10.1007/s10676-014-9343-8

Sharari, S., & Faqir, R. (2014). Protection of Individual Privacy under the Continental and Anglo-Saxon Systems: Legal and Criminal Aspects. *Beijing Law Review*, 5(3), 184–195. 10.4236/blr.2014.53018

Siman, E., Abiodun, J., Timothy, G. & Nandom, S. S. (2023). IoT-Driven Smart Cities: Enhancing Urban Sustainability and Quality of Life. 7th International Interdisciplinary Research & Development Conference.

Sinclair, B. (2017). *IoT Inc.: how your company can use the Internet of things to win in the outcome economy*. McGraw-Hill Education.

State of California, Department of Justice. (2018). California Consumer Privacy Act of 2018. https://oag.ca.gov/privacy/ccpa

Strban, G. (2022). Metodologija primerjalne razlage prava socialne varnosti. Pravne panoge in metodologija razlage prava. GV Založba.

Szabo, N. (1997). The Idea of Smart Contracts.

Thouvenin, F., & Tamò-Larrieux, A. (2021). Data Ownership and Data Access Rights: Meaningful Tools for Promoting the European Digital Single Market? In Burri, M. (Ed.), *Big Data and Global Trade Law* (pp. 316–339). Cambridge University Press. 10.1017/9781108919234.020

Ursić, H. (2018). The Failure of Control Rights in the Big Data Era: Does a Holistic Approach Offer a Solution? In Personal Data in Competition, Consumer Protection and Intellectual Property Law. MPI Studies on Intellectual Property and Competition Law, vol 28. Springer. 10.1007/978-3-662-57646-5_4

Van Ooijen, I., & Vrabec, H. U. (2019). Does the GDPR Enhance Consumers' Control over Personal Data? An Analysis from a Behavioural Perspective. *Journal of Consumer Policy*, 42(1), 91–107. 10.1007/s10603-018-9399-7

Verhulst, S. G. (2023). Operationalising digital self-determination. *Data & Policy*, 5, e14. 10.1017/dap.2023.11

Wang, W. S., Li, X., Qiu, X. Q., Zhang, X., Brusic, V., & Zhao, J. D. (2023). A privacy preserving framework for federated learning in smart healthcare systems. *Information Processing & Management*, 60(1), 23. 10.1016/j.ipm.2022.103167

Warren, S. D., & Brandeis, L. D. (1890). The Right to Privacy. *Harvard Law Review*, 4(5), 193–220. 10.2307/1321160

Winn, P. A. (2010). Older than the Bill of Rights: The Ancient Origins of the Right to Privacy. https://ssrn.com/abstract=1534309

Wong, J., Henderson, T., & Ball, K. (2022). Data protection for the common good: Developing a framework for a data protection-focused data commons. *Data & Policy*, 4, e3. 10.1017/dap.2021.40

Wooldridge, M. (1999). Intelligent agents. Multiagent systems: A modern approach to distributed artificial intelligence, 1, 27-73.

Your Europe. (2024). Prosti pretok neosebnih podatkov. https://europa.eu/youreurope/business/running-business/developing-business/free-flow-non-personal-data/index_sl.htm

Zhang, W., Sun, L. H., Wang, X. P., & Wu, A. B. (2022). The influence of AI word-of-mouth system on consumers' purchase behaviour: The mediating effect of risk perception. *Systems Research and Behavioral Science*, 39(3), 516–530. 10.1002/sres.2871

Zou, R. P., Lv, X. X., & Zhao, J. S. (2021). SPChain: Blockchain-based medical data sharing and privacy-preserving eHealth system. *Information Processing & Management*, 58(4), 18. 10.1016/j.ipm.2021.102604

ADDITIONAL READING

Craig, P., & de Bùrca, G. (2021). *The Evolution of EU Law*. Oxford University Press. 10.1093/oso/9780192846556.001.0001

Eckardt, M., & Kerber, W. (2024). Property Rights Theory, Bundles of Rights on IoT Data, and the EU Data Act. 10.1007/s10657-023-09791-8

Van Ooijen, I., & Vrabec, H. U. (2019). Does the GDPR Enhance Consumers' Control over Personal Data? An Analysis from a Behavioural Perspective. *Journal of Consumer Policy*, 42(1), 91–107. 10.1007/s10603-018-9399-7

KEY TERMS AND DEFINITIONS

Artificial Intelligence: Artificial Intelligence (AI) is an evolving technology that invokes advanced reasoning using machines based on underlying data. AI encompasses various subfields, including machine learning (ML) and deep learning, which allow systems to learn and adapt in novel ways from training data.

Data Collecting: Data collection is collecting from various stakeholders and sources and analysing data on relevant variables in a predetermined, organised way to reach data collectors' goals. Data collectors perform data governance, often excluding other stakeholders from the process.

Data Ownership: Data ownership can be understood as a form of ownership and as a form of control. The GDPR develops a framework for controlling data and protecting data users. Data ownership is still under debate. If data ownership is transferred to companies, data producers' (individuals) position would worsen, and providing ownership rights to data producers would require significant changes in data-sharing concepts.

Data Sharing: The concept of data sharing implies that multiple users can utilise the potential of a given data set. Improving data-sharing protocols is essential to achieve more significant societal benefits. The recently introduced EU legislation anticipates that incentives for data sharing will ensure that users and businesses have access to a greater quantity of data, thereby fostering the development of new products and services. In the context of the Data Act, data sharing can be understood as a process whereby data is shared for various purposes. A contract between the parties involved typically defines the terms of data sharing. In contrast, the data producers play an active role in the process, which means they decide who data is shared with and govern the data sharing conditions.

Internet of Things: The application of Internet of Things (IoT) technology allows for the automation of systems by integrating sensors and software. The Internet provides a platform for devices to interact with the external environment, facilitating a shift in the scope of IoT beyond mere physical objects. This has led to a transformation in numerous domains, with sensors transforming every aspect of our lives into computer-processed information. Consequently, the Internet of Things has significantly impacted business, industry, and infrastructure development, influencing consumer behaviour in the process.

Right to Data Privacy (or Correct to Data Protection): The European Charter of Fundamental Rights defines the right to privacy and data protection as two separate rights. Still, the substantive distinction between the two rights is not very clear. The right to privacy is often seen as a traditional right, while data protection is seen as a new right. This right ensures that personal data is protected from unauthorised access, misuse, and breaches, thereby preserving the confidentiality and integrity of an individual's private information. The GDPR has built a progressive legal framework for protecting personal data based on the right to control personal data.

Right to Data Self-Determination: The key rights guaranteed by the GDPR to individuals (referred to as "data subject") include the right to information (Articles 13 and 14), the right of access by the data subject (article 15), right to rectification (article 16), right to erasure ('right to be forgotten') (article 17), right to restriction of processing (article 18), right to data portability (article 20) and s right to object (article 21). These rights give individuals autonomy and power over decisions relating to their data, including deciding how it is collected, used and shared. Consequently, they can be grouped under a common denominator: the "right to data self-determination" or, more generally, the "right to control personal data". The right to control data belongs to individuals if the data are processed based on their explicit consent and if the processing is carried out on any other legitimate basis set out in Article 6 of the GDPR.

Chapter 7
Cherish Data Privacy and Human Rights in the Digital Age:
Harmonizing Innovation and Individual Autonomy

Bhupinder Singh
https://orcid.org/0009-0006-4779-2553
Sharda University, India

ABSTRACT

Data privacy encompasses the safeguarding and control individuals exercise over their personal information and data. It revolves around ensuring the confidentiality and security of sensitive data, including financial records, health information, and unique identifiers. In the era of extensive data collection, storage, and sharing in the digital landscape, preserving data privacy has become imperative to uphold individuals' rights and shield them from potential harm. From artificial intelligence and machine learning to internet of things (IoT) devices, innovative solutions have transformed our way of life and work. However, every innovation brings the responsibility to safeguard the privacy and security of individuals whose data is collected and processed. So, balancing innovation and personal security is a nuanced task. While innovation offers substantial benefits, it also poses risks to personal privacy without adequate regulation. This chapter dives into the diverse exploration of human rights protection concerning data and privacy of individuals in the digital arena.

DOI: 10.4018/979-8-3693-3334-1.ch007

Copyright © 2024, IGI Global. Copying or distributing in print or electronic forms without written permission of IGI Global is prohibited.

1. INTRODUCTION

The achieving of right balance necessitates collaboration between technology developers, policymakers and users (Ray et al., 2024). A pivotal aspect of achieving a balanced approach to data privacy involves empowering users with greater control over their personal information (Dinesh Arokia Raj et al., 2024). Innovation propels progress and propels technological advancements shaping our modern society (Yue & Shyu, 2024). This can be realized through transparent data collection practices, user-friendly privacy settings, and explicit consent mechanisms (Mithas et al., 2022). Granting individuals the ability to decide how their data is collected, used and shared fosters trust in digital platforms (Ivanov et al., 2019). The innovation and data privacy can coexist by incorporating robust security measures (Javaid et al., 2022). Technology companies and organizations should adopt industry best practices such as encryption, multi-factor authentication, and regular security audits (Fraga-Lamas et al., 2021). Prioritizing security at every stage of data processing minimizes the risks of data breaches and unauthorized access (Felsberger et al., 2022) (Rath et al., 2024). The preservation of any information pertaining to a known or identified natural (living) person, including names, dates of birth, photos, videos, email addresses, and phone numbers, is the focus of data protection (Asadollahi-Yazdi et al., 2020) (Meyendorf et al., 2023). It also includes additional information that is likewise regarded as personal data, such as IP addresses and communications content that is supplied by or connected to communication service end users (Zhong et al., 2017). The right to privacy is the foundation of the notion of data protection, and both are essential to upholding and advancing fundamental rights and values as well as facilitating the exercise of other freedoms and rights including the right to assemble and free expression (Angelopoulos et al., 2020) (Lu et al., 2020). The particular goals of data protection are to guarantee that personal data is processed fairly by the public and commercial sectors, including in terms of collection, use, and storage (Chander et al., 2022).

The globe has experienced unheard-of technological advancement, with digital technology permeating every facet of human existence (Kasowaki & Ahmet, 2024). Artificial intelligence (AI) and the Internet of Things (IoT) have significantly enhanced the collecting and use of personal data, starting with smartphones and social media (Sima et al., 2020). While there are many advantages to these developments, like increased efficiency, tailored experiences and connection, they also give rise to grave worries over data privacy and human rights (Tseng et al., 2021). In order to highlight the importance of valuing data privacy in the digital era, this chapter seeks to investigate the crucial subject of how to strike a balance between technical progress and individual liberty (Anastasi et al., 2021). The goal of digital rights is to encourage the use of data privacy-preserving technology (Dwivedi et al., 2021).

Cherish Data Privacy and Human Rights in the Digital Age

These technologies also need to defend human rights, democratic ideals, and intellectual property rights while stifling misinformation (Majstorovic & Mitrovic, 2019). New legislation designed for the internet era have been put into place in countries including Brazil, Chile, Spain, and Portugal as part of initiatives to enhance digital rights (Badri et al., 2018). The other nations have recognized rights connected to technology, particularly those concerning artificial intelligence including the United States, Colombia and Japan (Xu et al., 2018). The internet privacy index, which is based on information on international data privacy regulations, cybercrime legislation, freedom of speech, democratic metrics, and press freedom, is a crucial indicator of digital rights (Bhuiyan et al., 2020).

Figure 1. Landscapes of introduction split section (Original)

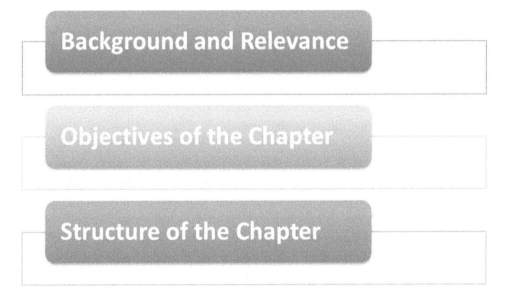

1.1 Background and Relevance

The digital technologies are not self-contained entities and they may be a powerful force for advancement in humankind and are essential in advancing and defending human rights (Reier Forradellas & Garay Gallastegui, 2021). But because to data-heavy technologies, such as those found in artificial intelligence apps, corporations and governments may now follow, analyze, forecast, and even control people's behavior to a never-before-seen degree in the digital world (Singh Rajawat et al., 2021). These technological developments pose serious hazards to privacy, autonomy, human dignity, and the wider exercise of human rights if they are implemented without adequate protections (Leng et al., 2020). The global exercise and violation of human rights have been altered by digital technology (Javaid et al., 2021). The Internet is now a vital instrument for both speeding economic development and achieving a wide range of human rights (Hassoun et al., 2023).

The mining and processing of personal data by businesses gives rise to two major problems as: (a) the violation of customers' rights and privacy; and (b) the widening gap between businesses that are able and unable to get consumer data (Aoun et al., 2021). The management of customer data in this digital economy becomes an issue that is covered by antitrust laws as the primary objectives of antitrust law are economic efficiency, consumer protection, and competitor protection (Arden et al., 2021) (Rane, 2023). But the primary focus of conventional competition analysis is on "price models." The several methods used by companies to determine the costs of their goods and services are referred to as pricing models (Awan et al., 2021). Client data is a "non-pricing model" which is often overlooked in conventional antitrust analyses (Kumar et al., 2020).

1.2 Objectives of the Chapter

This chapter has the following objectives to-

- explore the effects of digital technologies on personal autonomy, particularly the gathering and use of personal data (Li et al., 2017). In this analysis, the advantages and disadvantages of technical advancement in the digital era are examined, with special attention on how these factors affect data privacy and human rights (Kurniawan et al., 2022).
- analyzing the main difficulties brought on by the application of data-intensive technologies is the second goal (Jakka et al., 2022). This entails debating possible risks to privacy, human rights, and ethical issues in addition to investigating chances for ethical innovation that upholds individual autonomy and dignity (Xing et al., 2021).
- offer workable solutions for striking a balance between personal autonomy and technology advancement (Shi et al., 2020). This entails highlighting the significance of robust legal frameworks, moral data practices, and public awareness to guarantee that technology innovations do not jeopardize privacy or human rights (Finance et al., 2015).

Figure 2. Objectives of the chapter (Original)

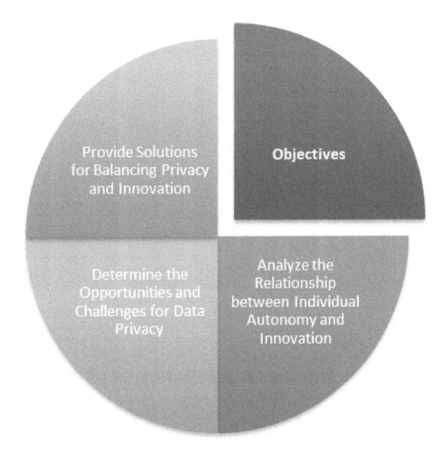

1.3 Structure of the Chapter

This chapter comprehensively discussed the diverse arena of Data Privacy and Human Rights in the Digital Age: Harmonizing Innovation and Individual Autonomy. Section 2 specifies the Digital Age and the Proliferation of Data. Section 3 explores the Human Rights and Data Privacy. Section 4 lays down the Legal and Ethical Landscape: Human Rights and Data Privacy in Digital Age. Section 5 highlights the Function of Authorities and Institutions. Section 6 conveys the Public Awareness and Societal Implications: Concerning Human Rights of Data Privacy. Finally, Section 7 Conclude the Chapter with Future Scope.

Figure 3. Flow of the chapter (Original)

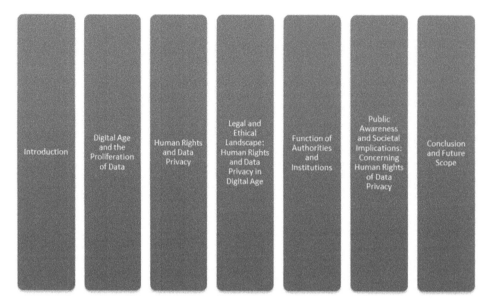

2. DIGITAL AGE AND THE PROLIFERATION OF DATA

The way data in which is gathered, handled and utilized has changed dramatically with the advent of the digital age (Saeed et al., 2023). The governments and organizations possess never-before-seen capacity to compile enormous volumes of personal data (Kumar & Mallipeddi, 2022). Humanity has greatly benefited from technological breakthroughs. But as technology advances, more and more of our liberties are under jeopardy (Bag & Pretorius, 2022). With more and more data being gathered and processed in the marketplace as we go deeper into the technology era, concerns about the right to privacy are mounting (Morgan et al., 2021). Data fraud, scam calls, cyber harassment, and other illegal acts have increased with the rise of digitalization (Singh et al., 2022). When users contribute their personal information to websites for businesses, online networking, analytics companies, government organizations, and other purposes, it is frequently exploited (Nica & Stehel, 2021). Clear national regulations governing data collecting, processing, sharing, monitoring, recording, access, and maintenance are lacking. Although this data-driven economy fosters innovation, privacy and individual liberty are at stake (Borowski, 2021). From financial transactions and health records to internet browsing patterns and location data, personal data may encompass a wide range of information. The potential dangers to human rights and privacy grow with the volume of data (Ahmad et al., 2021).

An essential component of the digital era is data. It is the engine that fuels the everyday gadgets we use as well as the digital economy (Wamba-Taguimdje et al., 2020). Data is used to make better decisions, develop new goods and services, increase operational effectiveness, customize interactions with customers, and even make money off of information (Popov et al., 2022). The term "Data is the new Gold" is frequently applied to data due to its great value as an asset for companies and organizations (Lu et al., 2019). It may be applied in a variety of ways to create value, frequently with artificial intelligence's assistance such as-

Improving Decision-Making: Information from data may be used to understand consumer behavior, industry trends, and operational effectiveness (Borowski, 2021). Decisions on new product development, advertising initiatives, and other key areas may then be made using this information (Lim et al., 2021).

Creating New Products and Services: Data may be used to identify new business prospects and assist in the creation of goods and services that are customized to meet the demands of clients (Vogt, 2021). For example, a business may examine consumer buying trends to find possible new product categories (Gadekar et al., 2022).

Increasing Operational Efficiency: Information may be used to identify inefficiencies and bottlenecks in company processes. Improving customer service and cutting costs can result from addressing these (Tao et al., 2021).

Customizing Customer Experience: With providing information, promotions and suggestions based on individual interests, the data may be utilized to tailor customer interactions (Chen et al., 2022).

Data Monetization: Data may be made money by utilizing it to develop customized advertising campaigns or by selling it to other companies (Mourtzis et al., 2022).

Figure 4. Digital age and the proliferation of data: "data is new gold" (Original)

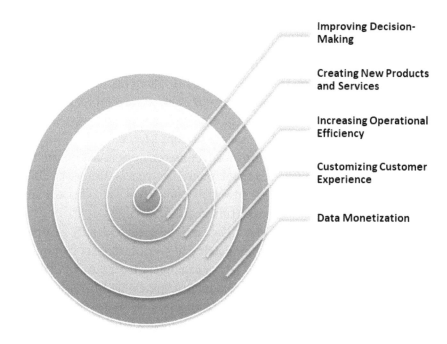

3. HUMAN RIGHTS AND DATA PRIVACY

The way that businesses and customers engage is changing dramatically as markets become more and more part of the "digital economy." The phrase 'digital economy' refers to marketplaces where the exchange of goods and services is facilitated by computer-based technologies (Wan et al., 2018). This industry's fundamental feature is the development of commercial strategies based on the exchange of "information" between businesses and their clientele (Cioffi et al., 2020). Customers' personal data is exchanged in a significant way in this data exchange. Customer data has essentially taken on the role of 'currency' in this online economy (Laskurain-Iturbe et al., 2021). Through the research that comes from mining sensitive data, businesses are able to promote their goods and services more successfully. Customers may generate demand and break free from the laws of supply and demand by taking advantage of their habits and purchase patterns (Brock & Von Wangenheim, 2019). In the digital era, safeguarding data privacy and human rights remains extremely difficult, even with laws in place and some of these difficulties are like-

Data Monitoring: Private and individual autonomy may be violated by corporate and governmental monitoring methods (Nguyen et al., 2020). Concerns regarding the exploitation and abuse of personal data are raised by widespread data collecting and monitoring (Hassoun et al., 2022).

Algorithmic Bias: The likelihood of algorithmic bias rises with the use of artificial intelligence (AI) and machine learning (ML) systems (Nahavandi, 2019). Human rights may be impacted by biased data that produces discriminatory results (Zhou et al., 2021).

Data Proliferation: The creation, gathering, and sharing of data have increased dramatically in the digital age (de la Pena Zarzuelo et al., 2020). People are producing enormous amounts of data every day due to the growth of social media, smart gadgets, and online transactions (Sanchez-Sotano et al., 2020). Companies and governments routinely collect this wealth of data, which raises questions about who may access our personal information (Bokhari & Myeong, 2023).

Lack of Knowledge: A lot of people are unaware of the full scope of the collection and usage of their personal data (Ali et al., 2022). Users may provide sensitive information without realizing it since data privacy regulations and terms of service agreements often contain intricate and difficult-to-understand elements (Kuzior, 2022).

Data Breach and Cybersecurity: Individual privacy is at risk due to the increasing number of data breaches (Bongomin et al., 2020). Cyberattacks have the potential to reveal private information, which might result in identity theft and other nefarious behavior (Pivoto et al., 2021).

Cross-Border Data Flows: Data often crosses national borders in today's global digital world, making it more difficult to police privacy laws (Gupta, 2023). Data privacy gaps may arise from differing degrees of protection offered by different governments (Dorfleitner & Braun, 2019).

Cyberattacks and High-Profile Data Breaches: Millions of people are now at danger of financial fraud and identity theft as a result of these more frequent cyberattacks (Nassiry, 2018). The continual development of innovative methods by cybercriminals to take advantage of flaws in digital systems highlights how difficult it is to maintain data security (Singh, 2024).

Figure 5. Difficulties in data privacy protection (Original)

4. LEGAL AND ETHICAL LANDSCAPE

There are some ethical and legal frameworks pertaining to data privacy that meet the intricacies of the digital era (Chen & Volz, 2021). These important laws that set standards for data protection and allow people more control over their personal information include the California Consumer Privacy Act (CCPA) in the US and the General Data Protection Regulation (GDPR) in Europe (Yang et al., 2021). These frameworks seek to achieve a balance between individual rights and technical advancement. They also draw attention to the difficulties in executing laws in a global digital environment (Zhang et al., 2018). Monitoring and protecting data assets at every stage of their existence, from acquisition to disposal, is known as data governance (Migliorelli & Dessertine, 2019). This process includes putting in place security measures to stop unwanted access, use, or disclosure of data in addition to developing rules and processes for gathering, using, storing, and sharing data (Naderi & Tian, 2022). Because it guarantees that data stays accurate, dependable, and secure, data governance is essential (Singh, 2023). It also guarantees that data is utilized in an ethical and responsible manner (Thomason 2018). There are fundamental solutions for protection of human rights via data safety and security as-

Strict Regulations on Data Privacy: The main actors in guaranteeing data privacy are the government and regulatory agencies (Liu et al., 2022). Regulations like the California Consumer Privacy Act (CCPA) and the General Data Protection Regulation (GDPR) of the European Union set down precise guidelines for data protection, requiring businesses to get users' express consent and to be open and honest about how they use their data (Yan et al., 2022).

Instruction for Users: It is essential to educate individuals about data privacy. People must to be aware of the dangers of disclosing private information online and how to control their privacy settings on various platforms (Bayram et al., 2022). Organizations may assist by providing simple-to-read privacy policies (Sachs et al., 2019).

Data Minimization: Businesses might adopt a "data minimization" strategy, collecting just the data necessary to provide their services and deleting the rest when they are no longer needed (Campbell-Verduyn, 2023). This strategy lessens the quantity of personal data that is out there and the possible consequences of data breaches (Singh, 2023).

Security and Encryption Measures: Using robust cybersecurity procedures and encryption techniques may significantly enhance data safety (Schulz & Feist, 2021). Even in the event that data is intercepted during transmission, end-to-end encryption guarantees that only authorized users may access the information (Kalaiarasi & Kirubahari, 2023).

Blockchain Technology: Blockchain technology offers a decentralized, unhackable approach to data management and storage (Chueca Vergara & Ferruz Agudo, 2021). With enabling users to limit access to their data and keep track of who has accessed it, it may provide them more control over the information they own (Singh, 2022).

Ethical Data Practices: User consent, accountability, and openness should be the main focuses of an organization's ethical data practices (Sharma & Singh, 2022). To do this, they must regularly examine privacy policies and request outside evaluations of their data management practices (Marke, 2018).

Figure 6. Solutions- data privacy and human rights protection in digital sphere (Original)

5. FUNCTION OF AUTHORITIES AND INSTITUTIONS: DATA PROTECTION, PRIVACY AND AI

Human rights and data privacy are important issues that are addressed by governments and organizations (Hoang et al., 2022). The governments are required to uphold privacy standards and implement data protection laws (Schloesser & Schulz, 2022). In order to handle cross-border data flows and create uniform privacy standards, international collaboration is also necessary (Singh, 2022). The notion of an individual's entitlement to privacy is multifaceted. It alludes to the unique right that internet users have over the gathering, storing, and sharing of their personal data (Harris, 2018). Identification information, preferences, interests, relationships, health, and financial records are examples of personal data (Shih et al., 2023). This data may be creatively utilized for a number of things, such as business profit-making and government spying. The organizations should put data protection first and follow moral guidelines (Ozili, 2023). This entails carrying out frequent audits, putting strong cybersecurity safeguards in place, and encouraging an employee-focused privacy culture (Bayram et al., 2022). Artificial intelligence (AI) is a field of computer science that seeks to create software that can carry out jobs that are normally completed by people (Singh, 2022). These activities, which are frequently considered intellectual, include seeing both visually and auditorily, learning and adapting, thinking, recognizing patterns, and making decisions (Micaroni, 2020). A variety of related methods and technologies, such as robots, machine learning, predictive analytics, and natural language processing, are together referred to as artificial intelligence (AI) (Dell'Erba, 2021).

The way individuals and businesses interact with data has significantly altered with the advent of artificial intelligence (AI) (Bin Amin et al., 2022). The advancement of AI technology has resulted in significant modifications to data protection laws throughout the globe (Singh, 2022). In the current digital era, data protection has emerged as a crucial concern. The best examples of how data protection laws have been modified to take into account the changing digital landscape is the General Data Protection Regulation (GDPR), which was established in 2018 (Puaschunder, 2023). A rule from the European Union (EU) called the GDPR specifies how personal information must be gathered, kept, and handled. Its main objectives are to protect people's rights and make data processing more transparent (Vikas et al., 2022). The right to data portability, the right to be forgotten, and the right to access one's personal information are just a few of the important new rules brought about by the GDPR that improve data protection (Macchiavello, 2023). The GDPR has been a significant turning point in the development of data protection regulations, spurring the adoption of comparable rules in nations including Brazil, India, and Japan (Marke & Sylvester, 2018).

6. PUBLIC AWARENESS AND SOCIETAL IMPLICATIONS: CONCERNING HUMAN RIGHTS OF DATA PRIVACY

Human rights and data privacy are more social concerns than only legal and regulatory ones (Singh, 2020). Educating the public and raising awareness about data privacy are essential to creating a culture that cherishes individual liberty (Udeagha & Ngepah, 2023). The promotion of privacy rights and awareness-raising may be greatly aided by the media, civil society, and advocacy organizations (Hassan et al., 2020). Privacy rights have been significantly impacted by the development of AI. AI has made it possible for businesses to gather, store, and analyze vast amounts of data at previously unheard-of rates (Fenwick & Vermeulen, 2020). There are now additional privacy problems as a result of the data being processed by AI algorithms. AI's ability to violate someone's right to privacy is a serious problem (Singh, 2019).

Artificial intelligence is capable of extracting sensitive information, such as sexual orientation, medical records, and religious convictions (Nenavath & Mishra, 2023). From data in order to identify specific people or groups, there is a risk to privacy rights when using AI for predictive analytics (Schulz et al., 2020). The advent of face recognition technology, which is frequently employed for monitoring and surveillance, has sparked debate over privacy issues and the place of AI in these kinds of applications (Moro-Visconti et al., 2020). It is been tough for people to comprehend how businesses exploit their data since AI algorithms are opaque and difficult to explain how they work (Rizzello & Kabli, 2020). Due to this ambiguity, there have been worries that people may be denied chances as a result of AI-based judgments without knowing why or how those decisions were reached (Jaiwant & Kureethara, 2023).

The UN Declaration of Human Rights, the International Covenant on Civil and Political Rights, and many other international and regional conventions recognize privacy as a basic human right (Bhowmik, 2022). Human dignity depends on privacy, which also supports other fundamental principles like freedom of expression and association. It is become one of the most important human rights concerns of the contemporary day (Radanliev, 2024).

7. CONCLUSION AND FUTURE SCOPE

Maintaining of the human rights and data privacy in the digital era necessitates striking a careful balance between innovation and personal freedom. Although technology has many advantages, there are serious hazards to human rights and privacy (Li et al., 2024). Societies must emphasize privacy, enact strict laws, and encourage moral data practices in order to balance these two factors. To ensure that personal liberty is not sacrificed in the name of technical advancement, governments, organizations, and people all have a part to play (Sreenivasan & Suresh, 2024). It can build a digital environment that respects and upholds human rights while encouraging innovation by embracing accountability, transparency, and ethical behavior. In the digital age, data privacy is a complicated challenge that necessitates cooperation between individuals, companies and legislators. It is critical that it prioritize protecting our personal data as we continue to adopt new technology (Anyanwu et al.,

2024). Through promoting stronger security protocols, raising public awareness, and advocating for more stringent laws, it can create a digital landscape in which people own their digital identities and data privacy is respected (Diro et al., 2024). It can only secure the benefits of the digital age while lowering the risks to privacy by working together (Sharma & Dwivedi, 2024).

REFERENCES

Ahmad, T., Zhang, D., Huang, C., Zhang, H., Dai, N., Song, Y., & Chen, H. (2021). Artificial intelligence in sustainable energy industry: Status Quo, challenges and opportunities. *Journal of Cleaner Production*, 289, 125834. 10.1016/j.jclepro.2021.125834

Ali, A., Septyanto, A. W., Chaudhary, I., Al Hamadi, H., Alzoubi, H. M., & Khan, Z. F. (2022, February). Applied Artificial Intelligence as Event Horizon Of Cyber Security. In *2022 International Conference on Business Analytics for Technology and Security (ICBATS)* (pp. 1-7). IEEE. 10.1109/ICBATS54253.2022.9759076

Anastasi, S., Madonna, M., & Monica, L. (2021). Implications of embedded artificial intelligence-machine learning on safety of machinery. *Procedia Computer Science*, 180, 338–343. 10.1016/j.procs.2021.01.171

Angelopoulos, A., Michailidis, E. T., Nomikos, N., Trakadas, P., Hatziefremidis, A., Voliotis, S., & Zahariadis, T. (2019). Tackling faults in the industry 4.0 era—A survey of machine-learning solutions and key aspects. *Sensors (Basel)*, 20(1), 109. 10.3390/s2001010931878065

Anyanwu, A., Olorunsogo, T., Abrahams, T. O., Akindote, O. J., & Reis, O. (2024). Data confidentiality and integrity: A review of accounting and cybersecurity controls in superannuation organizations. *Computer Science & IT Research Journal*, 5(1), 237–253. 10.51594/csitrj.v5i1.735

Aoun, A., Ilinca, A., Ghandour, M., & Ibrahim, H. (2021). A review of Industry 4.0 characteristics and challenges, with potential improvements using blockchain technology. *Computers & Industrial Engineering*, 162, 107746. 10.1016/j.cie.2021.107746

Arden, N. S., Fisher, A. C., Tyner, K., Lawrence, X. Y., Lee, S. L., & Kopcha, M. (2021). Industry 4.0 for pharmaceutical manufacturing: Preparing for the smart factories of the future. *International Journal of Pharmaceutics*, 602, 120554. 10.1016/j.ijpharm.2021.12055433794326

Asadollahi-Yazdi, E., Couzon, P., Nguyen, N. Q., Ouazene, Y., & Yalaoui, F. (2020). Industry 4.0: Revolution or Evolution? *American Journal of Operations Research*, 10(06), 241–268. 10.4236/ajor.2020.106014

Awan, U., Sroufe, R., & Shahbaz, M. (2021). Industry 4.0 and the circular economy: A literature review and recommendations for future research. *Business Strategy and the Environment*, 30(4), 2038–2060. 10.1002/bse.2731

Badri, A., Boudreau-Trudel, B., & Souissi, A. S. (2018). Occupational health and safety in the industry 4.0 era: A cause for major concern? *Safety Science*, 109, 403–411. 10.1016/j.ssci.2018.06.012

Bag, S., & Pretorius, J. H. C. (2022). Relationships between industry 4.0, sustainable manufacturing and circular economy: Proposal of a research framework. *The International Journal of Organizational Analysis*, 30(4), 864–898. 10.1108/IJOA-04-2020-2120

Bayram, O., Talay, I., & Feridun, M. (2022). Can FinTech promote sustainable finance? Policy lessons from the case of Turkey. *Sustainability (Basel)*, 14(19), 12414. 10.3390/su141912414

Bayram, O., Talay, I., & Feridun, M. (2022). Can Fintech Promote Sustainable Finance? Policy Lessons from the Case of Turkey. *Sustainability (Basel)*, 2022(14), 12414. 10.3390/su141912414

Bhowmik, D. (2022). An Introduction to Climate Fintech. *European Journal of Science. Innovation and Technology*, 2(4), 24–35.

Bhuiyan, A. B., Ali, M. J., Zulkifli, N., & Kumarasamy, M. M. (2020). Industry 4.0: Challenges, opportunities, and strategic solutions for Bangladesh. *International Journal of Business and Management Future*, 4(2), 41–56. 10.46281/ijbmf.v4i2.832

Bin Amin, S., Taghizadeh-Hesary, F., & Khan, F. (2022). Facilitating green digital finance in Bangladesh: Importance, prospects, and implications for meeting the SDGs. In *Green Digital Finance and Sustainable Development Goals* (pp. 143–165). Springer Nature Singapore. 10.1007/978-981-19-2662-4_7

Blobel, B. (2020, September). Application of industry 4.0 concept to health care. In *pHealth 2020: Proceedings of the 17th International Conference on Wearable Micro and Nano Technologies for Personalized Health* (Vol. 273, p. 23). IOS Press.

Bokhari, S. A. A., & Myeong, S. (2023). The influence of artificial intelligence on e-Governance and cybersecurity in smart cities: A stakeholder's perspective. *IEEE Access : Practical Innovations, Open Solutions*, 11, 69783–69797. 10.1109/ACCESS.2023.3293480

Bongomin, O., Gilibrays Ocen, G., Oyondi Nganyi, E., Musinguzi, A., & Omara, T. (2020). Exponential disruptive technologies and the required skills of industry 4.0. *Journal of Engineering*, 2020, 1–17. 10.1155/2020/8090521

Borowski, P. F. (2021). Innovative processes in managing an enterprise from the energy and food sector in the era of industry 4.0. *Processes (Basel, Switzerland)*, 9(2), 381. 10.3390/pr9020381

Borowski, P. F. (2021). Digitization, digital twins, blockchain, and industry 4.0 as elements of management process in enterprises in the energy sector. *Energies*, 14(7), 1885. 10.3390/en14071885

Brock, J. K. U., & Von Wangenheim, F. (2019). Demystifying AI: What digital transformation leaders can teach you about realistic artificial intelligence. *California Management Review*, 61(4), 110–134. 10.1177/1536504219865226

Campbell-Verduyn, M. (2023). Conjuring a cooler world? Imaginaries of improvement in blockchain climate finance experiments. *Environment and Planning C: Politics and Space*.

Chander, B., Pal, S., De, D., & Buyya, R. (2022). Artificial intelligence-based internet of things for industry 5.0. *Artificial intelligence-based internet of things systems*, 3-45.

Chen, Y., Lu, Y., Bulysheva, L., & Kataev, M. Y. (2022). Applications of blockchain in industry 4.0: A review. *Information Systems Frontiers*, 1–15. 10.1007/s10796-022-10248-7

Chen, Y., & Volz, U. (2021). Scaling up sustainable investment through blockchain-based project bonds. *ADB-IGF Special Working Paper Series "Fintech to Enable Development, Investment, Financial Inclusion, and Sustainability*.

Chueca Vergara, C., & Ferruz Agudo, L. (2021). Fintech and sustainability: Do they affect each other? *Sustainability (Basel)*, 13(13), 7012. 10.3390/su13137012

Cioffi, R., Travaglioni, M., Piscitelli, G., Petrillo, A., & De Felice, F. (2020). Artificial intelligence and machine learning applications in smart production: Progress, trends, and directions. *Sustainability (Basel)*, 12(2), 492. 10.3390/su12020492

de la Peña Zarzuelo, I., Soeane, M. J. F., & Bermúdez, B. L. (2020). Industry 4.0 in the port and maritime industry: A literature review. *Journal of Industrial Information Integration*, 20, 100173. 10.1016/j.jii.2020.100173

Dell'Erba, M. (2021). Sustainable digital finance and the pursuit of environmental sustainability. *Sustainable Finance in Europe: Corporate Governance, Financial Stability and Financial Markets*, 61-81.

Dinesh Arokia Raj, A., Jha, R. R., Yadav, M., Sam, D., & Jayanthi, K. (2024). Role of Blockchain and Watermarking Toward Cybersecurity. In *Multimedia Watermarking: Latest Developments and Trends* (pp. 103–123). Springer Nature Singapore. 10.1007/978-981-99-9803-6_6

Diro, A., Zhou, L., Saini, A., Kaisar, S., & Hiep, P. C. (2024). Leveraging zero knowledge proofs for blockchain-based identity sharing: A survey of advancements, challenges and opportunities. *Journal of Information Security and Applications*, 80, 103678. 10.1016/j.jisa.2023.103678

. Dorfleitner, G., & Braun, D. (2019). Fintech, digitalization and blockchain: possible applications for green finance. *The rise of green finance in Europe: opportunities and challenges for issuers, investors and marketplaces*, 207-237.

Dwivedi, Y. K., Hughes, L., Ismagilova, E., Aarts, G., Coombs, C., Crick, T., Duan, Y., Dwivedi, R., Edwards, J., Eirug, A., Galanos, V., Ilavarasan, P. V., Janssen, M., Jones, P., Kar, A. K., Kizgin, H., Kronemann, B., Lal, B., Lucini, B., & Williams, M. D. (2021). Artificial Intelligence (AI): Multidisciplinary perspectives on emerging challenges, opportunities, and agenda for research, practice and policy. *International Journal of Information Management*, 57, 101994. 10.1016/j.ijinfomgt.2019.08.002

Felsberger, A., Qaiser, F. H., Choudhary, A., & Reiner, G. (2022). The impact of Industry 4.0 on the reconciliation of dynamic capabilities: Evidence from the European manufacturing industries. *Production Planning and Control*, 33(2-3), 277–300. 10.1080/09537287.2020.1810765

Fenwick, M., & Vermeulen, E. P. (2020). Banking and regulatory responses to FinTech revisited-building the sustainable financial service 'ecosystems' of tomorrow. *Singapore Journal of Legal Studies*, (Mar), 165-189.

Finance, A. T. C. C. (2015). *Industry 4.0 Challenges and solutions for the digital transformation and use of exponential technologies. Finance, audit tax consulting corporate*. Swiss.

Fraga-Lamas, P., Lopes, S. I., & Fernández-Caramés, T. M. (2021). Green IoT and edge AI as key technological enablers for a sustainable digital transition towards a smart circular economy: An industry 5.0 use case. *Sensors (Basel)*, 21(17), 5745. 10.3390/s2117574534502637

Gadekar, R., Sarkar, B., & Gadekar, A. (2022). Key performance indicator based dynamic decision-making framework for sustainable Industry 4.0 implementation risks evaluation: Reference to the Indian manufacturing industries. *Annals of Operations Research*, 318(1), 189–249. 10.1007/s10479-022-04828-835910040

Gupta, R. (2023). Industry 4.0 adaption in indian banking Sector—A review and agenda for future research. *Vision (Basel)*, 27(1), 24–32. 10.1177/097226292199 682936977304

Harris, A. (2018). A conversation with masterminds in blockchain and climate change. In *Transforming climate finance and green investment with blockchains* (pp. 15–22). Academic Press. 10.1016/B978-0-12-814447-3.00002-1

Hassan, M. K., Rabbani, M. R., & Ali, M. A. M. (2020). Challenges for the Islamic Finance and banking in post COVID era and the role of Fintech. *Journal of Economic Cooperation & Development*, 41(3), 93–116.

Hassoun, A., Aït-Kaddour, A., Abu-Mahfouz, A. M., Rathod, N. B., Bader, F., Barba, F. J., Biancolillo, A., Cropotova, J., Galanakis, C. M., Jambrak, A. R., Lorenzo, J. M., Måge, I., Ozogul, F., & Regenstein, J. (2023). The fourth industrial revolution in the food industry—Part I: Industry 4.0 technologies. *Critical Reviews in Food Science and Nutrition*, 63(23), 6547–6563. 10.1080/10408398.2022.203473535114860

Hassoun, A., Prieto, M. A., Carpena, M., Bouzembrak, Y., Marvin, H. J., Pallares, N., Barba, F. J., Punia Bangar, S., Chaudhary, V., Ibrahim, S., & Bono, G. (2022). Exploring the role of green and Industry 4.0 technologies in achieving sustainable development goals in food sectors. *Food Research International*, 162, 112068. 10.1016/j.foodres.2022.11206836461323

Hoang, T. G., Nguyen, G. N. T., & Le, D. A. (2022). Developments in financial technologies for achieving the Sustainable Development Goals (SDGs): FinTech and SDGs. In *Disruptive technologies and eco-innovation for sustainable development* (pp. 1–19). IGI Global. 10.4018/978-1-7998-8900-7.ch001

Ivanov, D., Dolgui, A., & Sokolov, B. (2019). The impact of digital technology and Industry 4.0 on the ripple effect and supply chain risk analytics. *International Journal of Production Research*, 57(3), 829–846. 10.1080/00207543.2018.1488086

Jaiwant, S. V., & Kureethara, J. V. (2023). Green Finance and Fintech: Toward a More Sustainable Financial System. In *Green Finance Instruments, FinTech, and Investment Strategies: Sustainable Portfolio Management in the Post-COVID Era* (pp. 283–300). Springer International Publishing. 10.1007/978-3-031-29031-2_12

Jakka, G., Yathiraju, N., & Ansari, M. F. (2022). Artificial Intelligence in Terms of Spotting Malware and Delivering Cyber Risk Management. *Journal of Positive School Psychology*, 6(3), 6156–6165.

Javaid, M., Haleem, A., Singh, R. P., Khan, S., & Suman, R. (2021). Blockchain technology applications for Industry 4.0: A literature-based review. *Blockchain: Research and Applications*, 2(4), 100027.

Javaid, M., Haleem, A., Singh, R. P., Suman, R., & Gonzalez, E. S. (2022). Understanding the adoption of Industry 4.0 technologies in improving environmental sustainability. *Sustainable Operations and Computers*, 3, 203–217. 10.1016/j.susoc.2022.01.008

Kalaiarasi, H., & Kirubahari, S. (2023). Green finance for sustainable development using blockchain technology. In *Green Blockchain Technology for Sustainable Smart Cities* (pp. 167–185). Elsevier. 10.1016/B978-0-323-95407-5.00003-7

Kasowaki, L., & Ahmet, S. (2024). *Shielding the Virtual Ramparts: Understanding Cybersecurity Essentials* (No. 11700). EasyChair.

Kumar, S., & Mallipeddi, R. R. (2022). Impact of cybersecurity on operations and supply chain management: Emerging trends and future research directions. *Production and Operations Management*, 31(12), 4488–4500. 10.1111/poms.13859

Kumar, S. H., Talasila, D., Gowrav, M. P., & Gangadharappa, H. V. (2020). Adaptations of Pharma 4.0 from Industry 4.0. *Drug Invention Today*, 14(3).

Kurniawan, T. A., Maiurova, A., Kustikova, M., Bykovskaia, E., Othman, M. H. D., & Goh, H. H. (2022). Accelerating sustainability transition in St. Petersburg (Russia) through digitalization-based circular economy in waste recycling industry: A strategy to promote carbon neutrality in era of Industry 4.0. *Journal of Cleaner Production*, 363, 132452. 10.1016/j.jclepro.2022.132452

Kuzior, A. (2022). Technological unemployment in the perspective of Industry 4.0. *Virtual Economics*, 5(1), 7–23. 10.34021/ve.2022.05.01(1)

Laskurain-Iturbe, I., Arana-Landín, G., Landeta-Manzano, B., & Uriarte-Gallastegi, N. (2021). Exploring the influence of industry 4.0 technologies on the circular economy. *Journal of Cleaner Production*, 321, 128944. 10.1016/j.jclepro.2021.128944

Leng, J., Ye, S., Zhou, M., Zhao, J. L., Liu, Q., Guo, W., Cao, W., & Fu, L. (2020). Blockchain-secured smart manufacturing in industry 4.0: A survey. *IEEE Transactions on Systems, Man, and Cybernetics. Systems*, 51(1), 237–252. 10.1109/TSMC.2020.3040789

Li, B. H., Hou, B. C., Yu, W. T., Lu, X. B., & Yang, C. W. (2017). Applications of artificial intelligence in intelligent manufacturing: A review. *Frontiers of Information Technology & Electronic Engineering*, 18(1), 86–96. 10.1631/FITEE.1601885

Li, Z., Liang, X., Wen, Q., & Wan, E. (2024). The Analysis of Financial Network Transaction Risk Control based on Blockchain and Edge Computing Technology. *IEEE Transactions on Engineering Management*, 71, 5669–5690. 10.1109/TEM.2024.3364832

Lim, C. H., Lim, S., How, B. S., Ng, W. P. Q., Ngan, S. L., Leong, W. D., & Lam, H. L. (2021). A review of industry 4.0 revolution potential in a sustainable and renewable palm oil industry: HAZOP approach. *Renewable & Sustainable Energy Reviews*, 135, 110223. 10.1016/j.rser.2020.110223

Liu, H., Yao, P., Latif, S., Aslam, S., & Iqbal, N. (2022). Impact of Green financing, FinTech, and financial inclusion on energy efficiency. *Environmental Science and Pollution Research International*, 29(13), 1–12. 10.1007/s11356-021-16949-x34705207

Lu, C., Lyu, J., Zhang, L., Gong, A., Fan, Y., Yan, J., & Li, X. (2020). Nuclear power plants with artificial intelligence in industry 4.0 era: Top-level design and current applications—A systemic review. *IEEE Access: Practical Innovations, Open Solutions*, 8, 194315–194332. 10.1109/ACCESS.2020.3032529

Lu, Y. (2019). Artificial intelligence: A survey on evolution, models, applications and future trends. *Journal of Management Analytics*, 6(1), 1–29. 10.1080/23270012.2019.1570365

Macchiavello, E. (2023). *Sustainable Finance and Fintech. A Focus on Capital Raising. A Focus on Capital Raising*. CUP.

Majstorovic, V. D., & Mitrovic, R. (2019). Industry 4.0 programs worldwide. In *Proceedings of the 4th International Conference on the Industry 4.0 Model for Advanced Manufacturing: AMP 2019 4* (pp. 78-99). Springer International Publishing. 10.1007/978-3-030-18180-2_7

Marke, A. (Ed.). (2018). *Transforming climate finance and green investment with blockchains*. Academic Press.

Marke, A., & Sylvester, B. (2018). Decoding the current global climate finance architecture. In *Transforming climate finance and green investment with blockchains* (pp. 35–59). Academic Press. 10.1016/B978-0-12-814447-3.00004-5

Meyendorf, N., Ida, N., Singh, R., & Vrana, J. (2023). NDE 4.0: Progress, promise, and its role to industry 4.0. *NDT & E International*, 140, 102957. 10.1016/j.ndteint.2023.102957

Micaroni, M. (2020). *Sustainable Finance: Addressing the SDGs through Fintech and Digital Finance solutions in EU* (Doctoral dissertation, Politecnico di Torino).

Migliorelli, M., & Dessertine, P. (2019). The rise of green finance in Europe. *Opportunities and challenges for issuers, investors and marketplaces. Cham. Palgrave Macmillan*, 2, 2019.

Mithas, S., Chen, Z. L., Saldanha, T. J., & De Oliveira Silveira, A. (2022). How will artificial intelligence and Industry 4.0 emerging technologies transform operations management? *Production and Operations Management*, 31(12), 4475–4487. 10.1111/poms.13864

Morgan, J., Halton, M., Qiao, Y., & Breslin, J. G. (2021). Industry 4.0 smart reconfigurable manufacturing machines. *Journal of Manufacturing Systems*, 59, 481–506. 10.1016/j.jmsy.2021.03.001

Moro-Visconti, R., Cruz Rambaud, S., & López Pascual, J. (2020). Sustainability in FinTechs: An explanation through business model scalability and market valuation. *Sustainability (Basel)*, 12(24), 10316. 10.3390/su122410316

Mourtzis, D., Angelopoulos, J., & Panopoulos, N. (2022). A Literature Review of the Challenges and Opportunities of the Transition from Industry 4.0 to Society 5.0. *Energies*, 15(17), 6276. 10.3390/en15176276

Naderi, N., & Tian, Y. (2022). Leveraging Blockchain Technology and Tokenizing Green Assets to Fill the Green Finance Gap. *Energy Research Letters, 3*(3).

Nahavandi, S. (2019). Industry 5.0—A human-centric solution. *Sustainability (Basel)*, 11(16), 4371. 10.3390/su11164371

Nassiry, D. (2018). *The role of fintech in unlocking green finance: Policy insights for developing countries* (No. 883). ADBI working paper.

Nenavath, S., & Mishra, S. (2023). Impact of green finance and fintech on sustainable economic growth: Empirical evidence from India. *Heliyon*, 9(5), e16301. 10.1016/j.heliyon.2023.e1630137234625

Nguyen, T., Gosine, R. G., & Warrian, P. (2020). A systematic review of big data analytics for oil and gas industry 4.0. *IEEE Access : Practical Innovations, Open Solutions*, 8, 61183–61201. 10.1109/ACCESS.2020.2979678

Nica, E., & Stehel, V. (2021). Internet of things sensing networks, artificial intelligence-based decision-making algorithms, and real-time process monitoring in sustainable industry 4.0. *Journal of Self-Governance and Management Economics*, 9(3), 35–47.

Ozili, P. K. (2023). Assessing global interest in decentralized finance, embedded finance, open finance, ocean finance and sustainable finance. *Asian Journal of Economics and Banking*, 7(2), 197–216. 10.1108/AJEB-03-2022-0029

Pivoto, D. G., de Almeida, L. F., da Rosa Righi, R., Rodrigues, J. J., Lugli, A. B., & Alberti, A. M. (2021). Cyber-physical systems architectures for industrial internet of things applications in Industry 4.0: A literature review. *Journal of Manufacturing Systems*, 58, 176–192. 10.1016/j.jmsy.2020.11.017

Popov, V. V., Kudryavtseva, E. V., Kumar Katiyar, N., Shishkin, A., Stepanov, S. I., & Goel, S. (2022). Industry 4.0 and digitalisation in healthcare. *Materials (Basel)*, 15(6), 2140. 10.3390/ma15062140 35329592

Puaschunder, J. M. (2023). The Future of Resilient Green Finance. In *The Future of Resilient Finance: Finance Politics in the Age of Sustainable Development* (pp. 185–210). Springer International Publishing. 10.1007/978-3-031-30138-4_6

Radanliev, P. (2024). Cyber diplomacy: defining the opportunities for cybersecurity and risks from Artificial Intelligence, IoT, Blockchains, and Quantum Computing. *Journal of Cyber Security Technology*, 1-51.

Rane, N. (2023). Transformers in Industry 4.0, Industry 5.0, and Society 5.0: Roles and Challenges.

Rath, K. C., Khang, A., & Roy, D. (2024). The Role of Internet of Things (IoT) Technology in Industry 4.0 Economy. In *Advanced IoT Technologies and Applications in the Industry 4.0 Digital Economy* (pp. 1-28). CRC Press.

Ray, R. K., Chowdhury, F. R., & Hasan, M. R. (2024). Blockchain Applications in Retail Cybersecurity: Enhancing Supply Chain Integrity, Secure Transactions, and Data Protection. *Journal of Business and Management Studies*, 6(1), 206–214. 10.32996/jbms.2024.6.1.13

Reier Forradellas, R. F., & Garay Gallastegui, L. M. (2021). Digital transformation and artificial intelligence applied to business: Legal regulations, economic impact and perspective. *Laws*, 10(3), 70. 10.3390/laws10030070

Rizzello, A., & Kabli, A. (2020). Social finance and sustainable development goals: A literature synthesis, current approaches and research agenda. *ACRN Journal of Finance and Risk Perspectives*, 9.

Sachs, J. D., Woo, W. T., Yoshino, N., & Taghizadeh-Hesary, F. (2019). Importance of green finance for achieving sustainable development goals and energy security. *Handbook of green finance: Energy security and sustainable development*, 10, 1-10.

Saeed, S., Altamimi, S. A., Alkayyal, N. A., Alshehri, E., & Alabbad, D. A. (2023). Digital transformation and cybersecurity challenges for businesses resilience: Issues and recommendations. *Sensors (Basel)*, 23(15), 6666. 10.3390/s23156666 37571451

Sánchez-Sotano, A., Cerezo-Narváez, A., Abad-Fraga, F., Pastor-Fernández, A., & Salguero-Gómez, J. (2020). Trends of digital transformation in the shipbuilding sector. In *New Trends in the Use of Artificial Intelligence for the Industry 4.0*. IntechOpen. 10.5772/intechopen.91164

Schloesser, T., & Schulz, K. (2022). Distributed ledger technology and climate finance. In *Green Digital Finance and Sustainable Development Goals* (pp. 265–286). Springer Nature Singapore. 10.1007/978-981-19-2662-4_13

Schoenmaker, D., & Volz, U. (2022). *Scaling up sustainable finance and investment in the Global South*. CEPR Press.

Schulz, K., & Feist, M. (2021). Leveraging blockchain technology for innovative climate finance under the Green Climate Fund. *Earth System Governance*, 7, 100084. 10.1016/j.esg.2020.100084

Schulz, K. A., Gstrein, O. J., & Zwitter, A. J. (2020). Exploring the governance and implementation of sustainable development initiatives through blockchain technology. *Futures*, 122, 102611. 10.1016/j.futures.2020.102611

Sharma, A., & Singh, B. (2022). Measuring Impact of E-commerce on Small Scale Business: A Systematic Review. *Journal of Corporate Governance and International Business Law*, 5(1).

Sharma, S., & Dwivedi, R. (2024). A survey on blockchain deployment for biometric systems. *IET Blockchain*, 4(2), 124–151. 10.1049/blc2.12063

Shi, Z., Xie, Y., Xue, W., Chen, Y., Fu, L., & Xu, X. (2020). Smart factory in Industry 4.0. *Systems Research and Behavioral Science*, 37(4), 607–617. 10.1002/sres.2704

Shih, C., Gwizdalski, A., & Deng, X. (2023). *Building a Sustainable Future: Exploring Green Finance*. Regenerative Finance, and Green Financial Technology.

Sima, V., Gheorghe, I. G., Subić, J., & Nancu, D. (2020). Influences of the industry 4.0 revolution on the human capital development and consumer behavior: A systematic review. *Sustainability (Basel)*, 12(10), 4035. 10.3390/su12104035

Singh, B. (2019). Profiling Public Healthcare: A Comparative Analysis Based on the Multidimensional Healthcare Management and Legal Approach. *Indian Journal of Health and Medical Law*, 2(2), 1–5.

Singh, B. (2020). Global science and jurisprudential approach concerning healthcare and illness. *Indian Journal of Health and Medical Law*, 3(1), 7–13.

Singh, B. (2022). Understanding Legal Frameworks Concerning Transgender Healthcare in the Age of Dynamism. *Electronic Journal of Social and Strategic Studies*, 3(1), 56–65. 10.47362/EJSSS.2022.3104

Singh, B. (2022). Relevance of Agriculture-Nutrition Linkage for Human Healthcare: A Conceptual Legal Framework of Implication and Pathways. *Justice and Law Bulletin*, 1(1), 44–49.

Singh, B. (2022). COVID-19 Pandemic and Public Healthcare: Endless Downward Spiral or Solution via Rapid Legal and Health Services Implementation with Patient Monitoring Program. *Justice and Law Bulletin*, 1(1), 1–7.

Singh, B. (2023). Blockchain Technology in Renovating Healthcare: Legal and Future Perspectives. In *Revolutionizing Healthcare Through Artificial Intelligence and Internet of Things Applications* (pp. 177-186). IGI Global.

Singh, B. (2023). Federated Learning for Envision Future Trajectory Smart Transport System for Climate Preservation and Smart Green Planet: Insights into Global Governance and SDG-9 (Industry, Innovation and Infrastructure). *National Journal of Environmental Law*, 6(2), 6–17.

Singh, B. (2024). Legal Dynamics Lensing Metaverse Crafted for Videogame Industry and E-Sports: Phenomenological Exploration Catalyst Complexity and Future. *Journal of Intellectual Property Rights Law*, 7(1), 8–14.

. Singh, S. K., Sharma, S. K., Singla, D., & Gill, S. S. (2022). Evolving requirements and application of SDN and IoT in the context of industry 4.0, blockchain and artificial intelligence. *Software Defined Networks: Architecture and Applications*, 427-496.

Singh, V. K. (2022). Regulatory and Legal Framework for Promoting Green Digital Finance. In *Green Digital Finance and Sustainable Development Goals* (pp. 3–27). Springer Nature Singapore. 10.1007/978-981-19-2662-4_1

Singh Rajawat, A., Bedi, P., Goyal, S. B., Shukla, P. K., Zaguia, A., Jain, A., & Monirujjaman Khan, M. (2021). Reformist framework for improving human security for mobile robots in industry 4.0. *Mobile Information Systems*, 2021, 1–10. 10.1155/2021/4744220

Sreenivasan, A., & Suresh, M. (2024). Start-up sustainability: Does blockchain adoption drives sustainability in start-ups? A systematic literature reviews. *Management Research Review*, 47(3), 390–405. 10.1108/MRR-07-2022-0519

Tao, F., Akhtar, M. S., & Jiayuan, Z. (2021). The future of artificial intelligence in cybersecurity: A comprehensive survey. *EAI Endorsed Transactions on Creative Technologies*, 8(28), e3–e3. 10.4108/eai.7-7-2021.170285

Thomason, J., Ahmad, M., Bronder, P., Hoyt, E., Pocock, S., Bouteloupe, J., & Shrier, D. (2018). Blockchain—powering and empowering the poor in developing countries. In *Transforming climate finance and green investment with blockchains* (pp. 137–152). Academic Press. 10.1016/B978-0-12-814447-3.00010-0

Tseng, M. L., Tran, T. P. T., Ha, H. M., Bui, T. D., & Lim, M. K. (2021). Sustainable industrial and operation engineering trends and challenges Toward Industry 4.0: A data driven analysis. *Journal of Industrial and Production Engineering*, 38(8), 581–598. 10.1080/21681015.2021.1950227

Udeagha, M. C., & Ngepah, N. (2023). The drivers of environmental sustainability in BRICS economies: Do green finance and fintech matter? *World Development Sustainability*, 3, 100096. 10.1016/j.wds.2023.100096

Vikas, N., Venegas, P., & Aiyer, S. (2022). Role of Banks and Other Financial Institutions in Enhancing Green Digital Finance. In *Green Digital Finance and Sustainable Development Goals* (pp. 329–352). Springer Nature Singapore. 10.1007/978-981-19-2662-4_16

Vogt, J. (2021). Where is the human got to go? Artificial intelligence, machine learning, big data, digitalisation, and human–robot interaction in Industry 4.0 and 5.0: Review Comment on: Bauer, M.(2020). Preise kalkulieren mit KI-gestützter Onlineplattform BAM GmbH, Weiden, Bavaria, Germany. *AI & Society*, 36(3), 1083–1087. 10.1007/s00146-020-01123-7

Wamba-Taguimdje, S. L., Fosso Wamba, S., Kala Kamdjoug, J. R., & Tchatchouang Wanko, C. E. (2020). Influence of artificial intelligence (AI) on firm performance: The business value of AI-based transformation projects. *Business Process Management Journal*, 26(7), 1893–1924. 10.1108/BPMJ-10-2019-0411

Wan, J., Yang, J., Wang, Z., & Hua, Q. (2018). Artificial intelligence for cloud-assisted smart factory. *IEEE Access: Practical Innovations, Open Solutions*, 6, 55419–55430. 10.1109/ACCESS.2018.2871724

Xing, K., Cropley, D. H., Oppert, M. L., & Singh, C. (2021). Readiness for digital innovation and industry 4.0 transformation: studies on manufacturing industries in the city of salisbury. *Business Innovation with New ICT in the Asia-Pacific: Case Studies*, 155-176.

Xu, L. D., Xu, E. L., & Li, L. (2018). Industry 4.0: State of the art and future trends. *International Journal of Production Research*, 56(8), 2941–2962. 10.1080/00207543.2018.1444806

Yan, C., Siddik, A. B., Yong, L., Dong, Q., Zheng, G. W., & Rahman, M. N. (2022). A two-staged SEM-artificial neural network approach to analyze the impact of FinTech adoption on the sustainability performance of banking firms: The mediating effect of green finance and innovation. *Systems*, 10(5), 148. 10.3390/systems10050148

Yang, Y., Su, X., & Yao, S. (2021). Nexus between green finance, fintech, and high-quality economic development: Empirical evidence from China. *Resources Policy*, 74, 102445. 10.1016/j.resourpol.2021.102445

Yue, Y., & Shyu, J. Z. (2024). A paradigm shift in crisis management: The nexus of AGI-driven intelligence fusion networks and blockchain trustworthiness. *Journal of Contingencies and Crisis Management*, 32(1), e12541. 10.1111/1468-5973.12541

Zhang, X., Aranguiz, M., Xu, D., Zhang, X., & Xu, X. (2018). Utilizing blockchain for better enforcement of green finance law and regulations. In *Transforming climate finance and green investment with blockchains* (pp. 289–301). Academic Press. 10.1016/B978-0-12-814447-3.00021-5

Zhong, R. Y., Xu, X., Klotz, E., & Newman, S. T. (2017). Intelligent manufacturing in the context of industry 4.0: A review. *Engineering (Beijing)*, 3(5), 616–630. 10.1016/J.ENG.2017.05.015

Zhou, J., Zhang, S., Lu, Q., Dai, W., Chen, M., Liu, X., . . . Herrera-Viedma, E. (2021). A survey on federated learning and its applications for accelerating industrial internet of things. *arXiv preprint arXiv:2104.10501*.

Chapter 8
Digitalization in Corporations:
Integrating Utility of Digital Technology With Accessibility and Privacy of Data

Siddharth Kanojia
https://orcid.org/0000-0002-1479-5292
O.P. Jindal Global University, India

ABSTRACT

Corporations are quickly learning how to benefit from digitalization in various facets of their operations whilst placing a significant focus on the requirements and experiences of their clients. Furthermore, by utilizing sophisticated analytical models and effective risk-management techniques, digitalization is also being used as a potent tool to improve internal governance and decision-making processes. Digitalization has made it easier for companies to fetch and access the data of various stakeholders through surveys or customer interactions, purchasing it from third-party data brokers, or using public data sources such as social media or government records. Hence, it also raises important questions about data privacy and security, and the appropriate balance between technological innovation and the protection of legal rights and obligations. Accordingly, this chapter intends to analyze the efficacy and utility of the latest amendments in the provisions of corporate law which intends to take due cognizance of these technological advancements.

INTRODUCTION

The improvement of digital technology and its utility in the everyday lifestyle of people, communities, and artificial entities have marked the adaptation to an informative society through the digitalized economic system and advancements of various factors relating to public lifestyle. Accordingly, technology has been indulged into components of human life, inclusive of gaming, politics, development, transportation, law, education, science, and business. Nevertheless, it leads to cause a significant impact on the economy and the State and executive branches, and thus, they should move quickly to ensure that different aspects of the digital economy are governed by the law. Most of these efforts aimed toward enhancing company law withinside the context of emerging digitalization are presently focused on growing & selling a digital ecosystem that might lead to a powerful interplay among company actors and people immediately related to them. As a result of using digital technology through diverse topics, a change has happened to the economic relations present in society, such as 'companies' (Laptev & Feyzrakhmanova, 2021). This chapter attempts to analyze the incorporation of digitalization in the laws & regulations framed in the area of company law and discover the traits and instructions of the evolution of respective artificial establishments and applicable regulations related to privacy in digitalized communities. The analysis is pursued on the elements such as the integration of certain functions of corporations with digital technologies and the development of the legal & regulatory mechanisms of company governance followed by the emergence of various means of accessing the data of consumers and other key stakeholders.

The utility of digital technology for a company's governance and upkeeping of digital information has increased the performance of companies and optimized their enterprise procedures and helped in establishing smooth interactions among the members of the company & with the other stakeholders. Nonetheless, as conveyed by Ichak Adizes, companies may enjoy the acquisition of digital technology at some stage in their entire lifestyle. For instance, the technology may be consumed in the phase of incorporating a company and endowing it with legal personality. Thus, it seems affordable to include problems associated with the digital legal personalities of companies, digital company governance, and the operations of digital groups (networked and decentralized self-sustaining organizations) in discussions regarding the digitalization of company law (Adizes, I., 2004). Over the period, the legislators & experts have realized to transform the traditional provisions & regulations to bring them in line with the rapid technological advancements. Subsequently, various rules, regulations & provisions have been introduced & amended to embrace this advancement. Mentioned below are some of the major changes that one can observe in the present legislation dealing with the varied aspects of the companies.

Digital signature certificates (DSCs) and privacy concerns associated with their use have been the focus of international legal and technological discourse. For example, in the case of United States v. Ivanov (2001), Russian hacker Alexei Ivanov was found guilty of breaking into financial institutions in the United States. One important piece of evidence used by the prosecution was his use of digital signatures for transaction authentication. This case brought to light the significance of digital certificates in cybercrimes and the necessity of managing them securely. On the other hand, one of India's first successful cybercrime prosecutions, State of Tamil Nadu v. Suhas Katti, concerned the malicious use of a digital signature. The defendant was found guilty of pleading vulgar messages, purportedly written by a woman, on a Yahoo! message group. The electronic messages' authenticity was confirmed by the digital signature. Similarly, in the 2003 California Supreme Court case Intel Corp. v. Hamidi, Hamidi, a former employee, attacked Intel in numerous emails addressed to Intel staff members. Intel filed a lawsuit, claiming trespass to chattels, but Hamidi won the case in court. The case revealed the difficulties of digital interactions in legal contexts, even though it does not directly involve digital signatures. It does, however, touch on privacy and digital communications. Accordingly, Personal data, including names, email addresses, and occasionally even biometric information, is frequently needed in order to issue DSCs. Certificate Authorities (CAs), who store this data, have the potential to seriously violate privacy if they are compromised. One example of the potential risks associated with centralising sensitive information is the 2011 Sony PlayStation Network hack, which resulted in the exposure of millions of user records. A serious privacy risk is the possible misuse of this personal data, whether as a result of data breaches or unauthorised sharing.

Registration and Incorporation

To simplify and speed up the system for incorporating an artificial entity, the concerned Ministry of Corporate Affairs (MCA) has brought in Integrated Incorporation Form. This form handles a single application for name reservation, formation of a new entity, and assignment of DIN, greatly curtailing the time required to incorporate a corporation in India. The supporting materials for this e-Form provide details on the Directors and Subscribers, the MoA, and AoA, among other things. Once the e-Form is processed and discovered complete, a corporation might be registered (Agrawal, R. K., 2016). In addition to this, MCA 21 is a formidable e-Governance mission of the concerned ministry. The intent and aim of this mission are to transit the Ministry's mode of operation from conventional paper to electronic format. This project was started to provide citizens with more than 100 services online that are in addition to the Companies Act of 1956 and the current Companies Act of 2013. The services offered in this project are supplied smoothly and securely through Ministry's official portal. In current instances, the MCA Portal has been re-modeled and the re-modeled portal is extra consumer-friendly (Bhatnagar, S., 2009). In addition, a number of measures have been implemented by the Ministry of Corporate Affairs to guarantee the security and privacy of data submitted via the portal. Sensitive data encryption, two-factor login authentication, and frequent security audits to find and fix possible vulnerabilities are some of these precautions. The goal of these initiatives is to increase user trust and entice more companies to use the MCA's electronic services. Constant improvements to the MCA Portal show the Ministry's dedication to using technology to improve governance and make doing business in India easier.

The advent of online registration platforms is one of the biggest developments in corporate registration. Globally, regulatory agencies and governments have created online portals that let companies finish the registration process at their convenience. These platforms usually provide easy-to-use interfaces, sequential instructions, and immediate feedback, which greatly minimises the time and effort needed for integration. For instance, corporations can be register online in as little as 24 hours with the help of the Companies House website in the United Kingdom (Bamberger, K. A., & Mulligan 2015). With this system's significant improvements in accessibility and efficiency, business owners can now concentrate more on running their companies than on overcoming administrative challenges. Whereas, by offering a safe, transparent, and unchangeable ledger for storing corporate data, blockchain technology is completely changing the registration and incorporation process (Habib et al., 2022). Hence, ensuring that all transactions and modifications to the corporate structure are permanently recorded and unchangeable, this technology improves trust and lowers the possibility of fraud. In reference to this, Estonia's e-Residency programme provides non-residents with a safe digital identity (Tammpuu & Masso 2019). This takes advantage of Estonia's cutting-edge digital infrastructure and enables business owners from anywhere in the world to register and run an EU-based company fully online.

Similarly, Automation and artificial intelligence are essential for expediting the incorporation and registration process. AI-driven solutions can help users in real-time, fill out forms automatically, and confirm the legitimacy of documents. Automation expedites procedures, lowers the possibility of human error, and guarantees regulatory compliance. For example, AI is used in Singapore's BizFile+ system, which is run by the Accounting and Corporate Regulatory Authority (ACRA), to speed up and accurately process business registration. By automating numerous repetitive tasks, the system improves speed and reliability of the process (William & William 2019).

On the other hand, the blockchain technology and online registration platforms necessitate the gathering and archiving of enormous volumes of personal and business data. This includes private data like names, addresses, identifying numbers, and financial information (Chen & Zhu 2017). Significant privacy risks arise from the centralization of this data if it is not sufficiently protected. Further, the growing dependence on digital platforms exposes corporate data to security breaches and cyberattacks. Unauthorised access to confidential data can result in financial fraud, identity theft, and other privacy violations. Strong cybersecurity defences are essential to fending off these attacks. Whereas diverse jurisdictions have different laws protecting privacy of personal information. It can be difficult to ensure adherence to national and international data protection laws, such as the General Data Protection Regulation (GDPR) of the EU. Due to the strict regulations these laws place on data collection, storage, and processing, legal frameworks and business practices must be updated on a regular basis (Hoofnagle et al., 2019). The Modern technology has made it easier to register and incorporate corporations, which has major advantages in terms of efficiency, cost savings, transparency, and accessibility from around the world. These developments do, however, also present significant privacy risks that need to be properly addressed. It is imperative to tackle concerns pertaining to data collection, cybersecurity, regulatory compliance, and maintaining a delicate equilibrium between privacy and transparency in order to guarantee that the advantages of these technologies are completely realised while maintaining user privacy (Hoofnagle et al., 2019).

Digital Signatures Certificates

The Information Technology Act of 2000's Section 18 grants the required legal sanctity & acknowledgment to digital signatures that are principally based on asymmetric cryptosystems. Nowadays, the frequency of utilizing digital signatures is on par with that of handwriting signatures, and the treatment of digitally signed virtual files is on the level of that of paper files. Digital files must be manually signed in the same way that physical documents are, using a Digital Signature Certificate, as is the case with e-bureaucracy. An authentic digital signature gives the receiver an appropriate reason to confide that the communication was sent by a known sender and was not changed in transit (tamper-proof). Digital signatures are often used to distribute software, sign financial documents, and in other situations where it's important to deter fraud and tampering (Sharma, V., 2011). Throughout the year 2000, the Controller of Certifying Authorities office was established which intends to promote the usage and adaptability of digital signatures for authentication in the transactions relating to e-commerce. Thereon, the RCAI with its self-signed Root Certificate issues Public Key Certificates to the certified CAs, whilst those certified CAs in turn trouble DSCs to end users (Banday, M., 2011). Digital signature diffusion has greatly improved digital communications security and expedited administrative procedures. Businesses and governmental organisations can stop unwanted access and changes by making sure that documents are authenticated and securely signed. This legal framework has promoted trust in digital interactions by making the online environment for communications and transactions more effective and safer. Moreover, the continuous progress in cryptographic methods reinforces the resilience of digital signatures, rendering them an essential instrument in the contemporary digital economy.

Digital signature certificates (DSCs) and privacy concerns associated with their use have been the focus of international legal and technological discourse. For example, in the case of United States v. Ivanov (2001), Russian hacker Alexei Ivanov was found guilty of breaking into financial institutions in the United States (Al Hait 2014). One important piece of evidence used by the prosecution was his use of digital signatures for transaction authentication. This case brought to light the significance of digital certificates in cybercrimes and the necessity of managing them securely. On the other hand, one of India's first successful cybercrime prosecutions, State of Tamil Nadu v. Suhas Katti, concerned the malicious use of a digital signature. The defendant was found guilty of pleading vulgar messages, purportedly written by a woman, on a Yahoo! message group (Sirohi 2015). The electronic messages' authenticity was confirmed by the digital signature. Similarly, in the 2003 California Supreme Court case Intel Corp. v. Hamidi, Hamidi, a former employee, attacked Intel in numerous emails addressed to Intel staff members. Intel filed a lawsuit, claiming trespass to chattels, but Hamidi won the case in court (Kam 2004). The case revealed the difficulties of digital interactions in legal contexts, even though it does not directly involve digital signatures. It does, however, touch on privacy and digital communications. Accordingly, Personal data, including names, email addresses, and occasionally even biometric information, is frequently needed in order to issue DSCs. Certificate Authorities (CAs), who store this data, have the potential to seriously violate privacy if they are compromised. One example of the potential risks associated with centralising sensitive information is the 2011 Sony PlayStation Network hack, which resulted in the exposure of millions of user records. A serious privacy risk is the possible misuse of this personal data, whether as a result of data breaches or unauthorised sharing (Bonner 2012). Whereas, Although DSCs' non-repudiation feature is a plus, if a digital signature is used without the user's knowledge or consent, privacy concerns may arise (Kaltakis 2021). The unauthorised use of a digital signature can lead to deceptive promises and serious financial and legal consequences for the victim (Schneier 2015). Digital signatures can be abused, so it's important to make sure they're applied with full consent and understanding. This was demonstrated in the State of Tamil Nadu v. Suhas Katti (2004) case, where the signatures were used to defame an individual. Furthermore, the reliability of CAs is essential for DSCs (Sirohi 2015). The confidentiality and security of users' digital signatures are at risk if a CA is hacked or behaves maliciously. The 2011

DigiNotar breach, in which the CA was compromised and bogus certificates were issued, serves as a sobering reminder of the weaknesses in this trust framework (Van der Muelen 2013). Consumers entrust these third-party companies heavily with protecting their private keys and personal data, and any betrayal of this confidence may have far-reaching privacy repercussions. Hence, it is safe to conclude that digital signature certificates bring substantial privacy risks that need to be carefully managed, even though they offer many advantages in terms of security, authenticity, and non-repudiation. Strong regulatory frameworks, technological safeguards, and user awareness are necessary to address the issues posed by the reliance on personal data, trust in third parties, and technological vulnerabilities (Bibri & Bibri 2015).

Digital Corporate Governance

Digital India brings a new dimension to the field of information technology. It focuses on remodeling the country by comprehensively implementing digitalization. The initiative's additional objectives are to create world-class virtual infrastructure and communication, digitalization of services (government and different services), and promote generic digital literacy (Afsharipour, A., 2009). As a result of the liberalization measures, the governance of private sector firms and the financial sector, notably the banking sector and the stock markets, had experienced a considerable alteration. From the year 2013 onwards, there is an evident paradigm shift within the corporate governance landscape after the introduction of the exhaustive Companies Act. In 2015, SEBI added a brand new complete list of regulations (Listing Obligations and Disclosure Requirements, 2015) for the compliance of disclosure responsibility and reporting mechanisms. Furthermore, in 2017, SEBI formed the Kotak Committee to overtake the system of company governance practice (Goswami, O., 2002).

In this context, digital company governance will be interpreted because of the technique of handling the affairs of an agency using present virtual technology: digital disclosure of information, digital record control, participation in conferences through a recorded videoconference, remote e-balloting on agenda items, and the usage of artificial intelligence (AI) technology in company governance (Belloc, F., 2012). Depending on how much human input influences managerial decisions and how much of the control process is automated. Remote, Smart, and Artificial Intelligence management are three unique types of virtual firm control that may be found (governance). Hereto, the use of technology to enable remote involvement of a person, such as a shareholder, appointed executive officers, or authorized members of the board of directors in making and carrying out management decisions is referred to as remote management. Among the method of remote control may be distinguished, for instance, video conferencing, and digital balloting (Aguilera &

Cuervo-Cazurra, 2009). The Companies Act of 2013 and the latest committee on Corporate Governance i.e., the Kotak Committee have forwarded their recommendation for incorporating the adjustments to align and stability or synchronize with Listing Regulations (Veeramani at el., 2009). India has a strong legal foundation for digital corporate governance, with the Information Technology Act of 2000 giving digital signatures legal protection. Digital signatures based on asymmetric cryptosystems are recognised by Section 18 of the Act, which compares them to conventional handwritten signatures. Digital files signed with a Digital Signature Certificate (DSC) are accorded the same legal standing as paper documents thanks to this legal recognition (Thangavel 2014).

The year 2000 saw the creation of the Controller of Certifying Authorities (CCA) office, which was a major step in encouraging the use of digital signatures for authentication in electronic commerce. To establish a trustworthy and safe foundation for digital transactions, the CCA certifies Certifying Authorities (CAs) and issues Public Key Certificates to them (Bergström & Berghäll 2021). The CAs then issue DSCs to end users. The adoption of digital corporate governance practices has produced a number of advantageous effects. Digital signature adoption has improved the security of digital communications, decreased paperwork, and expedited administrative procedures. By preventing unauthorised access and modifications to confidential documents, businesses and government organisations can now promote trust in digital interactions. The MCA 21 project has made it much easier to do business in India by offering safe and easy online access to a variety of corporate services. Among this initiative's noteworthy accomplishments are the shortened processing times for different applications and the enhanced transparency of corporate governance procedures.

The incorporation of digital technologies into corporate governance has revolutionised organisational operations by providing increased accountability, efficiency, and transparency. Yet, this shift to digital also raises serious privacy issues. Contract Intelligence (COiN), a blockchain-based platform, was deployed by JP Morgan Chase to automate the examination of legal documents and extract important information (Byun 2022). This invention greatly increased accuracy while cutting down on the amount of time needed to review documents. Wherein, Data security and privacy are concerns raised by the large volume of sensitive data that COiN processes. Sensitive corporate data and private client information might be made public if the platform were compromised. To reduce these risks, strong cybersecurity measures and stringent access controls are crucial. Likewise, a significant data breach at the large credit reporting company Equifax in 2017 resulted in the exposure of 147 million people's personal data (Gaglione Jr 2019). Thus, inadequate cybersecurity measures led to the breach, which exposed weaknesses in digital governance frameworks. While, Due to Facebook's role in the Cambridge Analytica incident, significant weaknesses in its data governance procedures were made public. Without getting explicit permission, the social media behemoth gave Cambridge Analytica access to millions of users' personal information, which it then exploited for political advertising (Hinds et al., 2020). Therefore, the organisations can manage the challenges of digital corporate governance while protecting user privacy by putting strong cybersecurity procedures in place, making sure data privacy laws are followed, and encouraging a culture of openness and responsibility.

Electronic Voting

Corporations being driven through the stakeholder's democracy are hailing upon the usage of technology in essential decisions making. Companies with a massive shareholders base require active involvement and participation by the investors in principal decisions impacting the shareholders, which may be done effortlessly with the assistance of e-balloting infrastructure. E-balloting is not only effective, but it also hastens the entire balloting process, however. it reduces the possibility of errors. Meetings are possibly the exceptional practical expression of a democratic form of company function. For the actual proprietors of the company i.e. the shareholders it's a possibility for optimistic dialogue between the management and shareholders (Kanojia 2023). By being shareholders they have a say withinside the choice-making of the corporation which they exercise by casting their valuable votes.

It is considered that the shareholders' conferences are planned and prepared well and are carried out in a constructive, meaningful, and obvious manner. Exercising the balloting rights by the shareholders is an invariable tool to maintain the board liable for its activities (Strine Jr, L. E., 2010). In instances of technological advancement, numerous novel modes of verbal exchange which include the internet, video-conferencing, webcast, etc, and digital balloting, the boundaries have become baseless, as a result, shareholders located in different geographical regions can talk and cast their vote in conferences using new strategies of communication. Voting by the digital method is a facility given to the members of a corporation to cast their votes at the resolutions via digital mode. It presents a possibility to shareholders living in far-flung vicinity to participate in the decision-making process of the corporation. They might or might not attend the meeting physically. Boards in their fiduciary ability are looked upon for more responsibility and transparency for the effectiveness of their overall governance method. E-Voting is an additional step to inspire company democracy and promote excellent company governance.

An e-balloting service is an expedient device for executing voting at assigned conferences or meetings by the members, permitting them to check in for meetings and cast their vote by filing a digital ballot form online, comply with live broadcasts of conferences, and acquaint themselves with scheduled and associated files by gaining access to their verified accounts at the common electronic service via remote authentication. The Companies Act, of 2013 ushered in the idea of e-balloting to make certain wider shareholder participation within the decision-making procedure in corporations. The e-balloting procedure has been brought to ensure wider participation of Shareholders in the essential decisions of the corporation since the postal technique of balloting has its very own limitation. The procedure of seeking shareholders' approval via postal ballot—isn't always the simplest. It is time-consuming. However, it additionally includes large costs along with management and posting costs, paperwork, etc. To keep away from most of these hassles, the brand new Companies Act 2013, has delivered the idea of balloting via digital way. Passing suitable resolutions by shareholders of corporations and increasing shareholder participation therein is a crucial aspect of appropriate Corporate Governance. In keeping with Explanation to rule 20 to the Companies (Management and Administration) Rules, 2014 the expression "balloting by digital way" or "digital balloting system" means a 'secured system' primarily based procedure of display of digital ballots, recording of votes of the individuals and the range of votes polled in favor or against, such that the complete balloting exercised by the manner of digital means receives registered and counted in a digital registry in a centralized server with adequate 'cyber security' (Taylor et al, 2020).

There are serious privacy and security concerns with the use of electronic voting systems. One of the biggest automakers, Volkswagen AG, put in place an electronic voting system to make shareholder voting easier. The system's objectives were to boost attendance and expedite the voting procedure at annual general meetings (AGMs). It was revealed in the year 2019 that there were security vulnerabilities in Volkswagen's online voting system that might be used to obtain shareholder information and tamper with the outcome of votes (Kabeyi 2020). The event made people wonder if the electronic voting system was secure and reliable. In response, Volkswagen strengthened the platform's cybersecurity safeguards and carried out a comprehensive security audit to preserve shareholder privacy and guarantee the accuracy of voting results. Similarly, for shareholder's meetings, Telstra, an Australian telecommunications company, used an electronic voting system. When a privacy vulnerability in Telstra's electronic voting platform was found in 2017, the company was criticised (Gray 2020). Unauthorised access to the names and voting preferences of shareholders was made possible by the system. Thereby, stricter access controls, encryption of sensitive data, and routine security assessments were Telstra's solutions to the problem of preventing future breaches and safeguarding shareholder privacy. On the similar lines, an electronic voting system was established by Swiss Post, the country's postal service, for use in corporate governance and local and national elections. Researchers found serious security holes in the system in 2019, such openings that might let attackers tamper with votes and jeopardise voter anonymity.

In order to fix the vulnerabilities, Swiss Post temporarily halted the electronic voting system and started a thorough investigation (Germann & Serdült 2017). The case demonstrated that in order to safeguard voter's privacy and maintaining the integrity of the voting process, stringent security measures are essential.

Nevertheless, the incorporation of cutting-edge technologies keeps improving the efficacy, efficiency, and transparency of electronic voting procedures in corporate settings. For instance, Biometric authentication uses distinct biological characteristics to identify voters, improving the security and dependability of electronic voting systems (Ahmad et al., 2020). In this reference, Smartmatic's e-voting solutions aims to offer secure and effective process to verify voter identities (Kintu & Mohamed 2018). Besides, the confidentiality and integrity of voter data are guaranteed during transmission and storage by sophisticated encryption methods and secure transmission protocols. Here, organisations such as Follow My Vote offers to protect voter data and preserve privacy. Hence, the most recent technological advancements are improving efficacy, security, and transparency in corporate e-voting (McCorry et al., 2017). Modernising voting procedures to make them more accessible and safe requires a strong focus on blockchain, biometric authentication, artificial intelligence, and encryption. Even though these developments have many advantages, it is still crucial to address issues with cybersecurity, privacy protection, and technology adoption.

Board Meetings

In light of the restrictions put in place to stop the radical coronavirus infection from spreading further, the difficulty of shifting general conferences of company participants to a remote layout with the assistance of videoconferencing has turned out to be especially urgent. Video conferencing was used effectively to host a variety of meetings, such as general shareholder meetings (corporation participants), board of director meetings, meetings of collegial executive bodies (management board, directorate), meetings of the heads of parent and subsidiary corporations, organizing general shareholder meetings, members of limited liability corporations, and non-profit organization members. Accordingly, the Companies Act, 2013 now permits the convening of Board meetings by audio-visual means, such as videoconferencing. For the first time, the act happens to facilitate the idea of e-voting (Kanojia 2023). This in itself makes no provision for facilitating shareholders' conferences via video conferencing and different audiovisual means. In a similar context, Section 173 of the Act, read with Rules 3 & four of the Companies (Meetings of Board and its Powers) Rules, 2014 lay down the legal provisions concerning conducting Board conferences via video conferencing. Section 173 sub-section (2) gives that participation of directors in a meeting of the Board can be both in person or via video conferencing or other audio-visual means, as prescribed, that are capable of recording and recognizing the participation of the directors and recording and storing the proceedings of such conferences along with the date and time (Chaudhary & Taneja, 2018). It, in addition, provides that the Central Government might also additionally, with the aid of notification, specify such topics which shall not be handled in a meeting via video conferencing or other audiovisual methods

Data Accessibility and Privacy

Due to the advancements in digital technologies, corporations are easily able to obtain and use data collected from various sources, including consumers, to learn more about their trends, interests, and behaviors. This information and data are often used to improve products and services, personalize advertising and marketing, and make informed business decisions. Corporations can access data through several techniques, depending on the classification of data, its location, and the objective for accessing it. Accordingly, surveys are a common technique for accessing data, particularly for collecting information from large groups of people (Bose, R., 2009). Online, over-the-phone, or in-person, surveys can be used to gather both quantitative and qualitative data. Whereas, employing software to harvest data from websites or other online sources is known as data scraping. This technique can be used to collect an enormous volume of data quickly and can be particularly useful for analyzing trends or monitoring online activity. Followed by, using software to analyze massive volumes of data to find patterns or links is known as data mining. It can be used to extract insights from complex data sets and can be particularly useful for businesses or organizations looking to improve their operations or marketing strategies (Rygielski et al., 2002). Lastly, the monitoring of social media involves identifying the appropriate social media platforms for mentions of particular topics, brands, or keywords. It can be used to collect data on public sentiment, customer feedback, or brand reputation, and can be particularly useful for corporations who are looking to engage with their customers or improve their social media strategies.

However, the gathering and use of personal information by corporations has raised concerns about privacy, security, and ethics. Various studies have claimed that many stakeholders are worried about how their data is being used and whether it is being shared or sold without their knowledge or consent (Caudill & Murphy, 2000). There have been several instances related to abundant data access by corporations, some of which have raised concerns about privacy and security. In the year 2018, it came to light that Facebook (now Meta) had unwittingly sent millions of Facebook users' data to the political consultancy firm Cambridge Analytica. While, In the year 2021, WhatsApp updated its privacy policy, which required users to agree to share their data with the parent company Facebook. The policy was met with backlash from users and privacy advocates, who raised concerns about the collection and use of personal data (Fiesler & Hallinan, 2018). Considering the advent & extent of the issue, India has enforced several regulations related to the protection of data, which aim to safeguard personal information and prevent its unauthorized use or disclosure. One such important set of rules is the Information Technology (Reasonable Security Practices and Procedures and Sensitive Personal Data or Information) Rules, 2011 which were introduced under the Information Technology Act, 2000 and lay down the requirements for protecting sensitive personal data or information (SPDI) of individuals by companies and organizations. These rules require companies to implement reasonable security practices to protect the confidentiality and integrity of the SPDI and obtain consent from the individuals before collecting, using, or disclosing their SPDI (Lloyd, I., 2020). Followed by the drafting of the Personal Data Protection Bill, 2019 This seeks to establish a complete framework for the protection of personal data in India. This bill proposes an establishment of an independent authority i.e., Data Protection Authority, which should be inherited with the authority to oversee the implementation of data protection regulations and ensure compliance by companies and organizations. The bill also introduces several new requirements for the processing of personal data, including consent, purpose limitation, and data minimization. There are a few other sets of law & regulations that indirectly aims to protect the privacy of data such as The Aadhaar (Targeted Delivery of Financial and Other Subsidies, Benefits and Services) Act, 2016 which lays down the requirements for collecting, storing, and using Aadhaar data, and imposes penalties for unauthorized disclosure or misuse of the data. Similarly, the Indian Contract Act of 1872 lays down the legal requirements for contracts in India, including contracts concerning the collection and processing of personal & confidential data. In addition, it also requires that contracts be entered into voluntarily and with free consent and that the terms of the contract be fair and reasonable finally, The Right to Information Act, of 2005 requires public authorities to maintain records and provide access to information to citizens upon request, subject to certain exemptions.

Best Practices for Ensuring Privacy

It is often stated that the rules can be legislated to benefit society. Yet, society is required to adopt the practices & procedures which prevent instances of violation of law & hampering the legal rights of the citizens. Hence, in the presence of few regulations related to the privacy of data, Corporations are expected to implement some best practices to ensure the stringent protection of data accessed by them. Subsequently, to prevent unwanted access to sensitive data, businesses should put in place robust security measures including firewalls, intrusion detection systems, and encryption. Additionally, they should develop a comprehensive privacy policy that outlines the types of data collected, how it is used, and how it is protected. The privacy policy should also include instructions for customers to exercise their privacy rights and provide regular training to employees on the importance of data privacy and security. This training may include best practices for handling sensitive data, how to recognize and avoid phishing scams, and the appropriate use of passwords and other security measures. Furthermore, corporations should conduct regular security audits to identify potential vulnerabilities and threats. These audits should include penetration testing, vulnerability assessments, and security risk assessments and eventually, they should implement the protocol of data minimization, which means collecting and retaining only the minimum amount of data necessary to accomplish business objectives. This can help reduce the risk of data breaches and leaks (Patel et al., 2010). Implementing transparent data handling procedures is an additional paramount factor. Businesses have to make sure that all of their data processing operations are open and compliant with applicable data protection laws and regulations. This entails informing people in an understandable and accessible manner about the processing of their data, its goals, and its legal justification. Customers and other stakeholders are more likely to interact with businesses that are transparent about their data handling procedures, which contributes to the development of trust between these parties. Moreover, companies need to set up governance and accountability systems for data security. Organisations can maintain compliance with data protection regulations and best practices by designating a Data Protection Officer (DPO) or a dedicated team to oversee data privacy and security measures (Borisova 2019). Additionally, this group should be in charge of handling data breach incidents, including quickly and effectively notifying the appropriate authorities and impacted parties of breaches. A risk-based strategy for data protection is an additional best practice that businesses ought to think about implementing. Businesses can more efficiently allocate resources and put in place the necessary safeguards to reduce risks to individuals' privacy by identifying and prioritising data processing activities that pose the greatest risks to that privacy. This strategy not only strengthens data security but also shows how dedicated the company is to upholding people's right to privacy. Subsequently, it is critical to cultivate a privacy-conscious culture within the company. This entails making sure that each employee is aware of their responsibility for maintaining data privacy and incorporating data protection principles into the organization's basic values. Maintaining a high degree of data protection and lowering the possibility of data breaches and other privacy incidents can be accomplished by routinely updating training programmes and encouraging a privacy-conscious mindset. Organisations can protect their brand, keep customers' trust, and adhere to legal requirements by putting data privacy and security first.

CONCLUSION

In this era of digital transition, Corporations have taken up the baton to lead the optimum utilization of technology for augmenting their operation capacities and amplifying their reach to the end consumers & other key stakeholders and remaining competitive with rapidly evolving business ecosystems. In doing so, there is a possibility of unintentional or accidental disclosure of sensitive or confidential information to unauthorized parties which concurs with the privacy rights of the stakeholders. Accordingly, India has made efforts to strengthen its data protection laws, there are still several challenges that need to be addressed. The foremost challenge with data protection laws in India is the lack of effective enforcement mechanisms. Despite the existence of several regulations, there are plenty of instances related to data leaks, breaches & misuse of confidential information, indicating that the regulations are not being effectively enforced. Furthermore, the regulations and laws primarily focus on the protection of sensitive personal data or information (SPDI) and do not cover all types of personal data (Preethi & Satheesh, 2021). This limited scope leaves other types of personal data vulnerable to misuse or unauthorized access. Presently, India does not have a standardized legal or regulatory framework for data protection, which can make it difficult for companies and organizations to ensure compliance with different regulations and standards. Hence, indigenous corporations and sovereign governments in developing nations like India are required to collectively deliberate on the implementation of the measures to address these evident challenges to balance economic growth with the fundamental rights of the citizens.

REFERENCES

Adizes, I. (2004). *Managing corporate lifecycles*. The Adizes Institute Publishing.

Afsharipour, A. (2009). Corporate governance convergence: Lessons from the Indian experience. *Nw. J. Int'l L. & Bus.*, 29, 335.

Agrawal, R. K. (2016). A Comparative Study of Indian Companies Act, 2013 and Companies Act of Republic of Maldives, 1996. *Social Sciences*, 4(01), 2016.

Aguilera, R. V., & Cuervo-Cazurra, A. (2009). Codes of good governance. *Corporate Governance*, 17(3), 376–387. 10.1111/j.1467-8683.2009.00737.x

Ahmad, M., Rehman, A. U., Ayub, N., Alshehri, M. D., Khan, M. A., Hameed, A., & Yetgin, H. (2020). Security, usability, and biometric authentication scheme for electronic voting using multiple keys. *International Journal of Distributed Sensor Networks*, 16(7), 1550147720944025. 10.1177/1550147720944025

Al Hait, A. A. S. (2014). Jurisdiction in Cybercrimes: A comparative study. *JL Pol'y & Globalization*, 22, 75.

Anand, D., & Khemchandani, V. (2019). Study of e-governance in India: A survey. *International Journal of Electronic Security and Digital Forensics*, 11(2), 119–144. 10.1504/IJESDF.2019.098729

Bamberger, K. A., & Mulligan, D. K. (2015). *Privacy on the ground: driving corporate behavior in the United States and Europe*. MIT Press. 10.7551/mitpress/9905.001.0001

Banday, M. T. (2011). Easing PAIN with digital signatures. *International Journal of Computer Applications*, 975, 8887.

Belloc, F. (2012). Corporate governance and innovation: A survey. *Journal of Economic Surveys*, 26(5), 835–864. 10.1111/j.1467-6419.2011.00681.x

Bergström, A., & Berghäll, E. (2021). Public certificate management: An analysis of policies and practices used by CAs.

Bhatnagar, S. (2009). *Unlocking e-government potential: Concepts, cases and practical insights*. SAGE Publications India. 10.4135/9781446270202

Bibri, S. E., & Bibri, S. E. (2015). Ethical implications of AmI and the IoT: risks to privacy, security, and trust, and prospective technological safeguards. *The shaping of ambient intelligence and the Internet of Things: historico-epistemic, socio-cultural, politico-institutional and eco-environmental dimensions*, 217-238.

Bonner, L. (2012). Cyber risk: How the 2011 Sony data breach and the need for cyber risk insurance policies should direct the federal response to rising data breaches. *Wash. UJL & Pol'y*, 40, 257.

Borisova, A. (2019). The Data Protection Officer as an Instrument for Compliance with the Accountability Principle Under the GDPR. *Sustainable Development GOALS 2030: Challenges for South and Eastern European Countries and the Black Sea Region*, 346.

Bose, R. (2009). Advanced analytics: Opportunities and challenges. *Industrial Management & Data Systems*, 109(2), 155–172. 10.1108/02635570910930073

Byun, Y. C. (2022). *Knowledge Discovery and Cryptocurrency Price Prediction Based on Blockchain Framework* (Doctoral dissertation,   ).

Caudill, E. M., & Murphy, P. E. (2000). Consumer online privacy: Legal and ethical issues. *Journal of Public Policy & Marketing*, 19(1), 7–19. 10.1509/jppm.19.1.7.16951

Chaudhary, I., & Taneja, V. (2018). Analysis of Company Law Amendments: As a Progressive Step towards Better Corporate Governance. *RGNUL Fin. & Mercantile L. Rev.*, 5, 241.

Chen, Z., & Zhu, Y. (2017, June). Personal archive service system using blockchain technology: Case study, promising and challenging. In *2017 IEEE International Conference on AI & Mobile Services (AIMS)* (pp. 93-99). IEEE. 10.1109/AIMS.2017.31

Fiesler, C., & Hallinan, B. (2018, April). "We Are the Product" Public Reactions to Online Data Sharing and Privacy Controversies in the Media. In *Proceedings of the 2018 CHI conference on human factors in computing systems* (pp. 1-13).

Gaglione, G. S.Jr. (2019). The equifax data breach: An opportunity to improve consumer protection and cybersecurity efforts in America. *Buff. L. Rev.*, 67, 1133.

Germann, M., & Serdült, U. (2017). Internet voting and turnout: Evidence from Switzerland. *Electoral Studies*, 47, 1–12. 10.1016/j.electstud.2017.03.001

Goswami, O. (2002). Corporate governance in India. *Taking action against corruption in Asia and the Pacific*, 85-106.

Gray, H. (2020). *Congruent Regulation: Designing the Optimal Australian Utility Regime* (Doctoral dissertation, The Australian National University (Australia)).

Habib, G., Sharma, S., Ibrahim, S., Ahmad, I., Qureshi, S., & Ishfaq, M. (2022). Blockchain technology: Benefits, challenges, applications, and integration of blockchain technology with cloud computing. *Future Internet*, 14(11), 341. 10.3390/fi14110341

Hinds, J., Williams, E. J., & Joinson, A. N. (2020). "It wouldn't happen to me": Privacy concerns and perspectives following the Cambridge Analytica scandal. *International Journal of Human-Computer Studies*, 143, 102498. 10.1016/j.ijhcs.2020.102498

Hoofnagle, C. J., Van Der Sloot, B., & Borgesius, F. Z. (2019). The European Union general data protection regulation: What it is and what it means. *Information & Communications Technology Law*, 28(1), 65–98. 10.1080/13600834.2019.1573501

Kabeyi, M. J. B. (2020). Corporate governance in manufacturing and management with analysis of governance failures at Enron and Volkswagen Corporations. *Am J Oper Manage Inform Syst*, 4(4), 109–123.

Kaltakis, K., Polyzi, P., Drosatos, G., & Rantos, K. (2021). Privacy-preserving solutions in blockchain-enabled internet of vehicles. *Applied Sciences (Basel, Switzerland)*, 11(21), 9792. 10.3390/app11219792

Kam, S. (2004). Intel Corp. v. Hamidi: Trespass to chattels and a doctrine of cyber-nuisance. *Berkeley Tech. LJ, 19*, 427.

Kanojia, S. (2023). Application of blockchain in corporate governance: Adaptability, challenges and regulation in BRICS. *BRICS LJ*, 10(4), 53–67. 10.21684/2412-2343-2023-10-4-53-67

Kintu, N. B., & Mohamed, I. Z. (2018). A secure e-voting system using biometric fingerprint and crypt-watermark methodology. In *ASCENT International Conference Proceedings–Information Systems and Engineering* (pp. 1-18).

Laptev, V. A., & Feyzrakhmanova, D. R. (2021). Digitalization of institutions of corporate law: Current trends and future prospects. *Laws*, 10(4), 93. 10.3390/laws10040093

Lloyd, I. (2020). *Information technology law*. Oxford University Press. 10.1093/he/9780198830559.001.0001

McCorry, P., Shahandashti, S. F., & Hao, F. (2017). A smart contract for boardroom voting with maximum voter privacy. In *Financial Cryptography and Data Security:21st International Conference, FC 2017,Sliema, Malta,April 3-7, 2017, Revised Selected Papers 21* (pp. 357-375). Springer International Publishing. 10.1007/978-3-319-70972-7_20

Patel, A., Qassim, Q., & Wills, C. (2010). A survey of intrusion detection and prevention systems. *Information Management & Computer Security*, 18(4), 277–290. 10.1108/09685221011079199

Preethi, A., & Satheesh, N. (2021). Need for a Comprehensive Legislation on Employee's Privacy in India: Comparison with US and EU Models. *Issue 2 Int'l JL Mgmt. &. Human.*, 4, 1449.

Rygielski, C., Wang, J. C., & Yen, D. C. (2002). Data mining techniques for customer relationship management. *Technology in Society*, 24(4), 483–502. 10.1016/S0160-791X(02)00038-6

Schneier, B. (2015). *Secrets and lies: digital security in a networked world*. John Wiley & Sons. 10.1002/9781119183631

Sharma, V. (2011). *Information technology law and practice*. Universal Law Publishing.

Sirohi, M. N. (2015). *Transformational Dimensions of Cyber Crime*. Vij Books India Pvt Ltd.

Strine, L. E.Jr. (2010). One fundamental corporate governance question we face: Can corporations be managed for the long term unless their powerful electorates also act and think long term? *Business Lawyer*, 1–26.

Tammpuu, P., & Masso, A. (2019). Transnational digital identity as an instrument for global digital citizenship: The case of Estonia's E-residency. *Information Systems Frontiers*, 21(3), 621–634. 10.1007/s10796-019-09908-y

Taylor, P. J., Dargahi, T., Dehghantanha, A., Parizi, R. M., & Choo, K. K. R. (2020). A systematic literature review of blockchain cyber security. *Digital Communications and Networks*, 6(2), 147–156. 10.1016/j.dcan.2019.01.005

Thangavel, J. (2014). Digital Signature: Comparative study of its usage in developed and developing countries.

Van der Meulen, N. (2013). Diginotar: Dissecting the first dutch digital disaster. *Journal of Strategic Security*, 6(2), 46–58. 10.5038/1944-0472.6.2.4

Veeramani, S., Rong, R. P. N., & Singh, S. (2019). Digital Transformation and Corporate Governance in India: A Conceptual Analysis.

William, W., & William, L. (2019, November). Improving corporate secretary productivity using robotic process automation. In *2019 International Conference on Technologies and Applications of Artificial Intelligence (TAAI)* (pp. 1-5). IEEE. 10.1109/TAAI48200.2019.8959872

Chapter 9
Corporate Data Responsibility in East Africa:
Bridging the Gap Between Theory and Practice

Ripon Bhattacharjee
National Law University, Tripura, India

Aditya Agrawal
O.P. Jindal Global University, India

Madhulika Mishra
GLA University, Mathura, India

Akash Bag
https://orcid.org/0000-0001-8820-171X
Adamas University, India

ABSTRACT

East Africa's rapid digital connectivity growth without proper regulatory frameworks, notably in data protection, places regulatory obligations on technology actors. The chapter uses corporate social responsibility (CSR) and corporate digital responsibility (CDR) to examine how firms handle data privacy and ethical issues in the digital world. The research uses various case studies to illuminate the practical obstacles of data responsibility solutions. The lack of specific CSR and CDR techniques remains a major implementation gap despite actors' awareness and efforts to create complete policies. This emphasizes the need for context-specific guidelines to connect policy

DOI: 10.4018/979-8-3693-3334-1.ch009

Copyright © 2024, IGI Global. Copying or distributing in print or electronic forms without written permission of IGI Global is prohibited.

development and real-world application to ensure that the growing digital economy aligns with human rights and state goals.

INTRODUCTION

The chapter elucidates the profound implications of rapid digitalization and the exponential growth in internet usage on global socio-economic landscapes and human rights frameworks. In the face of such unprecedented digital expansion, highlighted by the increase from 1 billion to 4.9 billion internet users within 15 years, the necessity for a concerted, multi-stakeholder approach to managing the emergent risks—ranging from privacy breaches to cyberattacks—becomes paramount (Casella & Formenti, 2018). This necessity is underscored by António Guterres, the UN Secretary-General, in the United Nations Conference on Trade and Development's Digital Economy Report 2021, where he calls for collaborative efforts to address the potential for abuse and misuse of data by various actors, underlining the consequential nature of data usage not only for economic and trade development but also for human rights, peace, and security (Casella & Formenti, 2018). Simultaneously, the chapter reveals a striking discrepancy in global internet access, with a significant digital divide between developed and developing nations, where approximately 2.9 billion people, predominantly in the latter, remain offline. This gap not only underscores inequalities but also signifies a vast, untapped potential for economic growth and digital value creation in the global South, necessitating substantial investments to achieve Sustainable Development Goal number 9: affordable, reliable, and sustainable internet for all by 2030 (Hasan & Tucci, 2010).

Investments in the African ICT industry, particularly highlighted by the involvement of the private sector, African governments, international consortia, and notably, China, under its Belt and Road Initiative, underscore the global interest in developing Africa's digital infrastructure (Corrigan, 2020). Yet, this interest and the accompanying investments are not without controversy. Skepticism regarding the motivations behind these investments and their potential impacts—encompassing environmental, social, and governance considerations—points to a complex interplay of opportunities and risks, including the potential for telecommunication monopolies, mass surveillance, and data misuse. In light of these dynamics, the imperative for Corporate Social Responsibility (CSR) and Corporate Digital Responsibility (CDR) emerges as critical frameworks for managing the societal impacts of digitalization (Peltola et al., 2021). These frameworks advocate for a holistic approach to corporate accountability, extending beyond economic impacts to encompass environmental and societal considerations, with CDR specifically addressing the ethical challenges posed by the digital age (Herden et al., 2021). The evolving chapter on CDR, distinct yet related to traditional CSR, reflects the need for a nuanced understanding of the responsibilities of corporations in an increasingly digital world, emphasizing the importance of ethical considerations and corporate responsibility in the face of digital transformation (Lobschat et al., 2021). Thus, the material not only underscores the challenges and opportunities presented by the digital age but also highlights the critical role of multi-stakeholder engagement and responsible corporate practices in navigating the complex landscape of digitalization, emphasizing the need for ethical stewardship in harnessing digital technologies for economic growth, societal well-being, and the protection of human rights.

PROBLEMATISATION

The swift transformation of the digital environment underscores an imperative for methodologies of data stewardship that adapt with comparable agility. This evolving digital realm mandates a paradigm of transparency and accountability, compelling corporations to integrate strategies and frameworks to navigate digital proliferation effectively. The chapter surrounding Corporate Digital Responsibility (CDR) has predominantly unfolded within the theoretical and conceptual spheres, highlighting a notable dearth in empirical investigations, despite a burgeoning corpus of theoretical work suggesting industrial applications (Herden et al., 2021; Lobschat et al., 2021; Mueller, 2022). The scarcity of empirical research accentuates significant lacunae, particularly in the practical dimensions of CDR, despite recommendations proffered for industry praxis (Mueller, 2022).

Figure 1. Visualization of problematization and research gap for this research

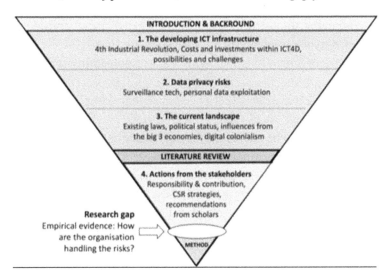

The extant literature reveals an exigency for a broader examination of CDR, advocating for a comprehensive chapter that transcends narrow disciplinary confines and embraces a societal perspective (Merwe & Achkar, 2022). Notably, the applicability of CDR within the context of the Global South's ICT sector remains uncharted territory, thereby furnishing a compelling rationale for this chapter's investigative focus. The employment of the funnel approach facilitates a methodical delineation of the research gap (Barker, 2014). This approach commences with a broad overview of the foundational issues, progressively honing in on specific challenges. Within this framework, the chapter initially addresses the universal challenge of facilitating equitable internet access in a manner that ensures safety and security. Subsequently, the discussion narrows to scrutinize the challenges and prospects within nascent ICT infrastructures, with a particular emphasis on data privacy and digital rights, before focusing more acutely on the East African digital landscape.

RESEARCH QUESTIONS AND SCOPE

The essence of this chapter is to delve empirically into the mechanisms through which various organizations navigate the concept of data responsibility. Leveraging the foundation laid by existing literature that situates data privacy within the ambit of Corporate Social Responsibility (CSR) and integrating the nascent notion of Corporate Digital Responsibility (CDR), this chapter embarks on an empirical exploration within the ICT sector of East Africa. This region, as identified by Kinuthia

(2020), is currently grappling with significant challenges related to data privacy and security, thereby necessitating a chapter on the obstacles hindering organizations from fully embracing digital responsibility. In light of these considerations, the chapter is guided by two pivotal research questions.

RQ1: How are organizations managing data responsibility when operating in East Africa?

RQ2: What do organizations experience as barriers to implementing higher data responsibility ambitions?

These questions are to be addressed through a qualitative multiple-case chapter approach, with data being garnered from semi-structured interviews conducted with sustainability professionals and counterparts within the ICT sector of East Africa. These individuals are deemed to possess sufficient knowledge and insight to contribute meaningful data to the chapter. The analysis of the collected data will be executed through thematic analysis, a method known for its efficacy in distilling and interpreting patterns within qualitative data. The research's scope is delineated by two principal parameters: its geographical focus on East Africa and its concentration on entities within the ICT sector. The constraints of the chapter, including its temporal limitations and the availability of data, as well as the researcher's interests informed by the extant literature, define these parameters. Additionally, the chapter will articulate a concise definition of data responsibility, thereby setting the stage for a comprehensive exploration of how this imperative is managed and the challenges that stymie its full implementation within the targeted region.

The geographical lens of this investigation is firmly set on East Africa, a region selected for its empirical richness and analytical potency. East Africa emerges as a distinctive locus of digital evolution within the African continent, characterized by its digital maturity relative to its continental counterparts. This maturity is exemplified by initiatives such as the One Network Area (ONA), a regional pact that streamlines cross-border telecommunications, reducing the financial burden of roaming charges, and signifying a collective stride towards technological integration (Arnold, 2022). Despite the progressive connectivity enhancements within the region, disparities in digital access persist, both across and within the East African countries. A gender gap is evident, with men enjoying significantly higher internet usage rates compared to women. Additionally, the almost universal adoption of national broadband policies across East African governments, excluding South Sudan, underscores the region's commitment to digital expansion. The region's varied governmental regimes further introduce a layer of complexity in how organizations navigate data responsibility and collaborate with state apparatuses, thus framing East Africa as an intriguing subject for scrutinizing the challenges organizations face in policy implementation and the broader ambition of democratizing internet access, aligning with Sustainable Development Goal 9 (SDG 9).

The analytical focus of this chapter narrows down to three distinct organizational entities within the ICT sector, chosen for their pivotal roles in data generation and ownership. These entities encompass:

1. **An International Telecommunication Company:** This entity boasts a global footprint, with operations in over 180 countries and a workforce exceeding 100,000. Its services span network and cloud software, positioning it as a significant player in the global telecommunications landscape.
2. **An East African Network Operator:** Operating across 14 African nations, including pivotal East African states such as Kenya, Uganda, Rwanda, and Tanzania, this operator delivers comprehensive telecommunications and mobile money services, directly influencing the region's digital ecosystem.
3. **A Consortium of European Donors:** This collective, with roots in Europe, is engaged in ICT and digitalization as focal areas of their developmental aid, reflecting a transcontinental commitment to fostering digital advancement.

The selection of these chapter objects is premised on their substantial influence on the ICT sector's dynamics in East Africa, offering a varied yet interconnected palette for examining data responsibility practices and the impediments to implementing efficacious digital policies and guidelines within this vibrant and complex region.

BACKGROUND AND LITERATURE REVIEW

Social and Digital Responsibility and CSR

Understanding the dynamics of social and digital responsibility in the context of rapid technological advancement is pivotal (Malan, 2018). The gap between the swift evolution of digital technologies and the slower pace of legislative development poses significant challenges. The crux of these challenges lies in crafting regulatory frameworks that are sufficiently adaptable to the fluid boundaries of markets and sectors reshaped by digital technologies (Erbacci et al., 2016). The absence of such regulations exerts increased pressure on tech companies to self-regulate, particularly concerning human rights, echoing the sentiments expressed by the World Economic Forum regarding the indispensability of foundational values in ensuring positive outcomes amidst the Fourth Industrial Revolution (M. Epstein et al., 2017). The chapter on Corporate Social Responsibility (CSR) presents a nuanced view of how organizations can navigate their societal, environmental, and economic impacts. The evolution of CSR as a concept, despite its widespread acknowledgment, remains mired in definitional ambiguities, complicating its operationalization within corporate strategies (M. Epstein et al., 2017).

Soumodip and Cory's analysis delineates six dimensions foundational to CSR, advocating for a holistic approach that encompasses ethical considerations, stakeholder impact, and a commitment to sustainability alongside core economic responsibilities (Sarkar & Searcy, 2016). This chapter is further enriched by contrasting perspectives on the role of CSR in business. Friedman's assertion that the sole social responsibility of business is to augment profits underscores a fundamental critique of CSR, suggesting that market mechanisms are better suited to address societal challenges (Fifka, 2009). Conversely, the contingent relationship between CSR performance and profit posits that organizational engagement with CSR is influenced by its perceived impact on financial outcomes. Within this framework, the Epstein model offers a comprehensive methodology for integrating CSR into corporate strategy, emphasizing the significance of understanding and managing the drivers of CSR performance. This model encapsulates a broad spectrum of influences ranging from external regulatory contexts to internal organizational dynamics and strategic considerations, underlining the complexity of implementing CSR in a manner that aligns with broader societal and environmental objectives.

Data Responsibility, CSR, CDR, and Emerging Markets

In emerging markets, Corporate Social Responsibility (CSR) emerges as a strategic response to governance deficiencies. Dartey-Baah and Amponsah-Tawiah articulate CSR as a mechanism to address the voids left by inadequate governance, a sentiment underscored by the public expectation for corporations to assume a more active role in societal welfare (Dartey-Baah & Amponsah-Tawiah, 2011). This perspective gains further validation through empirical studies, which indicate that governance quality and economic development are positively correlated with CSR performance in African contexts (Inekwe et al., 2020). Such findings challenge the prevailing practice of comparing African CSR outcomes directly with those of Western nations without contextual nuance, advocating instead for a region-specific approach to CSR (Idemudia, 2014). The narrative surrounding CSR in Africa points to a significant underexploitation of its potential, with calls for a more localized, "*South-centered*" CSR agenda. This chapter transitions into the realm of data privacy within CSR, as highlighted by Pollach's pioneering work, which revealed a conspicuous underrepresentation of data privacy concerns within corporate CSR initiatives (Pollach, 2011). Subsequent analyses by Allen and Peloza advocate for the integration of data privacy into CSR strategies as a means of bolstering corporate reputation and stakeholder relations (Allen & Peloza, 2015). The legal mandate for data privacy, exemplified by the European Union's General Data Protection Regulation (GDPR), underscores a shift towards obligatory corporate ethics in data handling.

The chapter extends into the domain of Corporate Digital Responsibility (CDR), introduced by Lobshart et al. to address ethical concerns specific to digital technology (Lobschat et al., 2021). CDR emphasizes the ethical imperatives in the development and deployment of digital technologies, advocating for a comprehensive, culturally embedded framework of values and norms. The construct of the CDR culture proposed encompasses shared values, specific norms, and tangible behaviors, illustrating the multi-layered nature of ethical digital engagement (Corrigan, 2020). Critical reflection on CDR suggests its differentiation from traditional CSR, due to the unique ethical considerations posed by digital innovation. However, there is an emerging viewpoint, as presented by Herden and colleagues, that considers CDR as an extension rather than a departure from CSR, incorporating it within the broader Environmental, Social, and Governance (ESG) framework (Herden et al., 2021). Despite the nascent stage of CDR in academic and practical arenas, its significance is increasingly acknowledged, as evidenced by initiatives like Germany's CDR Awards. The burgeoning interest in CDR, particularly within the German context, signals a broader, albeit initial, engagement with the concept. Future research directions are called to delve into the complexities introduced by globalization, emphasizing the importance of contextual sensitivity in the application of CDR principles across varying cultural landscapes (Merwe & Achkar, 2022). This evolving discussion highlights the necessity for an integrative approach to CSR and CDR, reflecting a comprehensive understanding of corporate ethics in the digital age against the backdrop of globalization and diverse governance environments (Mueller, 2022).

ICT's Double-Edged Sword in Africa

In the realm of global development, innovation, and technological advancements stand as cornerstones for economic growth and societal progress (Hasan & Tucci, 2010). Notably, the digital revolution has significantly enhanced the ease of accessing education, employment, and social connectivity, especially highlighted during the COVID-19 pandemic (Nchake & Shuaibu, 2022). Yet, this digital wave has not uniformly swept across the globe, with African nations markedly trailing behind their Western counterparts in both the breadth and quality of Information and Communication Technology (ICT) penetration (Awad & Albaity, 2022). The slow adaptation of technology in Africa, coupled with substantial variance across the continent, underscores a pivotal challenge. Furthermore, a distinct gap in empirical research exists, particularly concerning the correlation between economic growth and ICT infrastructure within African contexts. This absence of solid, data-driven insights into the economic implications of ICT advancements in developing regions leaves a critical area of understanding underexplored (Nchake & Shuaibu, 2022). Despite these hurdles, there's a burgeoning optimism for East Africa, poised on the cusp of an unprecedented economic surge, largely fueled by ICT innovations. This optimism is anchored in the African Union's ambitious Digital Transformation Strategy, aiming for universal internet access by 2030. Such initiatives reflect a significant commitment to harnessing digital technology for continental growth (Awad & Albaity, 2022). However, the realization of this potential is hindered by insufficient investments in science and technology, crucial for Africa's economic metamorphosis.

Private sector investments currently dominate financing in East Africa's ICT sector, overshadowing contributions from African governments and international stakeholders (Corrigan, 2020). This trend raises questions about the sustainability and equity of such funding sources, especially considering the looming specter of digital colonialism. The concern here is not just about the ownership and profit motives of large tech conglomerates and foreign governments but also about the implications for local ownership and the broader socio-economic impacts on African societies (Deshpande & Webster, 1989). Moreover, the European Union's Global Gateway strategy, while ambitious in its financial commitments to bridge global investment gaps, has faced criticism for potentially prioritizing European economic interests over sustainable development in non-European regions (Herden et al., 2021). This critique aligns with broader concerns regarding the neocolonial dynamics of foreign investments in Africa's ICT sector, which may exacerbate dependencies rather than foster equitable growth and innovation. The regulatory landscape in Africa, particularly concerning data privacy and digital rights, remains fragmented and underdeveloped. Despite the presence of privacy rights in national constitutions, comprehensive data privacy legislation is lacking, leaving a significant gap in protecting individuals' digital rights (Nchake & Shuaibu, 2022). This situation is further complicated by the diverse regulatory approaches to data governance seen across global powers, highlighting the need for coherent and effective data protection frameworks within the African context.

Assessing Democracy and Privacy in Africa

In the contemporary landscape, the Economist Intelligence Unit's annual assessment distinguishes "full democracies" based on a comprehensive scale, requiring a minimum score of eight out of ten. This evaluation hinges on five critical dimensions: electoral integrity, civil liberties, governance efficacy, civic engagement, and the maturity of political culture (Wike et al., 2021). Mauritius stands as a singular example of a "full democracy" within sub-Saharan Africa, as highlighted in the EIU's 2021 report, contrasting sharply with the rest of the region's classifications ranging from "flawed democracies" to "authoritarian regimes." The chapter extends to the surveillance practices prevalent across African nations, significantly impacting their democratic status. Despite international frameworks advocating for digital privacy rights, such as the UN's Resolution 42/15 on privacy in the digital age, African countries frequently curtail these rights under national security pretenses (Makulilo, 2016). This encroachment often manifests through invasive surveillance and data collection methods, lacking sufficient safeguards for individuals' privacy.

A noteworthy exploration in this context is the correlation between data protection laws and democratic indices, as depicted in a matrix by DLA Piper. This visualization aids in understanding the varied risks associated with digital privacy across different legal and democratic environments, emphasizing the need for international entities to adapt to diverse legal landscapes (Basimanyane, 2022). Further examination reveals the intrinsic risks associated with burgeoning ICT infrastructures in East Africa, necessitating a dialogue on digital responsibility alongside digital rights. The latter encompasses a spectrum of internet-related rights, pivotal among these being data privacy. Threats to digital rights are multifaceted, encompassing surveillance, censorship, and misuse of personal data, among others. Shoshana Zuboff's seminal work on "surveillance capitalism" introduces a critical perspective on the commodification of personal data within the digital economy, echoing neo-Marxist theories (Lobschat et al., 2021). This paradigm shift towards data as a capitalistic commodity underscores the burgeoning challenges of digital privacy and the manipulation of personal information for profit. Despite some African nations adopting privacy laws, the region remains significantly exposed to privacy infringements, underlining a dire need for enhanced privacy protections and the utilization of anonymization tools (Hutt, 2015). The prevalent issues of online impersonation, social engineering, and unregulated surveillance technologies further exacerbate privacy risks, especially in countries with inadequate data protection legislation (Makulilo, 2016). This scenario presents a complex intersection of human rights concerns, particularly in regions marked by disparities in governance, social equity, and legal frameworks, underscoring the imperative for a holistic approach to safeguarding digital rights in the face of evolving technological landscapes.

METHODOLOGY

Exploring the intricacies of digital responsibility management within East African organizations forms the crux of this examination (Yin, 2009). Employing a multiple-case chapter approach, augmented by an abductive research methodology, this investigation delves into the practicalities and obstacles encountered in the stewardship of data responsibility. Through the comparative lens of multiple case studies, this research not only seeks to comprehend the dynamics at play within organizations but also aspires to forge theoretical generalizations from the empirical evidence gathered (Bell et al., 2018). The abductive reasoning employed herein marries inductive and deductive approaches, drawing on existing corporate social responsibility (CSR) and corporate digital responsibility (CDR) research to enrich the analysis (Yin, 2009). Data gathering leveraged semi-structured interviews, distinguishing between respondent interviews aimed at answering the core research questions and informant interviews designed to furnish the contextual background (Lobschat et al., 2021). This bifurcation in data collection methods underscores the chapter's commitment to capturing a holistic view of digital responsibility practices (M. Epstein et al., 2017).

The selection of interview subjects was meticulously executed, incorporating purposive and snowball sampling techniques to ensure the relevance and expertise of participants, particularly from the ICT sector in East Africa (Yin, 2009). The thematic analysis of interview data, guided by the principles laid out by Nowell et al., facilitated a structured examination of the thematic undercurrents emerging from the chapter. This methodological choice underpins the chapter's ambition to systematically distill insights while acknowledging the inherent flexibility of thematic analysis as both a strength and a potential source of analytical divergence (Bell et al., 2018). This investigation acknowledges its methodological constraints, particularly the inherent challenges in case chapter comparisons and the potential for interviewer bias (Bryman, 2012). Despite these limitations, the research endeavors to contribute a nuanced understanding of how organizations navigate the complex terrain of digital responsibility, underpinned by a rigorous methodological framework that balances empirical chapter with theoretical reflection (Bryman, 2012).

This chapter centers on three primary actors within the information and communication technology (ICT) sector, identified due to their substantial influence on data generation and ownership. These actors, referred to as chapter objects 1, 2, and 3, offer a diverse yet interconnected glimpse into the complexities of the digital data ecosystem.

Study Object 1: represents a global telecommunication entity with operations spanning across 180 countries and a workforce exceeding 100,000 employees. This organization's vast global presence and its provision of network and cloud software services position it as a crucial node in the international telecommunication infrastructure, with significant implications for data generation and dissemination practices worldwide.

Study Object 2: is characterized by its regional focus, operating as a network provider across 14 African countries, including pivotal East African nations such as Kenya, Uganda, Rwanda, and Tanzania. Its services, encompassing telecommunications and mobile money, not only play a vital role in enhancing regional connectivity

Corporate Data Responsibility in East Africa

but also in advancing financial inclusion, thereby impacting data ownership and utilization patterns within the East African context.

Study Object 3: encompasses a consortium of European donors, unified by their commitment to supporting ICT and digitalization initiatives within their development assistance programs. This group's focus on digital advancement and its implications for data ownership and management highlights the role of international development efforts in shaping the global digital landscape, particularly in terms of fostering technological capacity and infrastructure in the developing world.

RESULTS

Organizations operating in East Africa are navigating the complexities of data responsibility, a challenge highlighted by the synchapter of interview findings detailed within the chapter. The research, structured into three distinct sections, delves into perspectives from both respondents and informants to shed light on organizational approaches to data management and the hurdles encountered in striving for elevated data responsibility standards. The analysis draws upon the Epstein model (2007) and the CDR-culture model by Lobshart et al. (2021) as theoretical frameworks to categorize the collected data into themes and sub-themes, enriching the narrative with a systematic examination of the empirical evidence. The methodology adopted for this exploration involved a meticulous coding process of the interview data, the specifics of which are outlined and further elaborated in the chapter. By employing a narrative technique to present the findings, the chapter assigns indices to interviewees to maintain confidentiality while ensuring the disclosure of their professional roles and the sectors in which their organizations operate. This approach enhances the transparency of the research, facilitating a deeper understanding of the intricacies associated with implementing higher data responsibility standards in the context of East Africa. Central to the investigation are the challenges and barriers organizations face in enhancing data responsibility, alongside the strategic management practices adopted to navigate these obstacles. The chapter's findings contribute significantly to the chapter on data responsibility, offering insights into the operational realities of organizations in East Africa and the impact of cultural and structural factors on their data management practices.

Table 1. List of interviewees

Index Role Field
Respondent Interviews

continued on following page

Table 1. Continued

Index	Role	Field
R1	Sustainability manager	Telecom
R2	Human Rights lawyer	Telecom
R3	Sustainability manager	Network operator
R4	Partnership Coordinator	Donor
R5	Policy Specialist	Donor
R6	Data privacy specialist	Donor
R7	Policy Specialist	Donor
Informant Interviews		
I1	Executive Director	NGO
I2	Chief Strategy Officer	NGO
I3	Researcher	Think tank
I4	Digital Transformation Advisor	NGO
I5	Policy Specialist	Think tank

Findings From Respondent Interviews

The analysis unfolds the complexities of managing data privacy within organizations, anchored in the theoretical constructs previously delineated. Initially, the exploration into organizations' overarching ambitions regarding data privacy issues is presented. This involves an examination of their motivations and overarching strategies for addressing data privacy, providing a broad perspective on organizational attitudes and approaches toward data management. Subsequently, attention shifts to the structural dimensions of organizations, which emerge as pivotal in influencing corporate social responsibility (CSR) performance. This segment underscores the significance of organizational structure in facilitating or impeding the achievement of data privacy objectives, suggesting that structural elements play a crucial role in an organization's ability to uphold data responsibility.

The transition from policy to practice represents a critical juncture in understanding how organizations operationalize data privacy commitments. Through the lens of the CDR (Corporate Data Responsibility) framework by Lobschart et al. (2021), this section delves into the tangible actions and cultural practices that embody organizations' commitments to data privacy, offering insights into the practical challenges and successes in embedding data responsibility into organizational culture. The discussion culminates in the identification of three primary barriers to the advancement of data responsibility ambitions: the dynamics of multi-stakeholder concerns, the pressures of a profit-driven market, and the influence of the external context. These factors, while distinct, collectively underscore the multifaceted challenges organizations encounter in striving for higher standards of data responsibility. Each barrier presents unique implications for organizations, necessitating tailored strategies to navigate the complexities of data privacy management effectively.

Data privacy and protection have become paramount in today's chapter, marking a significant shift in organizational priorities. Interviews across various entities reveal a unanimous elevation of data privacy within their sustainability frameworks. Specifically, respondents R3 and R6 highlight data privacy as not merely an operational necessity but a foundational pillar of their strategic orientation. This consensus among the chapter subjects reflects a broader, industry-wide reckoning with data privacy, spurred by high-profile incidents like the Facebook-Cambridge Analytica scandal and the Snowden affair. These events have thrust data privacy issues into the limelight, compelling a re-evaluation of privacy practices across the tech industry. Respondent R2 notes the critical juncture at which the sector finds itself, acknowledging that turning a blind eye to data privacy is no longer viable. R3 further elucidates this stance, asserting data privacy's centrality to their organizational ethos. This collective prioritization of data privacy signals a pivotal shift towards embedding these concerns into the very fabric of sustainability strategies, underscoring a commitment to ethical data management and protection. R3 explains: *"We hold a lot of customer information, and to be honest, mobile phones are a big thing in East Africa in terms of using them as payment, mobile wallets, etc. So, in our system, for example, you can find a lot of mobile transactions. Given this, [data privacy] is a key thing and a priority for us."*

The chapter on Corporate Digital Responsibility (CDR) reveals a notable gap in awareness and strategic alignment, with the majority of respondents unfamiliar with the CDR framework. Despite this, elements resonant with CDR principles—such as values, norms, and behaviors—are inherently present within their existing sustainability strategies, suggesting an intuitive integration of digital responsibility without formal acknowledgment. Respondent R1 says, *"If we look at 10 years ago, there wasn't so much focus on sustainability strategies, now it's a pull factor in all this. More and more people within [our organization] are getting their skills as well, so it's not just us at the corporate level."* This evolution from a peripheral to a central strategic focus illustrates the growing emphasis on sustainability as a compelling aspect of organizational identity and operation. R1's observations underscore the expansion of sustainability skills within their organization, indicating a broader, more ingrained commitment to CSR practices. R6 shares a personal evolution from skepticism to acknowledgment of substantive CSR efforts within industries. Initially critical of CSR as mere *"greenwashing"* or *"whitewashing,"* R6 now recognizes tangible actions undertaken by companies, particularly in environmental sustainability and worker welfare. This shift in perspective underscores a broader industry move towards authenticity in CSR initiatives.

In addressing data privacy beyond regulatory compliance, R2 points to industry best practices and collaboration with global organizations like GSMA and the Global Network Initiative (GNI). Membership in such initiatives represents a proactive approach to enhancing data privacy policies, emphasizing the importance of industry-wide standards and collective action. R3's organization, situated in Nairobi, Kenya, recently adopted European-influenced data privacy policies, reflecting a preference for the human-centered approach embodied by GDPR standards, as noted by R6. This alignment with European data privacy norms highlights a cross-continental influence on privacy standards, emphasizing the global nature of digital responsibility challenges and solutions. Within the landscape of corporate sustainability, those dedicated to data privacy and human rights constitute a notably small contingent, often embedded within broader sustainability teams. This configuration underscores a nuanced challenge in delineating the scope of data privacy vis-à-vis data security—a distinction that sometimes eludes even those deeply engaged in these discussions, as evidenced by respondents R3, R4, and R7. Their conflation of data privacy with data security, uncorrected during interviews, sheds light on varying interpretations and the broader conceptual ambiguity surrounding these critical issues.

In the context of chapter object 1, the sustainability group emerges as a pivotal yet compact team of experts charged with crafting strategies and guidelines for data privacy and human rights engagement. This strategic nucleus then entrusts implementation to other organizational arms, illustrating a decentralized execution model where, for instance, marketing departments play a key role in operationalizing data privacy principles. R3, associated with chapter object 2, articulates a centralized model of managing data privacy and security, underpinned by uniform technology and security measures across all markets. This approach underscores a centralized decision-making strategy, divergent from practices within chapter object 3, represented by donor agencies. These agencies aspire to infuse digitalization across organizational practices, though concrete mechanisms for embedding data protection remain less defined, as noted by respondents R5 and R6. R7, representing donor perspectives, emphasizes the utilization of local expertise in executing projects, suggesting a tailored approach to navigating the complexities of data privacy and security in diverse contexts. This strategy points to the recognition of localized knowledge as instrumental in aligning data privacy and security initiatives with specific regional nuances and challenges. R7 states, "*[for each project we do], we partner with local actors, and all our projects are based on research. We work agile and externally, so say that we were to do a project in Somalia on for example data security, then we use a Somali data protection expert*"

Corporate Data Responsibility in East Africa

Organizations are increasingly implementing robust measures to ensure data privacy and safeguard customer data, as illustrated through practical examples provided by the chapter subjects. Chapter object 1 adopts a meticulous approach to risk assessment prior to engaging in any sales opportunity. This process involves a comprehensive evaluation of the product, its intended use, the customer's track record, and the human rights context of the customer's country, including protection levels and historical surveillance incidents. The data for these assessments is sourced from third-party organizations with specialized databases on risk factors. Utilizing four key parameters, an aggregated score is derived to guide decision-making in sales opportunities, potentially leading to adjustments in business processes or the establishment of specific contractual requirements, as noted by respondents R1 and R2. Chapter object 2 emphasizes the importance of internal awareness and training to ensure the effective downstream implementation of their data privacy strategies. This is complemented by engaging external expertise for conducting risk assessments, echoing the practices of the chapter object 1 and underscoring a common industry approach to managing data privacy risks. R6 further adds that *"this project will be scaled up, as we identified that regulation is the base. We can't work on other issues if we don't have regulation. The goal is for the African regulators to identify for themselves what their needs are and then we believe that the European "offer" is somewhat more developed than the Chinese."*

R7, another donor representative, disagrees with this training program. He claims, *"I know people are now trying to push GDPR on Africa but that doesn't work at all because it's like a completely different landscape and scenario. Countries have very different levels of maturity. Kenya is very mature in terms of data protection. Rwanda too but that's also mostly for show, I believe."* R5 suggests that adopting European data protection standards in African countries might unintentionally benefit European companies, which are already familiar with these regulations, potentially giving them an advantage over local businesses. The consensus among chapter subjects underscores the complexity of addressing data privacy and protection, highlighting the necessity for concerted efforts across a spectrum of stakeholders. R2 articulates *"This is not a one-man-job, or a one-organization-job, but rather a multi-stakeholder concern. We as an individual company will not change the situation for the people in for example Ethiopia."* The dialogue identifies four pivotal actors: governments in East Africa, major global economies (the EU, the US, and China), companies managing data, and geopolitical dynamics, with R2 pinpointing government policies and international geopolitics as primary obstacles to progress. R3 contributes to the discussion. He mentioned, *"We ensure that the customer's data is kept within the business, and we don't share any information unless there is a request from the customer, or it has been legally allowed by the court of law or the government.*" This practice is situated within the broader context of navigating government and legal frameworks.

The conversation further explores the geopolitical implications of data privacy efforts. R2 raises concerns about the potential fragmentation of internet and communication technologies along geopolitical lines, advocating for a more inclusive approach that could prevent a divide between Western technologies (Ericsson and Nokia) and those prevalent in China and Africa (Huawei). R2 states *"Otherwise, we are going towards a world where the internet and communications will be fragmented between the US and EU where we will use Ericsson and Nokia and China and Africa where they will use Huawei, and that's not good either."* From the donor perspective, R6 and R5 provide insights into the public and private sector's roles in advancing data privacy initiatives. R6 calls for enhanced coordination and interaction between various actors, particularly emphasizing the value of direct dialogue with the tech industry for a deeper understanding of ICT complexities. Conversely, R5 describes their more integrated approach with the private sector as mutually beneficial, suggesting an impending shift towards aligning more closely with European legislation and values, particularly under Sweden's EU Presidency in 2023. This multifaceted discussion underscores the intricate interplay of local and international politics, corporate practices, and cross-sector collaboration in addressing the global challenge of data privacy and protection.

A critical challenge identified by chapter object 1 revolves around reconciling ethical norms and moral obligations with the drive for profit and efficiency. This dilemma manifests in the difficulty of adjusting business processes or setting contractual requirements to uphold digital responsibility. The concern is that by imposing such standards, they risk losing customers to competitors who may not enforce equivalent ethical constraints. This scenario highlights a broader market dynamic where the absence of a universal commitment to digital responsibility creates opportunities for entities with lesser ethical standards to capture market share, underscoring the tension between ethical commitment and competitive market pressures. R2 sums up this, *"In a way, we provide crucial infrastructures also from a human rights perspective, we enable people to communicate and to access and share information. To some extent also highlights problems in the country and so. Take Russia and Afghanistan, for example. In some cases, there are problems with withdrawing [from the countries] because that also has consequences for human rights. Then you might turn off certain functionalities and other operators will enter, who don't care about these issues at all come in instead. Should we say no to countries where the need may be very great from a socio-economic and human rights perspective, but bad for the business? So, it's a difficult balance. We have to be aware that we are providing something good, but we have to do it responsibly way."*

R5 highlights an intriguing trend among European companies striving to outdo one another in demonstrating ethical consciousness and digital responsibility. This competitive rush towards becoming "nicer" paradoxically may lead to losing bids in a market where alternatives, possibly less ethical, exist. This pursuit, while noble, blinds them to the realities of a profit-driven market, potentially jeopardizing their commercial viability. The challenge extends beyond data privacy, touching on broader issues within the ICT sector, such as digital inclusion. Chapter object 1's initiatives to enhance digital inclusion in African countries often falter, not due to a lack of effort but because of unsustainable business models. R2 elucidates this with the example of a project in Rwanda aimed at setting up rain gauging stations. Despite the potential multifaceted benefits of the project, from predicting malaria outbreaks to supporting agricultural insurance, it was ultimately abandoned. The withdrawal by both local and international partners due to the project's lack of commercial viability underscores the critical tension between pursuing non-commercial, socially responsible activities and the imperative of financial return on investment.

Respondents R3, R4, and R7 identify digital literacy as a significant challenge both within organizations and among East African citizens. R3 highlights that technology's effectiveness is contingent on the vigilance of its operators, pointing out that breaches often stem from employees' actions, borne out of ignorance rather than malintent. He advocates for educating customers on their digital rights to foster accountability and consciousness, noting a prevalent unawareness of these rights in East Africa. R4 echoes the concern about unawareness, adding to the complexity of emerging technologies like Artificial Intelligence (AI). She emphasizes the uncertain impact of AI on less affluent nations, suggesting a need for internal and external awareness and preparedness. R1 critiques the ICT industry's pace, comparing it unfavorably to the record industry's evolution and calling for innovative business models that do not compromise digital rights. The discussion reflects a consensus on the necessity of evolving business practices and legal frameworks to better accommodate digital rights and literacy. The call for new legislation is a recurring theme, with a push towards transforming voluntary principles into binding legal requirements to ensure that digital rights are taken more seriously. R2 concludes by framing digital responsibility as a legal compliance issue, urging companies to elevate their commitment from a social to a legal obligation, underlining the need for a holistic approach to address the challenges of digital literacy and rights in East Africa.

Findings From Informant Interviews

In the contemporary chapter surrounding data privacy and digital rights within East Africa, the emergent themes of policy implementation, legislation awareness, conflicting issues, and corporate responsibility underscore the complexities inherent in aligning legal frameworks with practical application. The crux of the analysis rests on informant interviews, yielding insights that underscore a pronounced dichotomy between theoretical policies and their practical execution. A consensus among respondents reveals a significant disparity in the region, characterized by a well-documented disconnect between high-level corporate policy formulation and grassroots implementation. Informant I2, leading a youth- and female-centric non-profit organization focused on internet governance and freedom, articulates a concern over the performative nature of policy-making and reporting. This sentiment is echoed by I1, an executive director at a non-governmental organization dedicated to digital rights, who critiques the prevalent rhetoric of commitment to data privacy which seldom translates into tangible actions. He states, *"In terms of inscription, in terms of privacy, in terms of freedom of expression, there is a lot of alignment with companies, not all, however. In terms of capacity building, the alignment is there from the government because it's in their interest to get more skills and increase in opportunities and all that."*

Nonetheless, I1 introduces a nuanced perspective, acknowledging instances of alignment between non-profit values and those of profit-driven entities, particularly in the realm of capacity building—a reflection of governmental interest in enhancing skill sets and opportunities. Further complicating the chapter, I3, a researcher at a think tank specializing in data privacy across East and South Africa, highlights the superficiality of legislative measures purported to safeguard privacy and counter-surveillance. I3 states, *"If we look at privacy and surveillance for example, there are countries that have legislations and they say that they do have data protection laws, but then at the end of the day there is not much going on."* The landscape of data privacy within East Africa is marred by the limitations of existing legal frameworks, which, as noted by informants I3 and I4, extend only as far as the statutes delineate. The real quandary emerges in the "grey areas" — domains inadequately addressed by regulations, thereby becoming hotbeds for data privacy breaches. The insufficiency of the regional data privacy regulations often leaves significant loopholes, exacerbating vulnerabilities to breaches. This issue is particularly evident in cases like online hate speech, which, while not directly a data privacy matter, lacks regulatory oversight and thus remains uncensored and unaddressed.

I3 points to the tendency within the private sector to attribute their limitations in ensuring data privacy to the confines of the law. Yet, this perspective overlooks the potential for industry-led initiatives to foster enhanced legal standards. A prime example of such proactive engagement is the Global Network Initiative (GNI), an international organization that advocates for the development of more robust legislative frameworks by uniting industry stakeholders. The participation of both private companies and non-governmental organizations, including those represented by respondents R1, R2, and I1, underscores the potential for collaborative efforts in shaping international standards for data privacy. I1's involvement with GNI, representing civil society, highlights a strategic alignment between non-profit objectives and corporate interests in the realm of data privacy. This partnership, facilitated through GNI, exemplifies a concerted effort to transcend legal limitations by fostering an ecosystem wherein entities collaborate toward the establishment of stronger, more comprehensive data privacy regulations.

A significant challenge in data privacy within East Africa, as identified by informant I4, lies in the widespread lack of awareness among individuals regarding the collection and breach of their data. This issue is vividly exemplified through observations from Kenya, where the majority of the populace remains uninformed about the breaches of their personal information. I4 states, *"If you go into a building, you will be asked by the security guard to sign in by putting in your name, ID number, phone number, and signature. With this, there are a lot of identity thefts that happen as the security guard does not understand the importance of safeguarding this data. We just had elections the other day, and a part of the requirement is to have a certain number of people registered under your party, so imagine the surprise when you look up yourself on the database and you find yourself registered for a party that you don't even know about, simply because someone picked up your details from a security guard somewhere."* The fundamental problem, as articulated by I4, is not just the breach itself but the pervasive ignorance of such breaches, highlighting an urgent need for comprehensive sensitization efforts.

Informant I2 echoes this sentiment, acknowledging existing initiatives aimed at elevating public consciousness about data privacy. However, these efforts fall short of achieving widespread awareness. I2 states, *"There are a lot of resources for capacity building. But after that, there is no support to implement, innovate, and create change. It does not exist. And you have to move so much more. You have to work hard to implement at the grassroots level."* Informant I1 articulates a significant challenge in the realm of digital rights within East Africa: the dichotomy between fulfilling basic human needs—such as access to food, water, and secure livelihoods—and addressing emerging human rights concerns spurred by the advancements of the Fourth Industrial Revolution. This "clash of civilizations" encapsulates the tension between traditional rights to survival and contemporary digital rights debates, underscoring a broader chapter on the prioritization of issues within the region.

The narrative extends to the conflicting priorities of government actions, particularly in crisis scenarios where the immediate response often involves measures like mass surveillance to ensure public safety, inadvertently compromising individual privacy and security. I1 states, *"For example when a bomb goes off in a city, the first thing the government wants is to find the predators and protect their citizens. They put up cameras to stabilize the situation, however that leads to mass surveillance.*" This response, while aimed at protecting citizens from immediate harm, raises concerns about the broader implications for digital rights and privacy. Furthermore, I1 delves into the intricate relationship between commercial enterprises and their societal responsibilities. The assertion that businesses primarily aim to generate profit, with social good being a secondary outcome, captures the essence of the dilemma facing digital rights advocacy. I1 states, *"Businesses are set up to make money, doing good is the byproduct. We have a problem with that doing good is the secondary thing. But at the same time, there would not be a business if doing good is their primary responsibility. And we need businesses."* The informant posits that while economic activity is essential for societal progress, there must be a balanced approach that incorporates ethical considerations and societal benefit. I4 reiterates this sentiment, acknowledging the dual pressures faced by profit-driven organizations to meet both their financial objectives and societal expectations.

The shift in corporate strategies towards recognizing the value of *"doing good"* as not only ethically right but also beneficial for business, marks a significant evolution in corporate responsibility. I1 states that *"it's a bit two-sided because the organizations have their priorities and they have their goals to meet, while at the same time pressure from the outside and their societal impact."* This change is driven by increased awareness and demands from civil society and consumers, highlighting a growing acknowledgment of the importance of ethical considerations in business operations. However, I4 also expresses concerns regarding the pace of technological evolution outstripping the development of regulatory frameworks, posing ongoing challenges to ensuring digital rights and privacy. The rapid advancement of technology, while offering myriad benefits, also introduces complexities in safeguarding individual rights within an ever-changing digital landscape. Over the last 10-20 years, there's been a significant shift in why companies focus on social issues, driven by increased demands from both society and consumers for ethical behavior. As people become more aware of their rights and the impact of businesses, they're pushing companies to act more responsibly. I3 states, *"Yes, the policies can be there, but the technology evolves fast, and that gives me a reason to worry."*

I2 criticizes the top-down approach companies often use to solve digital issues, advocating instead for empowering those directly affected to contribute to solutions, aiming for a more democratized internet. I2 states, *"What we see is that people that making decisions for others. It's too much top-down rather than bottom-up. The ones who are experiencing the most issues are not given the autonomy to solve their problem. An open, bottom-up approach to digital issues is the way I think the internet can be democratized."* I3 points out the significant role of economic power in influencing digital policy, noting the growing influence of countries like China, Russia, and Israel, and argues against a one-size-fits-all approach due to national differences. I3 states, "*I think is more about the economic muscle. We see that China, Russia, and Israel have increasingly more influence over the economy.*" I3 also states, "*It all depends on the economic influence. Some companies may want to apply the same approach and structure of doing things globally. But at a national level and country level, it is more of national restrictions. You must follow that.*" I4 emphasizes the need for legal reform and urges companies to advocate for changes to better address these digital challenges. I4 states, "*The laws need to change. The companies need to put pressure to change the laws.*"

DISCUSSION

Managing Data Responsibility in East Africa

The chapter elaborates upon the empirical findings derived from the preceding analysis, focusing on the operational dynamics of data responsibility within the context of East Africa, juxtaposed against the theoretical underpinnings and the articulated research inquiries. The exposition is structured to first illuminate the mechanisms through which data responsibility is managed in the region, followed by an examination of the impediments that hinder the enhancement of data stewardship practices. In the milieu of East Africa, the regulatory landscape for data protection is notably inadequate, a circumstance that is not only highlighted in the literature but also substantiated through narratives from informants who recount instances of regulatory failures impacting citizens (Makulilo, 2016). These narratives encompass the absence of safeguards against online harassment and prevalent incidents of identity theft, which are manipulated for various ends, including political machinations and illicit financial transactions. The insidious nature of these breaches, often unbeknownst to the individuals affected, exacerbates the complexity of addressing and mitigating these challenges.

This reality is consonant with Kinuthia's (2020) assertion that the primary infringement upon privacy in the region emanates from online impersonation and social engineering, aimed at financial exploitation, alongside social media compromises. Within this context, the imperative for a nuanced approach to digital responsibility emerges, especially in light of the rapid evolution and application of digital technologies in an environment bereft of robust regulatory frameworks. The concept of Corporate Digital Responsibility (CDR) is posited as a pivotal instrument for organizations, enabling them to navigate ethical quandaries and foster a conscientious digital presence, mindful of the ramifications, risks, and prospects of their digital endeavors (Herden et al., 2021). This framework assumes particular significance as a potential catalyst for fostering responsible technological advancement in a region characterized by significant regulatory gaps (Merwe & Achkar, 2022). Consequently, the chapter endeavors to scrutinize the application of CDR within the studied entities in East Africa, aiming to elucidate their strategies for managing data responsibility amidst these challenges.

The exploration at hand delves into the intricacies of how organizational culture serves as a foundational element in shaping and being shaped by Corporate Digital Responsibility (CDR). As elucidated by Lobschat et al (2021), the conceptual framework presented posits that the culture of CDR within organizations operates across three critical dimensions: shared values, specific norms, and manifested artifacts and behaviors. This tripartite model offers a comprehensive lens through which the dynamics of digital responsibility within organizational settings can be understood and assessed. Central to this discussion is the assertion that the essence of an organization's CDR culture is encapsulated within these layers. Shared values represent the collective beliefs and principles that underpin the organization's approach to digital responsibility. Specific norms, on the other hand, refer to the established standards and expectations that guide behavior within the context of digital engagements. Lastly, artifacts and behaviors are tangible expressions and actions that reflect the organization's commitment to digital responsibility, as discerned through both direct interviews and analysis of organizational documents such as annual and sustainability reports.

The chapter surrounding Corporate Digital Responsibility (CDR) remains nascent within the corporate realm, as evidenced by the unfamiliarity of most respondents with the concept, save for one. Despite the absence of a formal CDR framework within their operational paradigms, entities examined in this chapter inadvertently align with the ethos of digital stewardship. This alignment manifests through their ingrained organizational values, norms, and demonstrated artifacts and behaviors, echoing the foundational elements of a CDR culture. Yet, these organizations do not pioneer unique values and norms; instead, they pivot towards established sustainability frameworks such as the United Nations Guiding Principles (UNGP), Sustainable Development Goals (SDGs), and the UN Global Compact (Lobschat et al., 2021). These frameworks serve not merely as benchmarks but as reservoirs of inspiration, advocating for a harmonized approach toward sustainability and digital responsibility. The era of globalization and the subsequent international expansion of organizations complicate the tapestry of digital responsibility, introducing moral, normative, and cultural nuances into the equation. This complexity, as articulated by Mueller (2022) and Lobschart et al. (2021), necessitates a nuanced adoption of CDR cultures by international firms, contingent upon the socio-cultural and regulatory landscapes they navigate.

The criticality of contextual awareness in the formulation and implementation of CDR strategies cannot be overstated, with potential pitfalls and tensions lurking in the absence of such awareness (Barkema et al., 2015). Despite the recognized imperative for context-adjusted approaches to data responsibility, none of the entities under scrutiny exhibit a comprehensive strategy that thoroughly addresses the multifaceted nature of CDR. Nevertheless, certain organizational practices and artifacts hint at an implicit acknowledgment of context sensitivity. These include risk assessments and international training programs tailored to specific operational contexts (Lobschat et al., 2021). However, the overarching sustainability strategies of these entities remain conspicuously uniform across diverse geographical and cultural contexts, devoid of localized nuances. This uniformity extends to their approach to data safeguarding, underscoring a homogenized strategy that overlooks the distinct dynamics of operating in regions such as East Africa.

Table 2. Applying CDR to the findings, based on the framework by Lobschart et al (2021)

Category	SO1	SO2	SO3
Shared Values	Aligns with international sustainability and human rights principles.	Developed own conduct guidelines, inspired by international principles.	Focus on developing technology to support sustainable development, albeit without explicit values.
Specific Norms	Targets contribution to Sustainable Development Goals.	Follows GSMA's recommendations.	Activities based on EU-AU Digital Economy Task Force recommendations.
Artefacts and Behaviours	Implements sensitive business risk assessments.	Engages in training and development activities.	Provides an International Training Programme.

In the domain of Corporate Digital Responsibility (CDR), a nuanced exploration reveals that certain dimensions, namely the external and internal contexts within which organizations operate, elude seamless integration into the prevailing theoretical frameworks of CDR. Instead, a broader Corporate Social Responsibility (CSR) lens, particularly through the Corporate Sustainability Model posited by M. Epstein et al (2017), provides a more comprehensive understanding. This model elucidates the multifaceted drivers behind CSR performance, detailing the actions required for effective implementation and the resultant outcomes. It underscores the significance of organizational inputs and processes, encompassing the external and internal contexts, business landscape, and the allocation of human and financial resources, as pivotal to achieving substantive CSR performance (M. J. Epstein & Buhovac, 2014). The findings of this investigation underscore the complexity inherent in managing data responsibility. Despite the articulated policies, a tangible implementation shortfall is evident, highlighted by the modest size of teams dedicated to human rights-related issues.

These teams, although integral to the broader corporate responsibility framework, are often constrained by limited human resources, especially in large organizations where the proportion of staff allocated to digital rights remains minimal. This scenario not only reflects a resource allocation challenge but also indicates potential overlaps in responsibilities within larger corporate responsibility teams. Moreover, the chapter illuminates a critical knowledge gap among personnel tasked with navigating data privacy and security landscapes (M. J. Epstein & Buhovac, 2014). Despite being designated as suitable respondents, the conflation of data privacy and security terms among interviewees signals a pressing need for enhanced training. Data privacy and security, while related, address distinct aspects of data management, with the former focusing on the ethical handling of data and the latter on protecting data from unauthorized access and cyber threats. The chasm between the development of data responsibility policies and their pragmatic application emerges as a central theme. This gap can be attributed to a lack of context-adjusted approaches in data responsibility strategies and an overarching deficiency in both knowledge and dedicated personnel within organizations.

Barriers to Improve Data Responsibility

In the realm of Corporate Digital Responsibility (CDR), the analysis identifies two principal barriers impeding effective data privacy management and broader CDR initiatives. Firstly, the intrinsic complexity of data privacy as a multi-stakeholder issue emerges as a significant challenge (Sarkar & Searcy, 2016). The necessity for robust collaboration between private and public sectors is underscored, advocating for direct dialogue as a means to enhance mutual understanding and foster comprehensive solutions. Despite the potential for synergy, the diversity of actors within this sector often leads to fragmented efforts and a lack of cohesive action. The second barrier pertains to the voluntary nature of CSR/CDR commitments within a market primarily driven by profit motives (Merwe & Achkar, 2022). This dichotomy presents an enduring dilemma for organizations attempting to reconcile non-commercial activities with their commercial objectives, highlighting the inherent tension between ethical obligations and profit-driven imperatives (Merwe & Achkar, 2022). The absence of a unified global regulatory framework or universally endorsed standards further exacerbates these challenges. This fragmentation risks perpetuating piecemeal or ad hoc approaches to CDR, thereby underscoring the necessity for scalable solutions that transcend sectoral and geographical boundaries. The chapter echoes the sentiment that individual organizations cannot single-handedly address these pervasive issues, emphasizing the critical need for collective action across stakeholders.

Within this context, the Global Network Initiative (GNI) emerges as a pivotal platform for collaboration. Chapter Object 1 (SO1) and an informant from a Non-Governmental Organization (NGO), both members of the GNI, accentuate the initiative's role in facilitating proactive, long-term engagement with data privacy concerns. This affiliation is perceived as a vital conduit for advancing data privacy initiatives. Conversely, the absence of such collaborative frameworks among the donors of Chapter Object 3 (SO3) is noted with concern. This lack of engagement underscores the urgent call for enhanced public-private partnerships, as articulated by one donor interviewee. Another interviewee, involved in a separate donor initiative, affirms the mutual benefits derived from engaging more extensively with the private sector. The Global Network Initiative (GNI) embodies a concerted effort to address the multifaceted challenges of safeguarding digital rights worldwide. Its mission hinges on fostering a multi-stakeholder process that amalgamates the diverse perspectives, leverage, credibility, and expertise of various actors to navigate the evolving complexities of digital rights protection. Admission into the GNI necessitates a thorough self-assessment by organizations, demonstrating a commitment to integrate the GNI Principles into their operational ethos.

Beyond the multi-stakeholder dynamics, the chapter surrounding data privacy transcends into the geopolitical realm, underscored by the significant investments in the Information and Communication Technology (ICT) sector. Notably, a substantial portion of the $7.1 billion investment in ICT in 2018 emanated from the private sector, supplemented by contributions from African governments, China, and international coalitions including the Infrastructure Consortium for Africa (ICA), which boasts members like the World Bank, the African Development Bank, and G7 countries (Corrigan, 2020). This investment landscape is further complicated by the European Commission's ambitious EU Global Gateway strategy, promising a colossal investment in global infrastructure, thereby highlighting the intricate interplay of geopolitical interests and digital rights. The geopolitical undertones of data privacy are pronounced in the responses from Chapter Object 3 (SO3) interviewees, particularly from a European perspective. The contemplation of implementing a European-centric data privacy model, inspired by the General Data Protection Regulation (GDPR), in East Africa, evokes ambivalent reactions. On one hand, there exists a propensity to advocate for the GDPR model, perceived as more advanced compared to other frameworks. This inclination stems from a desire to elevate the standards of data privacy regulation in African contexts. On the other hand, the feasibility of transplanting the European data privacy model to East Africa is met with skepticism. Critics argue that the distinct digital and legal landscapes of East Africa render a direct replication of European standards impractical, highlighting the complexities of navigating digital rights across disparate geopolitical and regulatory environments.

The chapter on Corporate Social Responsibility (CSR) and Corporate Digital Responsibility (CDR) within the Global South illuminates the precarious nature of transplanting methodologies from the Global North without contextual adjustments. Such practices risk perpetuating digital colonialism, a critical concern highlighted by Inekwe et al. (2020) and Kwet (2019). The imperative for a South-centred CSR agenda underscores the need for a bottom-up approach, as posited by Idemudia (2014), ensuring that solutions are rooted in the realities of those most affected by digital inequities. An interviewee from a digital rights NGO in East Africa emphasized the misalignment of current corporate strategies with the actual needs and challenges faced at the grassroots level. This disconnect underscores the necessity for an open, bottom-up methodology to address digital privacy and democratize internet access, contrary to the predominantly top-down approaches observed among the studied entities. Notably, an exception was identified among the donors, where a policy of leveraging local expertise in projects signifies a nuanced understanding and respect for local contexts and knowledge systems.

The implementation of CSR and CDR initiatives, while instrumental in fostering trust and enhancing reputational capital, often incurs substantial costs. These financial implications, as Shuqin (2014) notes, can divert focus from economic productivity, presenting organizations with the challenging task of reconciling ethical obligations with the imperatives of profit maximization and efficiency. This dilemma is further exacerbated by the rapid pace of digitalization, which, according to Merwe & Achkar (2022), necessitates a higher threshold of responsibility. The prioritization of business as usual, absent robust accountability mechanisms and independent regulatory oversight, raises concerns regarding the potential for CDR to serve as a facade for corporate whitewashing or regulatory capture. Recent chapter within the realms of Corporate Social Responsibility (CSR) and Corporate Digital Responsibility (CDR) reflects a significant shift towards a more earnest engagement with sustainability issues (Hategan et al., 2018). Interviews with both respondents and informants reveal an increased commitment within organizations to address these concerns more authentically. This transformation is particularly notable when compared to attitudes prevalent 10-15 years ago, where a marked increase in the skills and genuine concern for sustainability issues among organizational members is observed.

Despite this positive shift, the perennial challenge of balancing profitability with ethical and social responsibilities persists. An illustrative example from Rwanda highlights this dilemma: a project with substantial potential for social impact was abandoned due to its lack of commercial viability. This underscores a critical tension within CSR and CDR initiatives—their justification often hinges on their ability to align with or contribute to, an organization's financial bottom line. Hategan et al (2018) research into the relationship between CSR and profit reinforces this notion, suggesting that organizational attitudes towards CSR are significantly influenced by the perceived impact of such initiatives on profitability, whether positive or negative. While the commitment to addressing data privacy and digital responsibility is unanimously affirmed by the chapter subjects, there emerges a consensus on the need for more stringent regulatory frameworks. The call for new legislation to impose stricter requirements on companies indicates a move towards formalizing commitments to data privacy and digital responsibility, potentially transcending the traditional voluntary nature of CSR initiatives. This perspective, however, presents a paradox to the definition of CSR as articulated by (Sarkar & Searcy, 2016), who advocate for organizations to voluntarily exceed legal minimums and embed ethical considerations into all aspects of their operations.

CONCLUSION AND SUGGESTIONS

Aim of the research—to investigate the management of data responsibility by organizations operating in East Africa and to identify barriers to enhanced data responsibility—provides valuable empirical insights into the intersection of digital advancement and corporate ethics. Organizations in East Africa find themselves at the crossroads of leveraging ICT for unprecedented economic growth while navigating the complex terrain of digital responsibility. The chapter underscores the importance of stakeholder awareness and action—spanning individuals, companies, and governments—toward the societal impacts of digital infrastructure. It posits a central inquiry: how can companies balance ethical norms and moral obligations with the drive for profit and efficiency? The investigation, informed by CSR and CDR theoretical frameworks, aims to bridge the research gap in the application of CDR within the Global South's ICT sector. Through semi-structured interviews across three distinct organizational types, the chapter dissects the modalities of managing data privacy in East Africa.

Key findings reveal a diversity of approaches to data responsibility among the examined entities. Despite the absence of explicit adherence to the CDR concept, certain practices and orientations of the chapter objects resonate with the guiding principles of CSR and CDR. These include the adoption of human rights-based values and norms, the allocation of resources, and the initiation of training programs aimed at fostering data privacy. The significance of collaborative efforts among various stakeholders is highlighted, showcasing instances of successful partnerships that aim to advance data responsibility. However, the chapter unveils a pronounced disparity between the formulation of data privacy policies and their pragmatic implementation.

This disconnect signals a critical area for further investigation and action, pointing to the need for a more cohesive and integrated approach to data responsibility that aligns theoretical frameworks with operational realities. The critical examination of Corporate Digital Responsibility (CDR) within the fast-evolving digital landscape of East Africa reveals a conspicuous gap between policy articulation and practical implementation. Despite a discernible shift towards a more conscientious approach to CSR and CDR among organizations, a persistent disparity between declared commitments and actionable strategies underscores the challenges of navigating the ethical complexities introduced by rapid technological advancements.

This gap, accentuated by the chapter's findings, signals a pressing need for organizations to not only verbalize their commitment to digital rights but to manifest such commitments through tangible, context-sensitive actions. The chapter underscores that while there is a nominal acknowledgment of the importance of digital responsibility, the actual commitment to these principles is not as robust as it could be, bordering on superficiality. This discrepancy between rhetoric and reality raises concerns about the sincerity of organizations' efforts in digital stewardship, potentially veering into the realm of whitewashing. The exploration into the operational dynamics of organizations in East Africa reflects a uniformity in approach, a one-size-fits-all strategy that neglects the unique socio-technological contexts of different regions. This homogenized strategy, though perhaps simpler to administer, fails to account for the nuanced demands of varying digital ecosystems, thereby lagging in responsiveness and relevance, much like state-based regulations have historically lagged behind technological innovation.

A notable observation from the chapter is the scarcity of resources—both in terms of personnel and focus—dedicated to the management of digital rights issues. Despite the genuine commitment of individual team members to digital ethics, the centralized decision-making processes and the limited size of teams devoted to these issues do not align with the organizations' purported dedication to digital responsibility. This misalignment suggests a substantial underestimation of the resources required to effectively navigate the digital rights landscape, especially in a region as diverse and rapidly changing as East Africa. Moreover, the chapter reveals a critical oversight in the lack of localized strategies for managing digital privacy and data responsibility. Despite the recognition of varying risk settings and the complex interplay of moral norms and cultural factors across different countries, organizations have yet to develop truly context-adjusted approaches to digital responsibility. This oversight not only undermines the efficacy of their CSR and CDR initiatives but also highlights a gap in understanding the intricacies of the settings they operate within.

This chapter delves into the evolving landscape of Corporate Digital Responsibility (CDR) in East Africa, juxtaposing organizational ambitions with the practicalities of implementing digital privacy and data management initiatives. It highlights the burgeoning acknowledgment of digital technologies' pivotal role in economic development, underscored by the imperative for universal internet access as espoused by the Sustainable Development Goals (SDGs). Despite the optimism surrounding Information and Communication Technology (ICT) investments, concerns persist regarding the equitable distribution of ownership within the ICT sector across the Global South and the exploitation by the Global North. This chapter aligns with critical observations about the lack of localized, South-centred CSR approaches among organizations operating in East Africa. The chapter's core revelation is the nascent state of the CDR framework among examined entities, reminiscent of early observations about the gradual incorporation of privacy programs into CSR strategies. This emergent phase of CDR underscores a critical juncture where the commitment to digital responsibility must transcend rhetoric to embrace comprehensive, actionable strategies. Moreover, the analysis emphasizes the dual need for funding and enhancing digital literacy to navigate the digital rights landscape effectively, advocating for heightened awareness among citizens and employees alike.

The research confronts several limitations, including its scope, constrained by the chapter's geographical and sectoral breadth and the methodological approach to selecting chapter subjects. These constraints underscore the challenges of extrapolating findings across the diverse and dynamically transforming ICT landscape of East Africa. Furthermore, the chapter grapples with varying levels of organizational transparency and the reluctance of potential interviewees to engage, which, in turn, has influenced the depth of the insights gathered. Looking forward, this chapter lays the groundwork for future inquiries into the practice of CDR in the Global South. It suggests avenues for broadening the empirical base through quantitative studies and more targeted qualitative analyses within specific East African contexts. The chapter advocates for a systematic exploration of contextual factors influencing CDR strategies and calls for the development of methodologies that resonate with the Global South's unique digital ecosystem. This approach not only promises to enrich the theoretical and practical understanding of CDR but also aims to contribute to the effective alignment of corporate strategies with sustainable development imperatives in the region.

REFERENCES

Allen, A., & Peloza, J. (2015). Someone to Watch Over Me: The Integration of Privacy and Corporate Social Responsibility. *Business Horizons*, 58(6), 635–642. Advance online publication. 10.1016/j.bushor.2015.06.007

Arnold, S. (2022). Drivers and Barriers of Digital Market Integration in East Africa: A Case Chapter of Rwanda and Tanzania. *Politics and Governance*, 10(2), 106–115. 10.17645/pag.v10i2.4922

Awad, A., & Albaity, M. (2022). ICT and economic growth in Sub-Saharan Africa: Transmission channels and effects. *Telecommunications Policy*, 46(8), 102381. 10.1016/j.telpol.2022.102381

Barkema, H., Chen, X.-P., George, G., Luo, Y., & Tsui, A. (2015). West Meets East: New Concepts and Theories. *Academy of Management Journal*, 58(2), 460–479. 10.5465/amj.2015.4021

Barker, M. (2014). In Vossler, A., & Moller, N. (Eds.), *Doing a literature review* (pp. 61–73). Sage. https://uk.sagepub.com/en-gb/eur/the-counselling-and-psychotherapy-research-handbook/book239261

Basimanyane, D. (2022). The Regulatory Dilemma on Mass Communications Surveillance and the Digital Right to Privacy in Africa: The Case of South Africa. *African Journal of International and Comparative Law*, 30(3), 361–382. 10.3366/ajicl.2022.0414

Bell, E., Bryman, A., & Harley, B. (2018). *Business Research Methods*. Oxford University Press.

Bryman, A. (2012). *Social Research Methods*. OUP Oxford.

Casella, B., & Formenti, L. (2018). FDI in the digital economy: A shift to asset-light international footprints. *Transnational Corporations*, 25(1), 101–130. 10.18356/cb688e94-en

Corrigan, T. (2020). *Africa's ICT Infrastructure: Its Present and Prospects*. https://policycommons.net/artifacts/1451480/africas-ict-infrastructure/2083288/

Dartey-Baah, K., & Amponsah-Tawiah, K. (2011). Exploring the limits of Western Corporate Social Responsibility Theories in Africa. *International Journal of Business and Social Science*, 2.

Deshpande, R., & Webster, F. E.Jr. (1989). Organizational Culture and Marketing: Defining the Research Agenda. *Journal of Marketing*, 53(1), 3–15. 10.1177/002224298905300102

Epstein, M., Buhovac, A., Elkington, J., & Leonard, H. (2017). Making Sustainability Work: Best Practices in Managing and Measuring Corporate Social, Environmental, and Economic Impacts. In *Making Sustainability Work: Best Practices in Managing and Measuring Corporate Social, Environmental and Economic Impacts* (p. 305). 10.4324/9781351276443

Epstein, M. J., & Buhovac, A. R. (2014). *Making Sustainability Work: Best Practices in Managing and Measuring Corporate Social, Environmental, and Economic Impacts*. Berrett-Koehler Publishers.

Erbacci, A., Deighton-Smith, R., & Kauffmann, C. (2016). Promoting inclusive growth through better regulation: The role of regulatory impact assessment. *OECD Regulatory Policy Working Papers, 2016*, 1–55. 10.1787/24140996

Fifka, M. (2009). Towards a More Business-Oriented Definition of Corporate Social Responsibility: Discussing the Core Controversies of a Well-Established Concept. *Journal of Service Science and Management -. Journal of Service Science and Management*, 02, 312–321. 10.4236/jssm.2009.24037

Hasan, I., & Tucci, C. (2010). The innovation-economic growth nexus: Global evidence. *Research Policy*, 39(10), 1264–1276. 10.1016/j.respol.2010.07.005

Hategan, C. D., Sirghi, N., Ruxandra Ioana, P., & Hategan, V. (2018). Doing Well or Doing Good: The Relationship between Corporate Social Responsibility and Profit in Romanian Companies. *Sustainability (Basel)*, 10(4), 1041. 10.3390/su10041041

Herden, C. J., Alliu, E., Cakici, A., Cormier, T., Deguelle, C., Gambhir, S., Griffiths, C., Gupta, S., Kamani, S. R., Kiratli, Y.-S., Kispataki, M., Lange, G., Moles de Matos, L., Tripero Moreno, L., Betancourt Nunez, H. A., Pilla, V., Raj, B., Roe, J., Skoda, M., ... Edinger-Schons, L. M. (2021). "Corporate Digital Responsibility." *Sustainability Management Forum | NachhaltigkeitsManagementForum, 29*(1), 13–29. 10.1007/s00550-020-00509-x

Hutt, R. (2015, November 13). *What are your digital rights?* World Economic Forum. https://www.weforum.org/agenda/2015/11/what-are-your-digital-rights-explainer/

Idemudia, U. (2014). Corporate Social Responsibility and Development in Africa: Issues and Possibilities. *Geography Compass*, 8(7), 421–435. 10.1111/gec3.12143

Inekwe, M., Hashim, F., & Yahya, S. (2020). CSR in developing countries – the importance of good governance and economic growth: Evidence from Africa. *Social Responsibility Journal*. 10.1108/SRJ-10-2019-0336

Kinuthia, D. (2020). *Exploring Data Anonymisation and Internet Safety in East Africa*. https://policycommons.net/artifacts/1445762/exploring-data-anonymisation-and-internet-safety-in-east-africa/2077526/

Lobschat, L., Mueller, B., Eggers, F., Brandimarte, L., Diefenbach, S., Kroschke, M., & Wirtz, J. (2021). Corporate Digital Responsibility. *Journal of Business Research*, 122, 875–888. 10.1016/j.jbusres.2019.10.006

Makulilo, A. (2016). *African Data Privacy Laws* (Vol. 33). 10.1007/978-3-319-47317-8

Malan, D. (2018, June 21). *Technology is changing faster than regulators can keep up—Here's how to close the gap*. World Economic Forum. https://www.weforum.org/agenda/2018/06/law-too-slow-for-new-tech-how-keep-up/

Mueller, B. (2022). Corporate Digital Responsibility. *Business & Information Systems Engineering*, 64(5), 689–700. 10.1007/s12599-022-00760-0

Nchake, M. A., & Shuaibu, M. (2022). Investment in ICT infrastructure and inclusive growth in Africa. *Scientific African*, 17, e01293. 10.1016/j.sciaf.2022.e01293

Peltola, M., Xue, G., & Yu, Z. (2021). *China-powered ICT Infrastructure: Lessons from Tanzania and Cambodia*. South African Institute of International Affairs.

Pollach, I. (2011). Online privacy as a corporate social responsibility: An empirical chapter. *Business Ethics (Oxford, England)*, 20(1), 88–102. 10.1111/j.1467-8608.2010.01611.x

Sarkar, S., & Searcy, C. (2016). Zeitgeist or chameleon? A quantitative analysis of CSR definitions. *Journal of Cleaner Production*, 135, 1423–1435. 10.1016/j.jclepro.2016.06.157

van der Merwe, J., & Achkar, Z. A. (2022). Data responsibility, corporate social responsibility, and corporate digital responsibility. *Data & Policy*, 4, e12. 10.1017/dap.2022.2

Wike, R., Fetterolf, J., Schumacher, S., & Moncus, J. J. (2021, October 21). Appendix A: Classifying democracies. *Pew Research Center's Global Attitudes Project*. https://www.pewresearch.org/global/2021/10/21/spring-2021-democracy-appendix-a-classifying-democracies/

Yin, R. K. (2009). Case Chapter Research: Design and Methods. *Sage (Atlanta, Ga.)*.

ADDITIONAL READING

Carl, K. (2021, September 28). *Corporate Digital Responsibility: Evaluating Privacy and Data Security Activities on Company-level*. 10.18420/informatik2021-065

Cheruiyot-Koech, R., & Reddy, C. D. (2022). Corporate Social Responsibility Preferences in South Africa. *Sustainability (Basel)*, 14(7), 7. Advance online publication. 10.3390/su14073792

Idemudia, U. (2014). Corporate Social Responsibility and Development in Africa: Issues and Possibilities. *Geography Compass*, 8(7), 421–435. Advance online publication. 10.1111/gec3.12143

Merwe, J., & Achkar, Z. (2022). Data responsibility, corporate social responsibility, and corporate digital responsibility. *Data & Policy*, 4, e12. Advance online publication. 10.1017/dap.2022.2

Napoli, F. (2023). Corporate Digital Responsibility: A Board of Directors May Encourage the Environmentally Responsible Use of Digital Technology and Data: Empirical Evidence from Italian Publicly Listed Companies. *Sustainability (Basel)*, 15(3), 3. Advance online publication. 10.3390/su15032539

Ndong Ntoutoume, A. G. (2023). Challenges of CSR in Sub-Saharan Africa: Clarifying the gaps between the regulations and human rights issues. *International Journal of Corporate Social Responsibility*, 8(1), 2. 10.1186/s40991-023-00079-3

KEY TERMS AND DEFINITIONS

Case Studies: In-depth investigations of particular instances or examples, used in research to illustrate broader trends or issues. The case studies within the research highlight practical obstacles to implementing data responsibility solutions in East Africa's ICT sector.

Corporate Digital Responsibility (CDR): An extension of CSR focused on ethical considerations specific to the digital age, including data privacy, security, and the ethical use of technology. CDR is pivotal for firms in addressing the nuances of digital transformation and its societal implications.

Corporate Social Responsibility (CSR): A business model where companies integrate social and environmental concerns in their operations and interactions with stakeholders. CSR in the digital context emphasizes how firms address data privacy and ethical issues, considering their societal impacts beyond profit.

Data Protection: Measures and policies aimed at safeguarding personal information from unauthorized access, use, or disclosure. Despite the digital expansion in East Africa, the lack of robust data protection regulations places individuals' privacy at risk.

Digital Connectivity: The extent to which different regions and populations can access and use the internet and other digital technologies. In East Africa, rapid digital connectivity growth has outpaced the development of regulatory frameworks, raising concerns about data protection and privacy.

Ethical Considerations: Moral principles guiding the conduct of individuals and organizations, especially pertinent to the handling of data and digital technologies. In East Africa, ethical considerations in digital expansion involve ensuring that technology use respects privacy, equality, and human rights.

Implementation Gap: The discrepancy between the theoretical frameworks or policies designed by organizations (including CSR and CDR policies) and their practical application or execution. This gap is a significant challenge in ensuring effective data protection and ethical digital practices in East Africa.

Sustainable Development Goal 9 (SDG 9): One of the 17 global goals set by the United Nations, focused on building resilient infrastructure, promoting inclusive and sustainable industrialization, and fostering innovation. The goal's relevance in East Africa underscores the importance of affordable and reliable internet access for all, highlighting the region's potential for digital value creation despite current disparities.

Conclusion

With the final part of our monograph *Balancing Human Rights, Social Responsibility and Digital Ethics*, we would like to reflect on the diverse and critical issues we have explored in this book. We have explored the complex relationship between digital technologies and ethical concerns, guided by the new guidelines of the EU Artificial Intelligence Act and the OECD AI recommendations.

This monograph addresses a wide range of issues, including the moral implications of AI, data protection, the digital divide, protection from online threats and the impact of digital transformation on sectors such as social media, e-commerce and digital healthcare. By combining ethical principles with legal perspectives, we have endeavored to provide readers of our monograph with a holistic and integrated understanding of digital ethics.

As digital technologies have profound implications for fundamental rights and social responsibility, particularly in relation to privacy, freedom of expression and access to information, the European Union's initiatives, such as the GDPR and the proposed AI Act, demonstrate its commitment to incorporating digital ethics into its legal framework. These efforts are in line with the OECD's AI Principles, which aim to ensure that AI systems uphold human rights and democratic values.

Our discussion goes beyond the technological challenges and addresses broader societal challenges. This book emphasizes the importance of adopting regulations and standards that protect human rights in an ever-evolving digital landscape. The European Commission's efforts to promote a European Union that properly addresses digital ethics and individual rights reflect a global trend that seeks responsible and ethical AI governance.

The chapters contributed by our authors have provided valuable insights and in-depth analyses of various aspects of digital ethics:

1. A Bibliometric Analysis of Digital Ethics and Human Rights - Rafiq Idris et al.
2. Artificial Intelligence: A New Tool to Protect Human Rights or an Instrument to Subdue Society by the State - Abhishek Kumar and Aparajita Mohanty
3. Interplay of Artificial Intelligence and Recruitment: The Gender Bias Effect - Shikha Saloni et al.
4. Socially Responsible Application of Artificial Intelligence in Human Resources Management - Ana Gricnik et al.

5. Legal Aspects of Digital Ethics in the Age of Artificial Intelligence - Hemendra Singh
6. Sovereignty Over Personal Data: Legal Frameworks and the Quest for Privacy in the Digital Age - Andreja Primec et al.
7. Cherish Data Privacy and Human Rights in the Digital Age: Harmonizing Innovation and Individual Autonomy - Bhupinder Singh
8. Digitalization in Corporations: Integrating Utility of Digital Technology with Accessibility and Privacy of Data - Siddharth Kanojia
9. Corporate Data Responsibility in East Africa: Bridging the Gap between Theory and Practice - Ripon Bhattacharjee et al.

Looking to the future, we believe it is essential to further understand the complex relationship between technological advances and fundamental human rights. This monograph aims to expand the discussion of digital ethics beyond traditional boundaries to include legal implications and regulatory norms to promote a deeper understanding of the topic.

To summarize, *Balancing Human Rights, Social Responsibility, and Digital Ethics* aims to be an important guide for navigating the ethical, legal, and societal dimensions of digital technologies. We hope that by exploring these critical issues, we can provide readers with the tools and insights necessary to contribute to a sustainable and just digital future. Our collective efforts can ensure that technological progress is not only innovative, but also ethical, equitable and consistent with the values that define our society.

We thank our authors for their invaluable insights and our readers for their participation in this important discourse. Together, let us continue to shine an important light on the principles of human rights, social responsibility and digital ethics in an ever-evolving digital world.

Compilation of References

. Dorfleitner, G., & Braun, D. (2019). Fintech, digitalization and blockchain: possible applications for green finance. *The rise of green finance in Europe: opportunities and challenges for issuers, investors and marketplaces*, 207-237.

. Singh, S. K., Sharma, S. K., Singla, D., & Gill, S. S. (2022). Evolving requirements and application of SDN and IoT in the context of industry 4.0, blockchain and artificial intelligence. *Software Defined Networks: Architecture and Applications*, 427-496.

Abascal, J., & Nicolle, C. (2005). Moving towards inclusive design guidelines for socially and ethically aware HCI. *Interacting with Computers*, 17(5), 484–505. 10.1016/j.intcom.2005.03.002

Abdul Rahman, N. A., Ahmi, A., Jraisat, L., & Upadhyay, A. (2022). Examining the trend of humanitarian supply chain studies: Pre, during and post COVID-19 pandemic. *Journal of Humanitarian Logistics and Supply Chain Management*, 12(4), 594–617. 10.1108/JHLSCM-01-2022-0012

Adizes, I. (2004). *Managing corporate lifecycles*. The Adizes Institute Publishing.

Afsharipour, A. (2009). Corporate governance convergence: Lessons from the Indian experience. *Nw. J. Int'l L. & Bus.*, 29, 335.

Agarwal, L. (2023). Defining Organizational AI Governance and Ethics. *SSRN*. https://ssrn.com/abstract=4553185

Agrawal, A., Gans, J. S., & Goldfarb, A. (2019). Exploring the impact of artificial intelligence: Prediction versus judgment. *Information Economics and Policy*, 47, 1–6. 10.1016/j.infoecopol.2019.05.001

Agrawal, A., Gans, J., & Goldfarb, A. (2018). *Prediction Machines: The Simple Economics of Artificial Intelligence*. Harvard Business Press.

Agrawal, R. K. (2016). A Comparative Study of Indian Companies Act, 2013 and Companies Act of Republic of Maldives, 1996. *Social Sciences*, 4(01), 2016.

Aguilera, R. V., & Cuervo-Cazurra, A. (2009). Codes of good governance. *Corporate Governance*, 17(3), 376–387. 10.1111/j.1467-8683.2009.00737.x

Ahmad, M., Rehman, A. U., Ayub, N., Alshehri, M. D., Khan, M. A., Hameed, A., & Yetgin, H. (2020). Security, usability, and biometric authentication scheme for electronic voting using multiple keys. *International Journal of Distributed Sensor Networks*, 16(7), 1550147720944025. 10.1177/1550147720944025

Ahmad, T., Zhang, D., Huang, C., Zhang, H., Dai, N., Song, Y., & Chen, H. (2021). Artificial intelligence in sustainable energy industry: Status Quo, challenges and opportunities. *Journal of Cleaner Production*, 289, 125834. 10.1016/j.jclepro.2021.125834

Ahmed, S., & Bajema, N. (2019). *Artificial Intelligence, China, Russia, and the Global Order Technological, Political, Global, and Creative Perspectives*. https://www.airuniversity.af.edu/Portals/10/AUPress/Books/B_0161_WRIGHT_ARTIFICIAL_INTELLIGENCE_CHINA_RUSSIA_AND_THE_GLOBAL_ORDER.PDF

Ahmed, O. (2018). Artificial intelligence in HR. *Int J Res Anal Rev*, 5(4), 971–978.

Ahmi, A. (2022). *Bibliometric Analysis using R for Non-Coders: A practical handbook in conducting bibliometric analysis studies using Biblioshiny for Bibliometrix R package*.

Ahn, M. J., & Chen, Y. C. (2022). Digital transformation toward AI-augmented public administration: The perception of government employees and the willingness to use AI in government. *Government Information Quarterly*, 39(2), 101664. 10.1016/j.giq.2021.101664

Al Hait, A. A. S. (2014). Jurisdiction in Cybercrimes: A comparative study. *JL Pol'y & Globalization*, 22, 75.

Alam, M. S., Dhar, S. S., & Munira, K. S. (2020). HR Professionals' intention to adopt and use of artificial intelligence in recruiting talents. *Business Perspective Review*, 2(2), 15–30.

Ali, A., Septyanto, A. W., Chaudhary, I., Al Hamadi, H., Alzoubi, H. M., & Khan, Z. F. (2022, February). Applied Artificial Intelligence as Event Horizon Of Cyber Security. In *2022 International Conference on Business Analytics for Technology and Security (ICBATS)* (pp. 1-7). IEEE. 10.1109/ICBATS54253.2022.9759076

Allen, A., & Peloza, J. (2015). Someone to Watch Over Me: The Integration of Privacy and Corporate Social Responsibility. *Business Horizons*, 58(6), 635–642. Advance online publication. 10.1016/j.bushor.2015.06.007

Almada, M., & Petit, N. (2022). The EU AI Act: Between product safety and fundamental rights. SSRN Electronic Journal. https://doi.org/10.2139/ssrn.4308072

Almendarez, L. (2013). Human Capital Theory: Implications for educational development in Belize and the Caribbean. *Caribbean Quarterly*, 59(3–4), 21–33. 10.1080/00086495.2013.11672495

Anand, D., & Khemchandani, V. (2019). Study of e-governance in India: A survey. *International Journal of Electronic Security and Digital Forensics*, 11(2), 119–144. 10.1504/IJESDF.2019.098729

Compilation of References

Anastasi, S., Madonna, M., & Monica, L. (2021). Implications of embedded artificial intelligence-machine learning on safety of machinery. *Procedia Computer Science*, 180, 338–343. 10.1016/j.procs.2021.01.171

Angelopoulos, A., Michailidis, E. T., Nomikos, N., Trakadas, P., Hatziefremidis, A., Voliotis, S., & Zahariadis, T. (2019). Tackling faults in the industry 4.0 era—A survey of machine-learning solutions and key aspects. *Sensors (Basel)*, 20(1), 109. 10.3390/s2001010931878065

Anyanwu, A., Olorunsogo, T., Abrahams, T. O., Akindote, O. J., & Reis, O. (2024). Data confidentiality and integrity: A review of accounting and cybersecurity controls in superannuation organizations. *Computer Science & IT Research Journal*, 5(1), 237–253. 10.51594/csitrj.v5i1.735

Aoun, A., Ilinca, A., Ghandour, M., & Ibrahim, H. (2021). A review of Industry 4.0 characteristics and challenges, with potential improvements using blockchain technology. *Computers & Industrial Engineering*, 162, 107746. 10.1016/j.cie.2021.107746

Arden, N. S., Fisher, A. C., Tyner, K., Lawrence, X. Y., Lee, S. L., & Kopcha, M. (2021). Industry 4.0 for pharmaceutical manufacturing: Preparing for the smart factories of the future. *International Journal of Pharmaceutics*, 602, 120554. 10.1016/j.ijpharm.2021.12055433794326

Aria, M., & Cuccurullo, C. (2017). bibliometrix: An R-tool for comprehensive science mapping analysis. *Journal of Informetrics*, 11(4), 959–975. 10.1016/j.joi.2017.08.007

Arlowski, M. (2021). Personal Data as a New Form of Intellectual Property. *Journal of the Patent and Trademark Office Society*, 102, 649.

Arnold, S. (2022). Drivers and Barriers of Digital Market Integration in East Africa: A Case Chapter of Rwanda and Tanzania. *Politics and Governance*, 10(2), 106–115. 10.17645/pag.v10i2.4922

Asadollahi-Yazdi, E., Couzon, P., Nguyen, N. Q., Ouazene, Y., & Yalaoui, F. (2020). Industry 4.0: Revolution or Evolution? *American Journal of Operations Research*, 10(06), 241–268. 10.4236/ajor.2020.106014

Attard-Frost, B., Brandusescu, A., & Lyons, K. (2024). The governance of artificial intelligence in Canada: Findings and opportunities from a review of 84 AI governance initiatives. *Government Information Quarterly*, 41(2), 101929–101929. 10.1016/j.giq.2024.101929

Awad, A., & Albaity, M. (2022). ICT and economic growth in Sub-Saharan Africa: Transmission channels and effects. *Telecommunications Policy*, 46(8), 102381. 10.1016/j.telpol.2022.102381

Awan, U., Sroufe, R., & Shahbaz, M. (2021). Industry 4.0 and the circular economy: A literature review and recommendations for future research. *Business Strategy and the Environment*, 30(4), 2038–2060. 10.1002/bse.2731

Ayşen, A., Gül, Y., Dilek, P., & Elmer. (2020). *Digital transformation in media & society*. https://westminsterresearch.westminster.ac.uk/download/cf99c72689c33eacd4e7cc8ad5680d1c0e939e2a2b9d3ffff86b6693c213a815/4794770/2B799F5AB7A14F628F90BFC458B38029.pdf#page=165

Badri, A., Boudreau-Trudel, B., & Souissi, A. S. (2018). Occupational health and safety in the industry 4.0 era: A cause for major concern? *Safety Science*, 109, 403–411. 10.1016/j.ssci.2018.06.012

Bag, S., & Pretorius, J. H. C. (2022). Relationships between industry 4.0, sustainable manufacturing and circular economy: Proposal of a research framework. *The International Journal of Organizational Analysis*, 30(4), 864–898. 10.1108/IJOA-04-2020-2120

Bamberger, K. A., & Mulligan, D. K. (2015). *Privacy on the ground: driving corporate behavior in the United States and Europe*. MIT Press. 10.7551/mitpress/9905.001.0001

Banday, M. T. (2011). Easing PAIN with digital signatures. *International Journal of Computer Applications*, 975, 8887.

Bankar, S., & Shukla, K. (2023). Performance Management and Artificial Intelligence: A Futuristic Conceptual Framework. In *Contemporary Studies of Risks in Emerging Technology, Part B* (pp. 341–360). Emerald Publishing Limited. 10.1108/978-1-80455-566-820231019

Bankins, S. (2021). The ethical use of artificial intelligence in human resource management: A decision-making framework. *Ethics and Information Technology*, 23(4), 841–845. 10.1007/s10676-021-09619-6

Bankins, S., & Formosa, P. (2023). The ethical implications of artificial intelligence (AI) for meaningful work. *Journal of Business Ethics*, 185(4), 1–16. 10.1007/s10551-023-05339-7

Barkema, H., Chen, X.-P., George, G., Luo, Y., & Tsui, A. (2015). West Meets East: New Concepts and Theories. *Academy of Management Journal*, 58(2), 460–479. 10.5465/amj.2015.4021

Barker, M. (2014). In Vossler, A., & Moller, N. (Eds.), *Doing a literature review* (pp. 61–73). Sage. https://uk.sagepub.com/en-gb/eur/the-counselling-and-psychotherapy-research-handbook/book239261

Basimanyane, D. (2022). The Regulatory Dilemma on Mass Communications Surveillance and the Digital Right to Privacy in Africa: The Case of South Africa. *African Journal of International and Comparative Law*, 30(3), 361–382. 10.3366/ajicl.2022.0414

Bayram, O., Talay, I., & Feridun, M. (2022). Can FinTech promote sustainable finance? Policy lessons from the case of Turkey. *Sustainability (Basel)*, 14(19), 12414. 10.3390/su141912414

Beattie, G., & Johnson, P. J. P. (2012). Possible unconscious bias in recruitment and promotion and the need to promote equality. *Perspectives*, 16(1), 7–13. 10.1080/13603108.2011.611833

Beaven, Z., & Laws, C. (2004). Principles and applications in ticketing and reservations management. In Yeoman, I., Robertson, M., Ali-Knight, J., Drummond, S., & McMahon-Beattie, U. (Eds.), *Festival and Events Management* (pp. 183–201). Butterworth-Heinemann. 10.1016/B978-0-7506-5872-0.50017-X

Bell, E., Bryman, A., & Harley, B. (2018). *Business Research Methods*. Oxford University Press.

Compilation of References

Belloc, F. (2012). Corporate governance and innovation: A survey. *Journal of Economic Surveys*, 26(5), 835–864. 10.1111/j.1467-6419.2011.00681.x

Beneduce, G. (2020). *Artificial intelligence in recruitment: just because it's biased, does it mean it's bad?* NOVA—School of Business and Economics.

Bennett, C. J. (2018). The European General Data Protection Regulation: An Instrument for the Globalization of Privacy Standards? *Information Polity*, 23(2), 239–246. 10.3233/IP-180002

Bergström, A., & Berghäll, E. (2021). Public certificate management: An analysis of policies and practices used by CAs.

Berryhill, J., Kok Heang, K., Clogher, R., & Mcbride, K. (2019). Hello, World: Artificial intelligence and its use in the public sector. *OECD Working Papers on Public Governance*, 36(19934351). 10.1787/19934351

Bharadiya, J. (2023). Artificial Intelligence in Transportation Systems A Critical Review. *American Journal of Computing and Engineering*, 6(1), 34–45. 10.47672/ajce.1487

Bhatia, T. (2018). *Artificial intelligence in HR*. HR Strategy and Planning Excellence Essentials.

Bhatnagar, S. (2009). *Unlocking e-government potential: Concepts, cases and practical insights*. SAGE Publications India. 10.4135/9781446270202

Bhowmik, D. (2022). An Introduction to Climate Fintech. *European Journal of Science. Innovation and Technology*, 2(4), 24–35.

Bhuiyan, A. B., Ali, M. J., Zulkifli, N., & Kumarasamy, M. M. (2020). Industry 4.0: Challenges, opportunities, and strategic solutions for Bangladesh. *International Journal of Business and Management Future*, 4(2), 41–56. 10.46281/ijbmf.v4i2.832

Bibri, S. E., & Bibri, S. E. (2015). Ethical implications of AmI and the IoT: risks to privacy, security, and trust, and prospective technological safeguards. *The shaping of ambient intelligence and the Internet of Things: historico-epistemic, socio-cultural, politico-institutional and eco-environmental dimensions*, 217-238.

Bienvenido-Huertas, D., Farinha, F., Oliveira, M. J., Silva, E. M. J., & Lança, R. (2020). Comparison of artificial intelligence algorithms to estimate sustainability indicators. *Sustainable Cities and Society*, 63, 102430. 10.1016/j.scs.2020.102430

Bin Amin, S., Taghizadeh-Hesary, F., & Khan, F. (2022). Facilitating green digital finance in Bangladesh: Importance, prospects, and implications for meeting the SDGs. In *Green Digital Finance and Sustainable Development Goals* (pp. 143–165). Springer Nature Singapore. 10.1007/978-981-19-2662-4_7

Birdi, K., Clegg, C., Patterson, M., Robinson, A., Stride, C. B., Wall, T. D., & Wood, S. J. (2008). The impact of human resource and operational management practices on company productivity: A longitudinal study. *Personnel Psychology*, 61(3), 467–501. 10.1111/j.1744-6570.2008.00136.x

Bissio, R. (2018). Vector of hope, source of fear. *Spotlight Sustain. Dev.*, 77-86.

Black & van Esch. (2020). AI-enabled recruiting: what is it and how should a manager use it? *Bus Horiz, 63*(2), 215–226. https://doi.org/.2019.12.00110.1016/j.bushor

Bleakley, C. (2020). Artificial Intelligence Emerges. *Poems That Solve Puzzles*, 75–92. 10.1093/oso/9780198853732.003.0005

Blobel, B. (2020, September). Application of industry 4.0 concept to health care. In *pHealth 2020:Proceedings of the 17th International Conference on Wearable Micro and Nano Technologies for Personalized Health* (Vol. 273, p. 23). IOS Press.

Blume, J. (2018). Contextual extraterritoriality analysis of the DPIA and DPO provisions in the GDPR. *Georgetown Journal of International Law*, 49(4), 1425–1460.

Bogen M, Rieke A (2018) Help wanted: an examination of hiring algorithms, equity, and bias. Academic Press.

Bokhari, S. A. A., & Myeong, S. (2023). The influence of artificial intelligence on e-Governance and cybersecurity in smart cities: A stakeholder's perspective. *IEEE Access : Practical Innovations, Open Solutions*, 11, 69783–69797. 10.1109/ACCESS.2023.3293480

Bolognini, L., & Balboni, P. (2019). *IoT and Cloud Computing: Specific Security and Data Protection Issues*. In the Internet of Things Security and Data Protection., 10.1007/978-3-030-04984-3_4

Bongomin, O., Gilibrays Ocen, G., Oyondi Nganyi, E., Musinguzi, A., & Omara, T. (2020). Exponential disruptive technologies and the required skills of industry 4.0. *Journal of Engineering*, 2020, 1–17. 10.1155/2020/8090521

Bonner, L. (2012). Cyber risk: How the 2011 Sony data breach and the need for cyber risk insurance policies should direct the federal response to rising data breaches. *Wash. UJL & Pol'y*, 40, 257.

Borisova, A. (2019). The Data Protection Officer as an Instrument for Compliance with the Accountability Principle Under the GDPR. *Sustainable Development GOALS 2030: Challenges for South and Eastern European Countries and the Black Sea Region*, 346.

Bornstein, S. (2018). Antidiscriminatory algorithms. *Alabama Law Review*, 70, 519.

Borowski, P. F. (2021). Digitization, digital twins, blockchain, and industry 4.0 as elements of management process in enterprises in the energy sector. *Energies*, 14(7), 1885. 10.3390/en14071885

Borowski, P. F. (2021). Innovative processes in managing an enterprise from the energy and food sector in the era of industry 4.0. *Processes (Basel, Switzerland)*, 9(2), 381. 10.3390/pr9020381

Bose, R. (2009). Advanced analytics: Opportunities and challenges. *Industrial Management & Data Systems*, 109(2), 155–172. 10.1108/02635570910930073

Bowen, D., & Ostroff, C. (2004). Understanding HRM–firm performance linkages: The role of the "strength of the HRM system". *Academy of Management Review*, 22(2), 203–221.

Compilation of References

Boxall, P. (2013). Mutuality in the management of human resources: Assessing the quality of alignment in employment relationships. *Human Resource Management Journal*, 1(23), 3–17. 10.1111/1748-8583.12015

Boxall, P. (2014). The future of employment relations from the perspective of human resource management. *The Journal of Industrial Relations*, 56(4), 578–593. 10.1177/0022185614527980

Boxall, P., & Purcell, J. (2011). *Strategy and Human Resource Management*. Palgrave Macmillan.

Brock, J. K. U., & Von Wangenheim, F. (2019). Demystifying AI: What digital transformation leaders can teach you about realistic artificial intelligence. *California Management Review*, 61(4), 110–134. 10.1177/1536504219865226

Brown, T., Mann, B., Ryder, N., Subbiah, M., Kaplan, J. D., Dhariwal, P., Neelakantan, A., Shyam, P., Sastry, G., & Askell, A. (2020). Language models are few-shot learners. *Advances in Neural Information Processing Systems*, 33, 1877–1901. 10.48550/arXiv.2005.14165

Brundtland, G. H. (1987). *Report of the World Commission on environment and development: our common future*. United Nations.

Bryman, A. (2012). *Social Research Methods*. OUP Oxford.

Brynjolfsson, E., & McAfee, A. (2014). *The Second Machine Age: Work, Progress, and Prosperity in a Time of Brilliant Technologies*. WW Norton & Company.

Bučiūnienė, I., & Kazlauskaitė, R. (2012). The linkage between HRM, CSR and performance outcomes. *Baltic Journal of Management*.

Buck, B., & Morrow, J. (2018). AI, performance management and engagement: Keeping your best their best. *Strategic HR Review*, 17(5), 261–262. 10.1108/SHR-10-2018-145

Bujold, A., Roberge-Maltais, I., Parent-Rocheleau, X., Boasen, J., Sénécal, S., & Léger, P. M. (2023). Responsible artificial intelligence in human resources management: A review of the empirical literature. *AI and Ethics*, 1–16. 10.1007/s43681-023-00325-1

Busuioc, M. (2020). Accountable Artificial Intelligence: Holding Algorithms to Account. *Public Administration Review*, 81(5), 825–836. Advance online publication. 10.1111/puar.1329334690372

Bygrave, L. A. (2017). Data protection by design and by default: Deciphering the EU's legislative requirements. *Oslo Law Review*, 4(2), 105–122. 10.18261/issn.2387-3299-2017-02-03

Byun, Y. C. (2022). *Knowledge Discovery and Cryptocurrency Price Prediction Based on Blockchain Framework* (Doctoral dissertation,).

Cain, G. G. (1986). The economic analysis of labor market discrimination: A survey. *Handbook Labor Econ*, 1, 693–785. 10.1016/S1573-4463(86)01016-7

Cain, J., & Romanelli, F. (2009). E-professionalism: A new paradigm for a digital age. *Currents in Pharmacy Teaching & Learning*, 1(2), 66–70. 10.1016/j.cptl.2009.10.001

Calegari, R., Ciatto, G., Mascardi, V., & Omicini, A. (2021). Logic-based technologies for multi-agent systems: A systematic literature review. *Autonomous Agents and Multi-Agent Systems*, 35(1), 1. 10.1007/s10458-020-09478-3

Cambronero, M.-E., Martínez, M. I., de la Vara, J. L., Cebrian, D., & Valero, V. (2022). GDPRValidator: A tool to enable companies using cloud services to be GDPR compliant. *PeerJ*, 8, e1171. 10.7717/peerj-cs.117136532816

Camilleri, M. (2023). Artificial Intelligence Governance: Ethical Considerations and Implications for Social Responsibility. *Expert Systems: International Journal of Knowledge Engineering and Neural Networks*. Advance online publication. 10.1111/exsy.13406

Campbell-Verduyn, M. (2023). Conjuring a cooler world? Imaginaries of improvement in blockchain climate finance experiments. *Environment and Planning C: Politics and Space*.

Carbado, D. W., Crenshaw, K. W., Mays, V. M., & Tomlinson, B. (2013). INTERSECTIONALITY: Mapping the movements of a theory. *Du Bois Review*, 10(2), 303–312. 10.1017/S1742058X1300034925285150

Carbonero, F., Davies, J., Ernst, E., Fossen, F. M., Samaan, D., & Sorgner, A. (2023). The impact of artificial intelligence on labor markets in developing countries: A new method with an illustration for Lao PDR and urban Viet Nam. *Journal of Evolutionary Economics*, 33(3), 707–736. Advance online publication. 10.1007/s00191-023-00809-736811092

Carroll, A. B. (1991). The pyramid of corporate social responsibility: Toward the moral management of organizational stakeholders. *Business Horizons*, 34(4), 39–48. 10.1016/0007-6813(91)90005-G

Casella, B., & Formenti, L. (2018). FDI in the digital economy: A shift to asset-light international footprints. *Transnational Corporations*, 25(1), 101–130. 10.18356/cb688e94-en

Caudill, E. M., & Murphy, P. E. (2000). Consumer online privacy: Legal and ethical issues. *Journal of Public Policy & Marketing*, 19(1), 7–19. 10.1509/jppm.19.1.7.16951

Chander, B., Pal, S., De, D., & Buyya, R. (2022). Artificial intelligence-based internet of things for industry 5.0. *Artificial intelligence-based internet of things systems*, 3-45.

Chang, Y. L., & Ke, J. (2023). Socially Responsible Artificial Intelligence Empowered People Analytics: A Novel Framework Towards Sustainability. *Human Resource Development Review*.

Chan-Olmsted, S. M. (2019). A Review of Artificial Intelligence Adoptions in the Media Industry. *International Journal on Media Management*, 21(3-4), 193–215. 10.1080/14241277.2019.1695619

Charles, V., Rana, N. P., & Carter, L. (2022). Artificial Intelligence for data-driven decision-making and governance in public affairs. *Government Information Quarterly*, 39(4), 101742. 10.1016/j.giq.2022.101742

Chaudhary, I., & Taneja, V. (2018). Analysis of Company Law Amendments: As a Progressive Step towards Better Corporate Governance. *RGNUL Fin. & Mercantile L. Rev.*, 5, 241.

Compilation of References

Chen, Y., & Volz, U. (2021). Scaling up sustainable investment through blockchain-based project bonds. *ADB-IGF Special Working Paper Series "Fintech to Enable Development, Investment, Financial Inclusion, and Sustainability.*

Cheng, L., & Liu, H. (2023). *Socially Responsible AI: Theories and Practices*. World Scientific. 10.1142/13150

Cheng, L., Varshney, K. R., & Liu, H. (2021). Socially Responsible AI Algorithms: Issues, Purposes, and Challenges. *Journal of Artificial Intelligence Research*, 71, 1137–1181. 10.1613/jair.1.12814

Cheng-Tek Tai, M. (2020). The impact of artificial intelligence on human society and bioethics. *Tzu Chi Medical Journal*, 32(4), 339–343. 10.4103/tcmj.tcmj_71_20

Chen, Y., Lu, Y., Bulysheva, L., & Kataev, M. Y. (2022). Applications of blockchain in industry 4.0: A review. *Information Systems Frontiers*, 1–15. 10.1007/s10796-022-10248-7

Chen, Z. (2023). Collaboration among recruiters and artificial intelligence: Removing human prejudices in employment. *Cognition Technology and Work*, 25(1), 135–149. 10.1007/s10111-022-00716-036187287

Chen, Z., & Zhu, Y. (2017, June). Personal archive service system using blockchain technology: Case study, promising and challenging. In *2017 IEEE International Conference on AI & Mobile Services (AIMS)* (pp. 93-99). IEEE. 10.1109/AIMS.2017.31

Chowdhury, S., Dey, P., Joel-Edgar, S., Bhattacharya, S., Rodriguez-Espindola, O., Abadie, A., & Truong, L. (2023). Unlocking the value of artificial intelligence in human resource management through AI capability framework. *Human Resource Management Review*, 33(1), 100899. 10.1016/j.hrmr.2022.100899

Christenhusz, G. M., Devriendt, K., & Dierickx, K. (2013). Disclosing incidental findings in genetics contexts: A review of the empirical ethical research. *European Journal of Medical Genetics*, 56(10), 529–540. 10.1016/j.ejmg.2013.08.00624036277

Chueca Vergara, C., & Ferruz Agudo, L. (2021). Fintech and sustainability: Do they affect each other? *Sustainability (Basel)*, 13(13), 7012. 10.3390/su13137012

Cioffi, R., Travaglioni, M., Piscitelli, G., Petrillo, A., & De Felice, F. (2020). Artificial intelligence and machine learning applications in smart production: Progress, trends, and directions. *Sustainability (Basel)*, 12(2), 492. 10.3390/su12020492

Cirillo, D., Catuara-Solarz, S., Morey, C., Guney, E., Subirats, L., Mellino, S., Gigante, A., Valencia, A., Rementeria, M. J., Chadha, A. S., & Nikolaos, M. (2020). Sex and gender differences and biases in artificial intelligence for biomedicine and healthcare. *Digital Medicine*, 8(3). https://www.nature.com/articles/s41746-020-0288-5

Clifton, J., Glasmeier, A., & Gray, M. (2020). When machines think for us: The consequences for work and place. *Cambridge Journal of Regions, Economy and Society*, 13(1), 3–23. 10.1093/cjres/rsaa004

Cognizant. (2024). *Human-centered artificial intelligence.* Retrieved from Cognizant: https://www.cognizant.com/us/en/glossary/human-centered-ai

Colbert, A., George, G., & Yee, N. (2016). The digital workforce and the workplace of the future. *Academy of Management Journal, 59*(3), 731-739.

Confederation of Indian Industry. (2023, May 17). *Artificial Intelligence in Governance.* CII Blog. https://ciiblog.in/artificial-intelligence-in-governance/

Considerati. (2021). European Data Act & Data Governance Act: the kids with gamechanging potential. https://www.considerati.com/publications/eu-data-act-data-governance-act.html

Coppola, L., Cianflone, A., Grimaldi, A. M., Incoronato, M., Bevilacqua, P., Messina, F., Baselice, S., Soricelli, A., Mirabelli, P., & Salvatore, M. (2019). Biobanking in health care: Evolution and future directions. *Journal of Translational Medicine*, 17(1), 1–18. 10.1186/s12967-019-1922-331118074

Corrigan, T. (2020). *Africa's ICT Infrastructure: Its Present and Prospects.* https://policycommons.net/artifacts/1451480/africas-ict-infrastructure/2083288/

Council of Europe. (2019). *Artificial intelligence and data protection.* Council of Europe Publishing; Council of Europe. https://edoc.coe.int/en/artificial-intelligence/8254-artificial-intelligence-and-data-protection.html

Craig, P., & de Bùrca, G. (2020). *EU Law. Text, Cases, and Materials.* Oxford University Press, 10.1093/he/9780198856641.001.0001

Craig, P., & de Bùrca, G. (2021). *The Evolution of EU Law.* Oxford University Press. 10.1093/oso/9780192846556.001.0001

Curzon, J., Kosa, T. A., Akalu, R., & El-Khatib, K. (2021). Privacy and Artificial Intelligence. *IEEE Transactions on Artificial Intelligence*, 2(2), 1–1. 10.1109/TAI.2021.3088084

Custers, B., Fosch-Villaronga, E., van der Hof, S., Schermer, B. W., Sears, A. M., & Tamò-Larrieux, A. (2022). The Role of Consent in an Algorithmic Society – Its Evolution, Scope, Failings and Re-conceptualisation. In Kostas, E., Leenes, R., & Kamara, I. (Eds.), *Research Handbook on EU Data Protection* (pp. 455–473). Edward Elgar Publishing. 10.4337/9781800371682.00027

Dalenberg, D. J. (2018). Preventing discrimination in the automated targeting of job advertisements. *Computer Law & Security Report*, 34(3), 615–627. 10.1016/j.clsr.2017.11.009

Damnjanović, M. (2018). Relation between corporate social responsibility reporting and financial indicators of Serbian companies. *the Proceedings of the 5th International Scientific Conference on Contemporary Issues in Economics, Business and Management (EBM 2018)*, 473-480.

Dartey-Baah, K., & Amponsah-Tawiah, K. (2011). Exploring the limits of Western Corporate Social Responsibility Theories in Africa. *International Journal of Business and Social Science*, 2.

Compilation of References

Dave, M., & Patel, N. (2023). Artificial intelligence in healthcare and education. *British Dental Journal*, 234(10), 761–764. 10.1038/s41415-023-5845-237237212

Davenport, T., Guha, A., Grewal, D., & Bressgott, T. (2020, October). How artificial intelligence will change the future of marketing. *Journal of the Academy of Marketing Science*, 48(1), 24–42. 10.1007/s11747-019-00696-0

de Almeida, P. G. R., dos Santos, C. D., & Farias, J. S. (2021). Artificial Intelligence Regulation: A Framework for Governance. *Ethics and Information Technology*, 23(3), 505–525. 10.1007/s10676-021-09593-z

de la Peña Zarzuelo, I., Soeane, M. J. F., & Bermúdez, B. L. (2020). Industry 4.0 in the port and maritime industry: A literature review. *Journal of Industrial Information Integration*, 20, 100173. 10.1016/j.jii.2020.100173

De Man, Y., Wieland-Jorna, Y., Torensma, B., de Wit, K., Francke, A. L., Oosterveld-Vlug, M. G., & Verheij, R. A. (2023). Opt-In and Opt-Out Consent Procedures for the Reuse of Routinely Recorded Health Data in Scientific Research and Their Consequences for Consent Rate and Consent Bias: Systematic Review. *Journal of Medical Internet Research*, 25, e42131. 10.2196/4213136853745

De Sousa, W. G., de Melo, E. P., Bermejo, P. D., Fairas, R. S., & Gomes, A. O. (2019). How and where is artificial intelligence in the public sector going? *A literature review and research agenda. Government Information Quarterly*, 36(4), 101392. 10.1016/j.giq.2019.07.004

de-Lima-Santos, M.-F., & Ceron, W. (2021). Artificial Intelligence in News Media: Current Perceptions and Future Outlook. *Journalism and Media*, 3(1), 13–26. 10.3390/journalmedia3010002

Dell'Erba, M. (2021). Sustainable digital finance and the pursuit of environmental sustainability. *Sustainable Finance in Europe: Corporate Governance, Financial Stability and Financial Markets*, 61-81.

Deshpande, R., & Webster, F. E.Jr. (1989). Organizational Culture and Marketing: Defining the Research Agenda. *Journal of Marketing*, 53(1), 3–15. 10.1177/002224298905300102

Diggelmann, O., & Cleis, M. N. (2014). How the Right to Privacy Became a Human Right. *Human Rights Law Review*, 14(3), 441–458. 10.1093/hrlr/ngu014

Dinesh Arokia Raj, A., Jha, R. R., Yadav, M., Sam, D., & Jayanthi, K. (2024). Role of Blockchain and Watermarking Toward Cybersecurity. In *Multimedia Watermarking: Latest Developments and Trends* (pp. 103–123). Springer Nature Singapore. 10.1007/978-981-99-9803-6_6

Directive 2002/58/EC of the European Parliament and of the Council of 12 July 2002 concerning the processing of personal data and the protection of privacy in the electronic communications sector (Directive on privacy and electronic communications),OJ L 201, 31.7.2002, 37–47.

Directive 95/46/EC of the European Parliament and of the Council of 24 October 1995 on the protection of individuals with regard to the processing of personal data and on the free movement of such data, OJ L 281, 23.11.1995, 31–50.

Dirican, C. (2015). The Impacts of Robotics, Artificial Intelligence On Business and Economics. *Procedia - Social and Behavioral Sciences, 195*, 564–573. 10.1016/j.sbspro.2015.06.134

Diro, A., Zhou, L., Saini, A., Kaisar, S., & Hiep, P. C. (2024). Leveraging zero knowledge proofs for blockchain-based identity sharing: A survey of advancements, challenges and opportunities. *Journal of Information Security and Applications*, 80, 103678. 10.1016/j.jisa.2023.103678

Dumas, J. (2019). General Data Protection Regulation (GDPR): Prioritizing resources. *Seattle University Law Review*, 42(3), 1115–1128.

Dwivedi, Y. K., Hughes, L., Ismagilova, E., Aarts, G., Coombs, C., Crick, T., Duan, Y., Dwivedi, R., Edwards, J., Eirug, A., Galanos, V., Ilavarasan, P. V., Janssen, M., Jones, P., Kar, A. K., Kizgin, H., Kronemann, B., Lal, B., Lucini, B., & Williams, M. D. (2021). Artificial Intelligence (AI): Multidisciplinary Perspectives on Emerging challenges, opportunities, and Agenda for research, Practice and Policy. *International Journal of Information Management*, 57(101994). Advance online publication. 10.1016/j.ijinfomgt.2019.08.002

Eckardt, M., & Kerber, W. (2024). Property Rights Theory, Bundles of Rights on IoT Data, and the EU Data Act. 10.1007/s10657-023-09791-8

Eckhaus, E. (2018). Measurement of organizational happiness. *Advances in Human Factors, Business Management and Leadership*, 266-278.

El Ioini, N., & Pahl, C. (2018). A review of distributed ledger technologies. On the Move to Meaningful Internet Systems. OTM 2018 Conferences: Confederated International Conferences: CoopIS, C&TC, and ODBASE 2018, Valletta, Malta, October 22-26, 2018, Proceedings, Part II, Springer. 10.1007/978-3-030-02671-4_16

Epstein, M., Buhovac, A., Elkington, J., & Leonard, H. (2017). Making Sustainability Work: Best Practices in Managing and Measuring Corporate Social, Environmental, and Economic Impacts. In *Making Sustainability Work: Best Practices in Managing and Measuring Corporate Social, Environmental and Economic Impacts* (p. 305). 10.4324/9781351276443

Epstein, M. J., & Buhovac, A. R. (2014). *Making Sustainability Work: Best Practices in Managing and Measuring Corporate Social, Environmental, and Economic Impacts*. Berrett-Koehler Publishers.

Erbacci, A., Deighton-Smith, R., & Kauffmann, C. (2016). Promoting inclusive growth through better regulation: The role of regulatory impact assessment. *OECD Regulatory Policy Working Papers, 2016*, 1–55. 10.1787/24140996

Erdos, D. (2021). Comparing Constitutional Privacy and Data Protection Rights within the EU University of Cambridge, Faculty of Law Research Paper No. 21/2021, Available at *SSRN*: https://ssrn.com/abstract=3843653 or http://dx.doi.org/10.2139/ssrn.3843653

Eubanks, B. (2022). *Artificial Intelligence for HR: Use AI to Support and Develop a Successful Workforce* (2nd ed.). Kogan Page Publishers.

Compilation of References

European Comission. (2001). *GREEN PAPER: Promoting a European framework for Corporate Social Responsibility*. Commission of the European Communities.

European Commission. (2020a). Communication From the Commission to the European Parliament, the Council, the European Economic and Social Committee and the Committee of the Regions. Shaping Europe's digital future, COM(2020) 67 final, 19. 2. 2020.

European Commission. (2020b). Communication From the Commission to the European Parliament, The Council, The European Economic and Social Committee and the Committee of the Regions a European Strategy for Data, COM/2020/66 final, 19. 2. 2020.

European Commission. (2023). Artificial Intelligence act. Available: https://eur-lex.europa.eu/legal-content/EN/TXT/?uri=CELEX%3A32023L2225&qid=1714133317618

European Commission. (2024). European Data Governance Act. https://digital-strategy.ec.europa.eu/en/policies/data-governance-act

European Data Protection Supervisor. (2021). TechDispatch #3/2020 - Personal Information Management Systems, https://www.edps.europa.eu/data-protection/our-work/publications/techdispatch/techdispatch-32020-personal-information_en

Federspiel, F., Mitchell, R., Asokan, A., Umana, C., & McCoy, D. (2023). Threats by artificial intelligence to human health and human existence. *BMJ Global Health*, 8(5), e010435. 10.1136/bmjgh-2022-01043537160371

Feige, I. (2019). *What is AI Safety? Towards a better understanding*. Retrieved from https://faculty.ai/blog/what-is-ai-safety/

Felsberger, A., Qaiser, F. H., Choudhary, A., & Reiner, G. (2022). The impact of Industry 4.0 on the reconciliation of dynamic capabilities: Evidence from the European manufacturing industries. *Production Planning and Control*, 33(2-3), 277–300. 10.1080/09537287.2020.1810765

Fenwick, M., & Vermeulen, E. P. (2020). Banking and regulatory responses to FinTech revisited-building the sustainable financial service 'ecosystems' of tomorrow. *Singapore Journal of Legal Studies*, (Mar), 165-189.

Fernandes, T., & Oliveira, E. (2021). Understanding consumers' acceptance of automated technologies in service encounters: Drivers of digital voice assistants adoption. *Journal of Business Research*, 122, 180–191. 10.1016/j.jbusres.2020.08.058

Feuerriegel, S., Dolata, M., & Schwabe, G. (2020). Fair AI: Challenges and opportunities. *Business & Information Systems Engineering*, 62(4), 379–384. 10.1007/s12599-020-00650-3

Fiesler, C., & Hallinan, B. (2018, April). "We Are the Product" Public Reactions to Online Data Sharing and Privacy Controversies in the Media. In *Proceedings of the 2018 CHI conference on human factors in computing systems* (pp. 1-13).

Fifka, M. (2009). Towards a More Business-Oriented Definition of Corporate Social Responsibility: Discussing the Core Controversies of a Well-Established Concept. *Journal of Service Science and Management -. Journal of Service Science and Management*, 02, 312–321. 10.4236/jssm.2009.24037

Figueroa-Armijos, M., Clark, B. B., & da Motta Veiga, S. P. (2022). Ethical perceptions of AI in hiring and organizational trust: The role of performance expectancy and social influence. *Journal of Business Ethics*, 1–19.

Filho, W. L., Kovaleva, M., Tsani, S., îrcă, D., Shiel, C., Dinis, M. P., Nicolau, M., Sima, M., Fritzen, B., Salvia, A. L., Minhas, A., Kozlova, V., Doni, F., Spiteri, J., Gupta, T., Wakunuma, K., Sharma, M., Barbir, J., Shulla, K., & Tripathi, S. (2022). Promoting gender equality across the sustainable development goals. *Environment, Development and Sustainability*, 25(12), 14177–14198. 10.1007/s10668-022-02656-1

Finance, A. T. C. C. (2015). *Industry 4.0 Challenges and solutions for the digital transformation and use of exponential technologies. Finance, audit tax consulting corporate.* Swiss.

Fraga-Lamas, P., Lopes, S. I., & Fernández-Caramés, T. M. (2021). Green IoT and edge AI as key technological enablers for a sustainable digital transition towards a smart circular economy: An industry 5.0 use case. *Sensors (Basel)*, 21(17), 5745. 10.3390/s2117574534502637

Francis, L. P., & Francis, J. G. (2017). Data Reuse and the Problem of Group Identity. Studies in Law, Politics, and Society, 73. University of Utah College of Law Research Paper No. 311. 10.2139/ssrn.3371134

Frank, M. R., Autor, D., Bessen, J. E., Brynjolfsson, E., Cebrian, M., Deming, D. J., Feldman, M., Groh, M., Lobo, J., Moro, E., Wang, D., Youn, H., & Rahwan, I. (2019). Toward Understanding the Impact of Artificial Intelligence on Labor. *Proceedings of the National Academy of Sciences of the United States of America*, 116(14), 6531–6539. 10.1073/pnas.1900949116309 10965

Fritz, G., & Dannhausen, E. (2024). When GDPR and Data Act clash: what Businesses need to know. https://www.lexology.com/library/detail.aspx?g=6b3f477e-ab49-415d-bfb4-1d27e8b8db53

Gadekar, R., Sarkar, B., & Gadekar, A. (2022). Key performance indicator based dynamic decision-making framework for sustainable Industry 4.0 implementation risks evaluation: Reference to the Indian manufacturing industries. *Annals of Operations Research*, 318(1), 189–249. 10.1007/s10479-022-04828-835910040

Gaglione, G. S.Jr. (2019). The equifax data breach: An opportunity to improve consumer protection and cybersecurity efforts in America. *Buff. L. Rev.*, 67, 1133.

Gasser, U., Ienca, M., Scheibner, J., Sleigh, J., & Vayena, E. (2020). Digital tools against COVID-19: Taxonomy, ethical challenges, and navigation aid. *The Lancet. Digital Health*, 2(8), e425–e434. 10.1016/S2589-7500(20)30137-032835200

Compilation of References

Gen, H. (2023). *Implementing Artificial Intelligence in the Indian Military.* Www.delhipolicygroup.org. https://www.delhipolicygroup.org/publication/policy-briefs/implementing-artificial-intelligence-in-the-indian-military.html

Germann, M., & Serdült, U. (2017). Internet voting and turnout: Evidence from Switzerland. *Electoral Studies*, 47, 1–12. 10.1016/j.electstud.2017.03.001

Ghatak, R. (2022). *People Analytics: Data to Decisions.* Springer. 10.1007/978-981-19-3873-3

Ghosh, S., Majumder, S., & Peng, S. L. (2023). Aritificial Intelligence Techniques in Human Resource Management. In *An empirical study on adoption of Aritificial Intelligence in Human Resource Management* (pp. 29-85). CRC Press.

Giudici, P., & Raffinetti, E. (2023). SAFE Artificial Intelligence in finance. *Finance Research Letters*, 56, 104088. 10.1016/j.frl.2023.104088

Go, E., & Sundar, S. (2019). Humanizing chatbots: The effects of visual, identity and conversational cues on humanness perceptions. *Computers in Human Behavior*, 97, 304–316. 10.1016/j.chb.2019.01.020

Gömann, M. (2017). The new territorial scope of EU data protection law: Deconstructing a revolutionary achievement. *Common Market Law Review*, 54(2), 567–590. 10.54648/COLA2017035

Goswami, O. (2002). Corporate governance in India. *Taking action against corruption in Asia and the Pacific*, 85-106.

Government of Canada. (2019). Personal Information Protection and Electronic Documents Act. https://laws-lois.justice.gc.ca/eng/acts/p-8.6/

Graef, I., & Gellert, R. (2021). The European Commission's proposed Data Governance Act: some initial reflections on the increasingly complex EU regulatory puzzle of stimulating data sharing. TILEC Discussion Paper No. DP2021-006. 10.2139/ssrn.3814721

Graef, I., Gellert, R., & Husovec, M. (2018). Towards a Holistic Regulatory Approach for the European Data Economy: Why the Illusive Notion of Non-Personal Data is Counterproductive to Data Innovation. TILEC Discussion Paper No. 2018-029. 10.2139/ssrn.3256189

Graßmann, C., & Schermuly, C. C. (2021). Coaching with artificial intelligence: Concepts and capabilities. *Human Resource Development Review*, 20(1), 106–126. 10.1177/1534484320982891

Gray, H. (2020). *Congruent Regulation: Designing the Optimal Australian Utility Regime* (Doctoral dissertation, The Australian National University (Australia)).

Gray, M. L., & Suri, S. (2019). *Ghost work: How to stop Silicon Valley from building a new global underclass.* Eamon Dolan Books.

Greenwood, M. (2007). Stakeholder engagement: Beyond the myth of corporate responsibility. *Journal of Business Ethics*, 74(4), 315–327. 10.1007/s10551-007-9509-y

Guest, D. E. (1997). Human resource management and performance: A review and research agenda. *International Journal of Human Resource Management*, 8(3), 263–276. 10.1080/095851997341630

Guest, D. E., Michie, J., Sheehan, M., & Conway, N. (2000). *Employee Relations, HRM and Business Performance: An analysis of the 1998 Workplace Employee Relations Survey*. CIPD.

Guo, P., Xiao, K., Ye, Z., Zhu, H., & Zhu, W. (2022). Intelligent career planning via stochastic subsampling reinforcement learning. *Scientific Reports*, 12(1).35585154

Gupta, R. (2023). Industry 4.0 adaption in indian banking Sector—A review and agenda for future research. *Vision (Basel)*, 27(1), 24–32. 10.1177/097226292199682936977304

Habib, G., Sharma, S., Ibrahim, S., Ahmad, I., Qureshi, S., & Ishfaq, M. (2022). Blockchain technology: Benefits, challenges, applications, and integration of blockchain technology with cloud computing. *Future Internet*, 14(11), 341. 10.3390/fi14110341

Hacker, P., Engel, A., & Mauer, M. (2023). Regulating ChatGPT and other large generative AI models. *Proceedings of the 2023 ACM Conference on Fairness, Accountability, and Transparency*. 10.1145/3593013.3594067

Harris, A. (2018). A conversation with masterminds in blockchain and climate change. In *Transforming climate finance and green investment with blockchains* (pp. 15–22). Academic Press. 10.1016/B978-0-12-814447-3.00002-1

Harris, R. (2020). Forging path towards meaningful digital privacy: Data monetization and the CCPA. *Loyola of Los Angeles Law Review*, 54(1), 197–234.

Hasan, I., & Tucci, C. (2010). The innovation-economic growth nexus: Global evidence. *Research Policy*, 39(10), 1264–1276. 10.1016/j.respol.2010.07.005

Hassan, M. K., Rabbani, M. R., & Ali, M. A. M. (2020). Challenges for the Islamic Finance and banking in post COVID era and the role of Fintech. *Journal of Economic Cooperation & Development*, 41(3), 93–116.

Hassoun, A., Aït-Kaddour, A., Abu-Mahfouz, A. M., Rathod, N. B., Bader, F., Barba, F. J., Biancolillo, A., Cropotova, J., Galanakis, C. M., Jambrak, A. R., Lorenzo, J. M., Måge, I., Ozogul, F., & Regenstein, J. (2023). The fourth industrial revolution in the food industry—Part I: Industry 4.0 technologies. *Critical Reviews in Food Science and Nutrition*, 63(23), 6547–6563. 10.1080/10408398.2022.203473535114860

Hassoun, A., Prieto, M. A., Carpena, M., Bouzembrak, Y., Marvin, H. J., Pallares, N., Barba, F. J., Punia Bangar, S., Chaudhary, V., Ibrahim, S., & Bono, G. (2022). Exploring the role of green and Industry 4.0 technologies in achieving sustainable development goals in food sectors. *Food Research International*, 162, 112068. 10.1016/j.foodres.2022.11206836461323

Hategan, C. D., Sirghi, N., Ruxandra Ioana, P., & Hategan, V. (2018). Doing Well or Doing Good: The Relationship between Corporate Social Responsibility and Profit in Romanian Companies. *Sustainability (Basel)*, 10(4), 1041. 10.3390/su10041041

Compilation of References

Hayes, P., Poel, I. V. D., & Steen, M. (2020). Algorithms and values in justice and security. *AI & Society*, 35(3), 533–555. 10.1007/s00146-019-00932-9

Helbing, D., Frey, B. S., Gigerenzer, G., Hafen, E., Hagner, M., Hofstetter, Y., . . . Zwitter, A. (2019). Will democracy survive big data and artificial intelligence? *Towards digital enlightenment: Essays on the dark and light sides of the digital revolution*, 73-98.

Helbing, D., & Pournaras, E. (2015). Society: Build digital democracy. *Nature*, 527(7576), 33–34. 10.1038/527033a26536943

Hemalatha, A., Kumari, P. B., Nawaz, N., & Gajenderan, V. (2021). Impact of artificial intelligence on recruitment and selection of information technology companies. In *2021 international conference on artificial intelligence and smart systems (ICAIS)* (pp. 60–66). IEEE. 10.1109/ICAIS50930.2021.9396036

Henriksson, E. A. (2018). Data protection challenges for virtual reality applications. *Interactive Entertainment Law Review*, 1(1), 57–61. 10.4337/ielr.2018.01.05

Herden, C. J., Alliu, E., Cakici, A., Cormier, T., Deguelle, C., Gambhir, S., Griffiths, C., Gupta, S., Kamani, S. R., Kiratli, Y.-S., Kispataki, M., Lange, G., Moles de Matos, L., Tripero Moreno, L., Betancourt Nunez, H. A., Pilla, V., Raj, B., Roe, J., Skoda, M., ... Edinger-Schons, L. M. (2021). "Corporate Digital Responsibility." *Sustainability Management Forum | NachhaltigkeitsManagementForum*, 29(1), 13–29. 10.1007/s00550-020-00509-x

Hinds, J., Williams, E. J., & Joinson, A. N. (2020). "It wouldn't happen to me": Privacy concerns and perspectives following the Cambridge Analytica scandal. *International Journal of Human-Computer Studies*, 143, 102498. 10.1016/j.ijhcs.2020.102498

Hmoud, B., & Laszlo, V. (2019). Will artificial intelligence take over human resources recruitment and selection? *NetwIntell Stud*, 7(13), 21–30.

Hoang, T. G., Nguyen, G. N. T., & Le, D. A. (2022). Developments in financial technologies for achieving the Sustainable Development Goals (SDGs): FinTech and SDGs. In *Disruptive technologies and eco-innovation for sustainable development* (pp. 1–19). IGI Global. 10.4018/978-1-7998-8900-7.ch001

Hoofnagle, C. J., Van Der Sloot, B., & Borgesius, F. Z. (2019). The European Union general data protection regulation: What it is and what it means. *Information & Communications Technology Law*, 28(1), 65–98. 10.1080/13600834.2019.1573501

Hrast, A. (2018). Usmeritve Evropske Unije na področju družbene odgovornosti in trajnostnega razvoja. In *Uvod v politično ekonomijo družbeno odgovorne družbe* (pp. 285-294). Maribor: Kulturni center, zavod za umetniško produkcijo in založništvo.

Hrast, A., & Mulej, M. (2009). *Družbena odgovornost in izzivi časa 2009: Delo -odnosi do zaposlenih in različnih starostnih generacij. Zbornik prispevkov*. IRDO.

Hsieh, R. (2014). Improving HIPAA enforcement and protecting patient privacy in digital healthcare environment. Loyola University Chicago Law Journal, 46(1), 175-224.

Huang, C., Zhang, Z., Mao, B., & Yao, X. (2023). An overview of artificial intelligence ethics. *IEEE Transactions on Artificial Intelligence*, 4(4), 799–819. 10.1109/TAI.2022.3194503

Hui, Alamri, & Toghraie. (2023). Greening Smart Cities: an Investigation of the Integration of Urban Natural Resources and Smart City Technologies for Promoting Environmental Sustainability. *Sustainable Cities and Society, 99*, 104985–104985. 10.1016/j.scs.2023.104985

Hunt, E. B. (2014). Artificial Intelligence. In *Google Books*. Academic Press. https://books.google.co.in/books?hl=en&lr=&id=9y2jBQAAQBAJ&oi=fnd&pg=PP1&dq=artificial+intelligence&ots=uQSx0laUyD&sig=nHVwQiR88VrgtdCsKvev3yZwB74#v=onepage&q=artificial%20intelligence&f=false

Hunt, P. A., Greaves, I., & Owens, W. A. (2006). Emergency thoracotomy in thoracic trauma - A Review. *Injury*, 37(1), 1–19. 10.1016/j.injury.2005.02.01416410079

Hutt, R. (2015, November 13). *What are your digital rights?* World Economic Forum. https://www.weforum.org/agenda/2015/11/what-are-your-digital-rights-explainer/

Idemudia, U. (2014). Corporate Social Responsibility and Development in Africa: Issues and Possibilities. *Geography Compass*, 8(7), 421–435. 10.1111/gec3.12143

Illes, J., Moser, M. A., McCormick, J. B., Racine, E., Blakeslee, S., Caplan, A., Hayden, E. C., Ingram, J., Lohwater, T., McKnight, P., Nicholson, C., Phillips, A., Sauvé, K. D., Snell, E., & Weiss, S. (2010). Neurotalk: Improving the communication of neuroscience research. *Nature Reviews. Neuroscience*, 11(1), 61–69. 10.1038/nrn277319953102

Imik Tanyildizi, N., & Tanyildizi, H. (2022). Estimation of voting behavior in election using support vector machine, extreme learning machine and deep learning. *Neural Computing & Applications*, 34(20), 17329–17342. Advance online publication. 10.1007/s00521-022-07395-y

Inekwe, M., Hashim, F., & Yahya, S. (2020). CSR in developing countries – the importance of good governance and economic growth: Evidence from Africa. *Social Responsibility Journal*. 10.1108/SRJ-10-2019-0336

Informacijski pooblaščenec. (2022). Letno poročilo informacijskega pooblaščenca. https://www.ip-rs.si/fileadmin/user_upload/Pdf/porocila/LP2022.pdf

International Energy Agency. (2017). *Digitalisation and Energy*. Retrieved from International Energy Agency: https://www.iea.org/reports/digitalisation-and-energy

ISO. (2010). *ISO 26000, Social Responsibility*. Retrieved from ISO: https://www.iso.org/iso/discovering_iso_26000.pdf

Ivanov, D., Dolgui, A., & Sokolov, B. (2019). The impact of digital technology and Industry 4.0 on the ripple effect and supply chain risk analytics. *International Journal of Production Research*, 57(3), 829–846. 10.1080/00207543.2018.1488086

Jackson, M. C. (2021). Artificial intelligence & algorithmic bias: The issues with technology reflecting history & humans. *J Bus Technol Law*, 16, 299.

Compilation of References

Jaiwant, S. V., & Kureethara, J. V. (2023). Green Finance and Fintech: Toward a More Sustainable Financial System. In *Green Finance Instruments, FinTech, and Investment Strategies: Sustainable Portfolio Management in the Post-COVID Era* (pp. 283–300). Springer International Publishing. 10.1007/978-3-031-29031-2_12

Jakka, G., Yathiraju, N., & Ansari, M. F. (2022). Artificial Intelligence in Terms of Spotting Malware and Delivering Cyber Risk Management. *Journal of Positive School Psychology*, 6(3), 6156–6165.

Jarrahi, M. H., Askay, D., Eshraaghi, A., & Smith, P. (2023). Artificial intelligence and knowledge management: A partnership between human and AI. *Business Horizons*, 66(1), 87–99. 10.1016/j.bushor.2022.03.002

Jatobá, M. N., Ferreira, J. J., Fernandes, P. O., & Teixeira, J. P. (2023). Intelligent human resources for the adoption of artificial intelligence: A systematic literature review. *Journal of Organizational Change Management*, 36(7), 1099–1124. 10.1108/JOCM-03-2022-0075

Javaid, M., Haleem, A., Singh, R. P., Khan, S., & Suman, R. (2021). Blockchain technology applications for Industry 4.0: A literature-based review. *Blockchain: Research and Applications*, 2(4), 100027.

Javaid, M., Haleem, A., Singh, R. P., Suman, R., & Gonzalez, E. S. (2022). Understanding the adoption of Industry 4.0 technologies in improving environmental sustainability. *Sustainable Operations and Computers*, 3, 203–217. 10.1016/j.susoc.2022.01.008

Jean, N., Burke, M., Xie, M., Davis, W. M., Lobell, D. B., & Ermon, S. (2016). Combining satellite imagery and machine learning to predict poverty. *Science*, 353(6301), 790–794. 10.1126/science.aaf789427540167

Jobin, A., Ienca, M., & Vayena, E. (2019). The global landscape of AI ethics guidelines. *Nature Machine Intelligence*, 1(9), 389–399. 10.1038/s42256-019-0088-2

Johansson, J., & Herranen, S. (2019) The application of artificial intelligence (AI) in human resource management: current state of AI and its impact on the traditional recruitment process. Bachelor thesis, Jonkoping University.

Johnson, K. N. (2019). Automating the risk of bias. *George Washington Law Review*, 87(6). https://www.gwlr.org/wp-content/uploads/2020/01/87-Geo.-Wash.-L.-Rev.-1214.pdf

Johnson, R. D., Stone, D. L., & Lukaszewski, K. M. (2020). The benefits of eHRM and AI for talent acquisition. *J Tour Futur*, 7(1), 40–52. 10.1108/JTF-02-2020-0013

Jones, D., Molitor, D., & Reif, J. (2019). What do workplace wellness programs do? Evidence from the Illinois workplace wellness study. *The Quarterly Journal of Economics*, 134(4), 1747–1791. 10.1093/qje/qjz02331564754

Kabeyi, M. J. B. (2020). Corporate governance in manufacturing and management with analysis of governance failures at Enron and Volkswagen Corporations. *Am J Oper Manage Inform Syst*, 4(4), 109–123.

Kalaiarasi, H., & Kirubahari, S. (2023). Green finance for sustainable development using blockchain technology. In *Green Blockchain Technology for Sustainable Smart Cities* (pp. 167–185). Elsevier. 10.1016/B978-0-323-95407-5.00003-7

Kaltakis, K., Polyzi, P., Drosatos, G., & Rantos, K. (2021). Privacy-preserving solutions in blockchain-enabled internet of vehicles. *Applied Sciences (Basel, Switzerland)*, 11(21), 9792. 10.3390/app11219792

Kam, S. (2004). Intel Corp. v. Hamidi: Trespass to chattels and a doctrine of cyber-nuisance. *Berkeley Tech. LJ, 19*, 427.

Kamalov, F., Calonge, D. S., & Gurrib, I. (2023). New Era of Artificial Intelligence in Education: Towards a Sustainable Multifaceted Revolution. *Sustainability (Basel)*, 15(16), 12451. 10.3390/su151612451

Kanojia, S. (2023). Application of blockchain in corporate governance: Adaptability, challenges and regulation in BRICS. *BRICS LJ*, 10(4), 53–67. 10.21684/2412-2343-2023-10-4-53-67

Kasowaki, L., & Ahmet, S. (2024). *Shielding the Virtual Ramparts: Understanding Cybersecurity Essentials* (No. 11700). EasyChair.

Kataev, M., Bulysheva, L., & Mosiaev, A. (2022). Application of artificial intelligence methods in the school educational process. *Systems Research and Behavioral Science*, 39(3), 531–541. 10.1002/sres.2873

Kaur, M., Rekha, A., & Resmi, A. (2021). Research landscape of artificial intelligence in human resource management: A bibliometric overview. In Yadav, R., Yadav, R., & Gupta, S. B. (Eds.), *Artificial Intelligence and Speech Technology* (pp. 463–477)., 10.1201/9781003150664-51

Khajuria, A. (2024). The EU AI Act: A landmark in AI regulation. Observer Research Foundation. https://www.orfonline.org/research/the-eu-ai-act-a-landmark-in-ai-regulation

Kinder, T., Stenvall, J., Koskimies, E., Webb, H., & Janenova, S. (2023). Local public services and the ethical deployment of artificial intelligence. *Government Information Quarterly*, 40(4), 101865. 10.1016/j.giq.2023.101865

Kintu, N. B., & Mohamed, I. Z. (2018). A secure e-voting system using biometric fingerprint and crypt-watermark methodology. In *ASCENT International Conference Proceedings–Information Systems and Engineering* (pp. 1-18).

Kinuthia, D. (2020). *Exploring Data Anonymisation and Internet Safety in East Africa*. https://policycommons.net/artifacts/1445762/exploring-data-anonymisation-and-internet-safety-in-east-africa/2077526/

Klinger, E., Wiesmaier, A., & Hinzmann, A. (2023). A review of existing GDPR solutions for citizens and SMEs. arXiv.org. https://doi.org//arXiv.2302.03581 10.48550

Compilation of References

Klumpp, M. (2017). Automation and artificial intelligence in business logistics systems: Human reactions and collaboration requirements. *International Journal of Logistics*, 21(3), 224–242. 10.1080/13675567.2017.1384451

Köchling, A., Riazy, S., Wehner, M. C., & Simbeck, K. (2021). Highly accurate, but still discriminatory: A fairness evaluation of algorithmic video analysis in the recruitment context. *Business & Information Systems Engineering*, 63(1), 39–54. 10.1007/s12599-020-00673-w

Kolvart, M., Poola, M. & Rull, A. (2016). "Smart contracts." The Future of Law and eTechnologies, 133-147. 10.1007/978-3-319-26896-5

Kong, Y., Xie, C., Wang, J., Jones, H., & Ding, H. (2021). AI-assisted recruiting technologies: Tools, challenges, and opportunities. In *The 39th ACM international conference on design of communication* (pp. 359–361). ACM.

König, P. D. (2022). Fortress Europe 4.0? An analysis of EU data governance through the lens of the resource regime concept. *European Policy Analysis*, 8(4), 484–504. 10.1002/epa2.1160

Kop, M. (2021). EU Artificial Intelligence Act: The European approach to AI. Stanford - Vienna Transatlantic Technology Law Forum, Transatlantic Antitrust and IPR Developments, Stanford University, Issue No. 2/2021. https://law.stanford.edu/publications/eu-artificial-intelligence-act-the-european-approach-to-ai/

Kossek, E. E., & Buzzanell, P. M. (2018). Women's career equality and leadership in organizations: Creating an evidence-based positive change. *Human Resource Management*, 57(4), 813–822. 10.1002/hrm.21936

Kovalenko, Y. (2022). The Right to Privacy and Protection of Personal Data: Emerging Trends and Implications for Development in Jurisprudence of European Court of Human Rights. *Masaryk University Journal of Law and Technology*, 16(1), 37–57. 10.5817/MUJLT2022-1-2

Kramar, R. (2022). Sustainable human resource management: Six defining characteristics. *Asia Pacific Journal of Human Resources*, 60(1), 146–170. 10.1111/1744-7941.12321

Kretschmer, M., Pennekamp, J., & Wehrle, K. (2021). Cookie Banners and Privacy Policies: Measuring the Impact of the GDPR on the Web. *ACM Transactions on the Web*, 15(4), 20. Advance online publication. 10.1145/3466722

Krzywdzinski, M., Gerst, D., & Butollo, F. (2023). Promoting human-centred AI in the workplace. Trade unions and their strategies for regulating the use of AI in Germany. *Transfer: European Review of Labour and Research*, 29(1), 53–70. 10.1177/10242589221142273

Kulik, J. A., Mahler, H. I., & Moore, P. J. (1996). Social comparison and affiliation under threat: Effects on recovery from major surgery. *Journal of Personality and Social Psychology*, 71(5), 967–979. 10.1037/0022-3514.71.5.9678939044

Kumar, S. H., Talasila, D., Gowrav, M. P., & Gangadharappa, H. V. (2020). Adaptations of Pharma 4.0 from Industry 4.0. *Drug Invention Today*, 14(3).

Kumar, S., & Mallipeddi, R. R. (2022). Impact of cybersecurity on operations and supply chain management: Emerging trends and future research directions. *Production and Operations Management*, 31(12), 4488–4500. 10.1111/poms.13859

Kundu, S. C., & Mor, A. (2017). Workforce diversity and organizational performance: A study of IT industry in India. *Employee Relations*, 39(2), 160–183. 10.1108/ER-06-2015-0114

Kuo, T. C., & Smith, S. (2018). A systematic review of technologies involving eco-innovation for enterprises moving towards sustainability. *Journal of Cleaner Production*, 192, 207–220. 10.1016/j.jclepro.2018.04.212

Kurniawan, T. A., Maiurova, A., Kustikova, M., Bykovskaia, E., Othman, M. H. D., & Goh, H. H. (2022). Accelerating sustainability transition in St. Petersburg (Russia) through digitalization-based circular economy in waste recycling industry: A strategy to promote carbon neutrality in era of Industry 4.0. *Journal of Cleaner Production*, 363, 132452. 10.1016/j.jclepro.2022.132452

Kuzior, A. (2022). Technological unemployment in the perspective of Industry 4.0. *Virtual Economics*, 5(1), 7–23. 10.34021/ve.2022.05.01(1)

Kyriazanos, D. M., Thanos, K. G., & Thomopoulos, S. C. A. (2019). Automated decisions making in airports checkpoints: Bias detection toward smarter security and fairness. *IEEE Security and Privacy*, 17(2), 8–16. 10.1109/MSEC.2018.2888777

Laine, J. (2021). There is no decision: design of cookie consent banner and its effect on user consent (Publication Number URN:NBN:fi:tuni-202111208544) Tampere University). https://trepo.tuni.fi/handle/10024/135598

Lambrecht, A., & Tucker, C. (2019). Algorithmic bias? An empirical study of apparent gender-biased discrimination in the display of STEM career ads. *Management Science*, 65(7), 2947–3448. 10.1287/mnsc.2018.3093

Langer, M., König, C. J., & Hemsing, V. (2020). Is anybody listening? The impact of automatically evaluated job interviews on impression management and applicant reactions. *Journal of Managerial Psychology*, 35(4), 271–284. 10.1108/JMP-03-2019-0156

Laptev, V. A., & Feyzrakhmanova, D. R. (2021). Digitalization of institutions of corporate law: Current trends and future prospects. *Laws*, 10(4), 93. 10.3390/laws10040093

Laskurain-Iturbe, I., Arana-Landín, G., Landeta-Manzano, B., & Uriarte-Gallastegi, N. (2021). Exploring the influence of industry 4.0 technologies on the circular economy. *Journal of Cleaner Production*, 321, 128944. 10.1016/j.jclepro.2021.128944

Latonero, M. (2018). *Governing Artificial Intelligence: Upholding human rights & dignity*. Data & Society.

Laurim, V., Arpaci, S., Prommegger, B., & Krcmar, H. (2021). Computer, whom should I hire?–acceptance criteria for artificial intelligence in the recruitment process. In *Proceedings of the 54th Hawaii international conference on system sciences* (p. 5495). 10.24251/HICSS.2021.668

Compilation of References

Lee, N. T. (2018). Detecting racial bias in algorithms and machine learning. *Journal of Information, Communication, and Ethics in Society*, 16(3), 252–260. 10.1108/JICES-06-2018-0056

Leenes, R., & De Conca. (2018). Artificial intelligence and privacy—AI enters the house through the Cloud. *Edward Elgar Publishing EBooks*. 10.4337/9781786439055.00022

Leng, J., Ye, S., Zhou, M., Zhao, J. L., Liu, Q., Guo, W., Cao, W., & Fu, L. (2020). Blockchain-secured smart manufacturing in industry 4.0: A survey. *IEEE Transactions on Systems, Man, and Cybernetics. Systems*, 51(1), 237–252. 10.1109/TSMC.2020.3040789

Leong, C. (2018). Technology & recruiting 101: How it works and where it's going. *Strategic HR Review*, 17(1), 50–52. 10.1108/SHR-12-2017-0083

Levin, A., & Nicholson, M. J. (2023). Privacy Law in the United States, the EU and Canada: The Allure of the Middle Ground. Toronto Metropolitan University. 10.32920/22227775.v1

Levine, M., Philpot, R., Nightingale, S. J., & Kordoni, A. (2024). Visual digital data, ethical challenges, and psychological science. *The American Psychologist*, 79(1), 109–122. 10.1037/amp000119238236219

Li, B. H., Hou, B. C., Yu, W. T., Lu, X. B., & Yang, C. W. (2017). Applications of artificial intelligence in intelligent manufacturing: A review. *Frontiers of Information Technology & Electronic Engineering*, 18(1), 86–96. 10.1631/FITEE.1601885

Libby, K. (2019). This Bill Hader Deepfake video is Amazing. It's also Terrifying for our Future. Popular Mechanics. Retrieved from https://www.popularmechanics.com/technology/security/a28691128/deepfaketechnology/

Li, F. (2020). The digital transformation of business models in the creative industries: A holistic framework and emerging trends. *Technovation*, 92-93, 92. 10.1016/j.technovation.2017.12.004

Lim, C. H., Lim, S., How, B. S., Ng, W. P. Q., Ngan, S. L., Leong, W. D., & Lam, H. L. (2021). A review of industry 4.0 revolution potential in a sustainable and renewable palm oil industry: HAZOP approach. *Renewable & Sustainable Energy Reviews*, 135, 110223. 10.1016/j.rser.2020.110223

Lin, X. L., & Wang, X. Q. (2023). Following too much on Facebook brand page: A concept of brand overload and its validation. *International Journal of Information Management*, 73, 19. 10.1016/j.ijinfomgt.2023.102682

Liu, H., Yao, P., Latif, S., Aslam, S., & Iqbal, N. (2022). Impact of Green financing, FinTech, and financial inclusion on energy efficiency. *Environmental Science and Pollution Research International*, 29(13), 1–12. 10.1007/s11356-021-16949-x34705207

Li, Z., Liang, X., Wen, Q., & Wan, E. (2024). The Analysis of Financial Network Transaction Risk Control based on Blockchain and Edge Computing Technology. *IEEE Transactions on Engineering Management*, 71, 5669–5690. 10.1109/TEM.2024.3364832

Lloyd, I. (2020). *Information technology law*. Oxford University Press. 10.1093/he/9780198830559.001.0001

Lobschat, L., Mueller, B., Eggers, F., Brandimarte, L., Diefenbach, S., Kroschke, M., & Wirtz, J. (2021). Corporate Digital Responsibility. *Journal of Business Research*, 122, 875–888. 10.1016/j.jbusres.2019.10.006

Lu, C., Lyu, J., Zhang, L., Gong, A., Fan, Y., Yan, J., & Li, X. (2020). Nuclear power plants with artificial intelligence in industry 4.0 era: Top-level design and current applications—A systemic review. *IEEE Access : Practical Innovations, Open Solutions*, 8, 194315–194332. 10.1109/ACCESS.2020.3032529

Luo, X., Tong, S., Fang, Z., & Qu, Z. (2019). Frontiers: Machines vs. humans: The impact of artificial intelligence chatbot disclosure on customer purchases. *Marketing Science*, •••, 937–947. 10.1287/mksc.2019.1192

Lu, Y. (2019). Artificial intelligence: A survey on evolution, models, applications and future trends. *Journal of Management Analytics*, 6(1), 1–29. 10.1080/23270012.2019.1570365

Macchiavello, E. (2023). *Sustainable Finance and Fintech. A Focus on Capital Raising. A Focus on Capital Raising*. CUP.

Machin, M., Sanguesa, J. A., Garrido, P., & Martinez, F. J. (2018). *On the use of artificial intelligence techniques in intelligent transportation systems | IEEE Conference Publication | IEEE Xplore*. Ieeexplore.ieee.org. https://ieeexplore.ieee.org/abstract/document/8369029

Majstorovic, V. D., & Mitrovic, R. (2019). Industry 4.0 programs worldwide. In *Proceedings of the 4th International Conference on the Industry 4.0 Model for Advanced Manufacturing: AMP 2019 4* (pp. 78-99). Springer International Publishing. 10.1007/978-3-030-18180-2_7

Majumder, S., Ghosh, S., & Polkowski, Z. (2023). A brief introduction to Human Resource Management. In *Artificial Intelligence Techniques in Human Resource Management* (pp. 1-25). CRC Press. 10.1201/9781003328346-1

Makridakis, S. (2017). The forthcoming Artificial Intelligence (AI) revolution: Its impact on society and firms. *Futures*, 90, 46–60. 10.1016/j.futures.2017.03.006

Makulilo, A. (2016). *African Data Privacy Laws* (Vol. 33). 10.1007/978-3-319-47317-8

Malan, D. (2018, June 21). *Technology is changing faster than regulators can keep up—Here's how to close the gap*. World Economic Forum. https://www.weforum.org/agenda/2018/06/law-too-slow-for-new-tech-how-keep-up/

Manroop, L., Malik, A., & Milner, M. (2024). The ethical implications of big data in human resource management. *Human Resource Management Review*, 34(2), 101012. 10.1016/j.hrmr.2024.101012

Maon, F., Lindgreen, A., & Swaen, V. (2008). Thinking of the organization as a system: The role of managerial perceptions in developing a corporate social responsibility strategic agenda. *Systems Research and Behavioral Science*, 25(3), 413–426. 10.1002/sres.900

Compilation of References

Marabelli, M., Newell, S., & Handunge, V. (2021). The lifecycle of algorithmic decision-making systems: Organizational choices and ethical challenges. *The Journal of Strategic Information Systems*, 30(3), 101683. Advance online publication. 10.1016/j.jsis.2021.101683

Martinez, C. F., & Fernandez, A. (2020). AI and recruiting software: Ethical and legal implications. *Paladyn : Journal of Behavioral Robotics*, 11(1), 199–216. Advance online publication. 10.1515/pjbr-2020-0030

Martins Gonçalves, R., Mira da Silva, M., & Rupino da Cunha, P. (2023). Implementing GDPR-Compliant Surveys Using Blockchain. *Future Internet*, 15(4), 143. 10.3390/fi15040143

Masterson, V. (2023, December 14). *AI beats top weather forecasting computers*. World Economic Forum. https://www.weforum.org/agenda/2023/12/ai-weather-forecasting-climate-crisis/

Matsa, P., & Gullamajji, K. (2019). To Study Impact of Artificial Intelligence on Human Resource Management. *International Research Journal of Engineering and Technology*, 6(8), 1229–1238.

McCarthy, J., Minsky, M. L., Rochester, N., & Shannon, C. E. (1955). A proposal for the Dartmouth summer research project on artificial intelligence. *AI Magazine*, 27(4).

McCorry, P., Shahandashti, S. F., & Hao, F. (2017). A smart contract for boardroom voting with maximum voter privacy. In *Financial Cryptography and Data Security:21st International Conference, FC 2017,Sliema, Malta,April 3-7, 2017, Revised Selected Papers 21* (pp. 357-375). Springer International Publishing. 10.1007/978-3-319-70972-7_20

Mckinsey & Company. (2024, April 30). *What's the future of AI? | McKinsey*. Www.mckinsey.comMckinsey & Company. https://www.mckinsey.com/featured-insights/mckinsey-explainers/whats-the-future-of-ai

McWilliams, A., Siegel, D. S., & Wright, P. M. (2006). Corporate social responsibility: Strategic implications. *Journal of Management Studies*, 43(1), 1–18. 10.1111/j.1467-6486.2006.00580.x

Mehrabi, N., Morstatter, F., Saxena, N., Lerman, K., & Galstyan, A. (2021). A survey on bias and fairness in machine learning. *ACM Computing Surveys*, 54(6), 1–35. 10.1145/3457607

Mendoza, S. (2019). GDPR Compliance - It Takes a Village. *Seattle University Law Review*, 42(3), 1155–1162.

Meyendorf, N., Ida, N., Singh, R., & Vrana, J. (2023). NDE 4.0: Progress, promise, and its role to industry 4.0. *NDT & E International*, 140, 102957. 10.1016/j.ndteint.2023.102957

Meyer, D. (2018). Amazon reportedly killed an AI recruitment system because it couldn't stop the tool from discriminating against women. *Fortune*. fortune.com/2018/10/10/amazon-ai-recruitment-bias-women-sexist/.

Miasato, A., & Silva, F. R. (2019). Artificial intelligence as an instrument of discrimination in workforce recruitment. ActaUnivSapientiae. *Legal Studies*, 8(2), 191–212.

Micaroni, M. (2020). *Sustainable Finance: Addressing the SDGs through Fintech and Digital Finance solutions in EU* (Doctoral dissertation, Politecnico di Torino).

Michell, M. J., Iqbal, A., Wasan, R. K., Evans, D. R., Peacock, C., Lawinski, C. P., Douiri, A., Wilson, R., & Whelehan, P. (2012). A comparison of the accuracy of film-screen mammography, full-field digital mammography, and digital breast tomosynthesis. *Clinical Radiology*, 67(10), 976–981. 10.1016/j.crad.2012.03.00922625656

Migliorelli, M., & Dessertine, P. (2019). The rise of green finance in Europe. *Opportunities and challenges for issuers, investors and marketplaces. Cham. Palgrave Macmillan*, 2, 2019.

Mikaelf, P., Framnes, V. A., Danielsen, F., Krogstie, J., & Olsen, D. (2017). *Big data analytics capability: antecedents and business value.*

Milaj-Weishaar, J. (2020). Safeguarding Privacy by Regulating the Processing of Personal Data – An EU Illusion? *European Journal of Law and Technology*, 11(2).

Miller, K. (2024, March 18). *Privacy in an AI Era: How Do We Protect Our Personal Information?* https://hai.stanford.edu/news/privacy-ai-era-how-do-we-protect-our-personal-information

Miller, G. (2012). The Smartphone Psychology Manifesto. *Perspectives on Psychological Science*, 7(3), 221–237. 10.1177/174569161244121526168460

Miller, T. (2019). Explanation in artificial intelligence: Insights from the social sciences. *Artificial Intelligence*, 267, 1–38. 10.1016/j.artint.2018.07.007

Milosavljević, N. (2023). The limitation of the data protection right for the purpose of intellectual property rights protection. In Ćalić, R. (Ed.), *Uporednopravni izazovi u savremenom pravu, In memoriam dr* (pp. 587–617). Stefan Andonović. 10.56461/ZR_23.SA.UPISP_NM

Miron, M., Tolan, S., Gomez, E., & Castillo, C. (2020). Evaluating causes of algorithmic bias in juvenile criminal recidivism. *Artificial Intelligence and Law*, 29(2), 111–147. 10.1007/s10506-020-09268-y

Mitchell, T. M. (1997). *Machine Learning*. McGraw-Hill.

Mithas, S., Tafti, A., & Mitchell, W. (2013). How a firm's competitive environment and digital strategic posture influence digital business strategy. *MIS Quarterly*, 511-536.

Mithas, S., Chen, Z. L., Saldanha, T. J., & De Oliveira Silveira, A. (2022). How will artificial intelligence and Industry 4.0 emerging technologies transform operations management? *Production and Operations Management*, 31(12), 4475–4487. 10.1111/poms.13864

Mittelstadt, B. (2019). Principles Alone Cannot Guarantee Ethical AI. *Nature Machine Intelligence*, 1(11), 501–507. 10.1038/s42256-019-0114-4

Miyachi, K., & Mackey, T. K. (2021). HOCBS: A privacy-preserving blockchain framework for healthcare data leveraging an on-chain and off-chain system design. *Information Processing & Management*, 58(3), 24. 10.1016/j.ipm.2021.102535

Compilation of References

Montasari, R. (2022). Artificial Intelligence and National Security. In *Artificial Intelligence and National Security.* 10.1007/978-3-031-06709-9

Morandini, S., Fraboni, F., De Angelis, M., Puzzo, G., Giusino, D., & Pietrantoni, L. (2023). The Impact of Artificial Intelligence on Workers' Skills: Upskilling and Reskilling in Organisations. *Informing Science*, 26(1), 39–68. 10.28945/5078

Morgan, J., Halton, M., Qiao, Y., & Breslin, J. G. (2021). Industry 4.0 smart reconfigurable manufacturing machines. *Journal of Manufacturing Systems*, 59, 481–506. 10.1016/j.jmsy.2021.03.001

Morley, J., Cowls, J., Taddeo, M., & Floridi, L. (2020). Ethical guidelines for COVID 19. *Nature*, 582(7810), 29–31. 10.1038/d41586-020-01578-032467596

Moro-Visconti, R., Cruz Rambaud, S., & López Pascual, J. (2020). Sustainability in FinTechs: An explanation through business model scalability and market valuation. *Sustainability (Basel)*, 12(24), 10316. 10.3390/su122410316

Morshidi, A. Bin, Satar, N. S. M., Azizan, A. A. D. A., Idris, R. Z., Idris, R., Radzi, M. S. M., Yusoff, S. M., & Sarjono, F. (2023). A Bibliometric Analysis of Artificial Intelligence and Human Resource Management Studies. In *Exploring the Intersection of AI and Human Resources Management* (pp. 85–117). IGI Global. 10.4018/979-8-3693-0039-8.ch006

Morshidi, A., b, K. Y. S., Yussof, K. M., Idris, R. Z., Idris, R., & Abas, A. (2022). Online work adjustment in the sustainable development context during COVID-19 pandemic: A systematic literature review. *International Journal of Work Innovation*, 3(3), 269–288. 10.1504/IJWI.2022.127668

Mourtzis, D., Angelopoulos, J., & Panopoulos, N. (2022). A Literature Review of the Challenges and Opportunities of the Transition from Industry 4.0 to Society 5.0. *Energies*, 15(17), 6276. 10.3390/en15176276

Mueller, B. (2022). Corporate Digital Responsibility. *Business & Information Systems Engineering*, 64(5), 689–700. 10.1007/s12599-022-00760-0

Mujtaba, D. F., & Mahapatra, N. R. (2019, November). Ethical considerations in AI-based recruitment. In *2019 IEEE International Symposium on Technology and Society (ISTAS)* (pp. 1-7). IEEE.

Mulej, M. (2018). Politična ekonomija in družbena odgovornost. In *Uvod v politično ekonomijo družbeno odgovorne družbe* (pp. 197-284). Maribor: Kulturni center, zavod za umetniško produkcijo in založništvo.

Mulej, M., & Mihec, M. (2020). Social Responsibility as a Precondition of Innovation in Higher Education. In *Handbook of Research on Enhancing Innovation in Higher Education Institutions* (pp. 49-74). IGI Global. 10.4018/978-1-7998-2708-5.ch003

Mulej, M., Hrast, A., & Mihec, M. (2024a). *Inovativna trajnostna družbeno odgovorna družba, Knj. 2: Nekaj informacijskih spodbud za obstoj človeštva.* Ljubljana: IRDO-Inštitut za razvoj družbene odgovornosti.

Mulej, M., Hrast, A., & Šarotar Žižek, S. (2024b). *Bases of the innovative sustainable socially responsible society (ISSR Society). Volume 2: some information management viewpoints of ISSR SOCIETY*. Ljubljana: IRDO - Inštitut za razvoj družbene odgovornosti.

Mulej, M., Hrast, A., & Štrukelj, T. (2024c). *Inovativna trajnostna družbeno odgovorna družba, Knj. 1: Nekaj ekonomskih spodbud zoper propad človeštva*. Ljubljana: IRDO – Inštitut za razvoj družbene odgovornosti.

Mulej, M., Hrast, A., & Štrukelj, T. (2024d). *Inovativna trajnostna družbeno odgovorna družba, Knj. 3: Skrb za starejše*. Ljubljana: IRDO - Inštitut za razvoj družbene odgovornosti.

Mulej, M., Hrast, A., Štrukelj, T., Likar, B., & Šarotar Žižek, S. (2024e). *Bases of an innovative sustainable socially responsible (ISSR) society. Volume 1: Social responsibility - a non-technical innovation process toward ISS SOCIETY*. Ljubljana: IRDO - Inštitut za razvoj družbene odgovornosti.

Mulej, M., Merhar, V., & Žakelj, V. (2016a). Prihodnost - preživetje brez ekosocializma in družbene odgovornosti? In *Nehajte sovražiti svoje otroke in vnuke. Knj. 1, Družbenoekonomski okvir in osebne lastnosti družbeno odgovornih* (pp. 11-23). Maribor: Kulturni center, zavod za umetniško produkcijo in založništvo.

Mulej, M., Ženko, Z., Hrast, A., & Merhar, V. (2016b). Prihodnost ekonomije v obdobju izobilja in krize izobilja brez zdravih ambicij. In *Nehajte sovražiti svoje otroke in vnuke. Knj. 1, Družbenoekonomski okvir in osebne lastnosti družbeno odgovornih* (pp. 35-85). Maribor: Kulturni center, zavod za umetniško produkcijo in založništvo.

Muller, C. (2020). *The Impact of Artificial Intelligence on Human Rights, Democracy and the Rule of Law*. https://allai.nl/wp-content/uploads/2020/06/The-Impact-of-AI-on-Human-Rights-Democracy-and-the-Rule-of-Law-draft.pdf

Murdoch, B. (2021). Privacy and Artificial Intelligence: Challenges for Protecting Health Information in a New Era. *BMC Medical Ethics*, 22(1), 122. Advance online publication. 10.1186/s12910-021-00687-334525993

Musch, S., Borrelli, M., & Kerrigan, C. (2023). The EU AI Act as global artificial intelligence regulation. *SSRN*. https://doi.org/10.2139/ssrn.4549261

Myll, U.M. (2024). Trade Secrets and the Data Act. IIC - International Review of Intellectual Property and Competition Law, 55, 368-393. 10.1007/s40319-024-01432-0

Naderi, N., & Tian, Y. (2022). Leveraging Blockchain Technology and Tokenizing Green Assets to Fill the Green Finance Gap. *Energy Research Letters, 3*(3).

Nagano, A. (2018). Economic Growth and Automation Risks in Developing Countries Due to the Transition Toward Digital Modernity. *Proceedings of the 11th International Conference on Theory and Practice of Electronic Governance*, 42-50. 10.1145/3209415.3209442

Nagler, J., van den Hoven, J., & Helbing, D. (2019). Towards Digital Enlightenment. *Springer International Publishing*, 41-46.

Compilation of References

Nahavandi, S. (2019). Industry 5.0—A human-centric solution. *Sustainability (Basel)*, 11(16), 4371. 10.3390/su11164371

Naim, M. F. (2023). Reinventing workplace learning and development: Envisaging the role of AI. In *The adoption and Effect of artificial intelligence on human resources management* (pp. 215–227). Part A. 10.1108/978-1-80382-027-920231011

Nam, K., Dutt, C. S., Chathoth, P., Daghfous, A., & Khan, M. S. (2020). The adoption of artificial intelligence and robotics in the hotel industry: Prospects and challenges. *Electronic Markets*, 31(3), 553–574. https://link.springer.com/article/10.1007/s12525-020-00442-3. 10.1007/s12525-020-00442-3

Nassar, A., & Kamal, M. (2021). Ethical dilemmas in AI-powered decision-making: A deep dive into big data-driven ethical considerations. *International Journal of Responsible Artificial Intelligence*, 11(8), 1–11.

Nassiry, D. (2018). *The role of fintech in unlocking green finance: Policy insights for developing countries* (No. 883). ADBI working paper.

Nchake, M. A., & Shuaibu, M. (2022). Investment in ICT infrastructure and inclusive growth in Africa. *Scientific African*, 17, e01293. 10.1016/j.sciaf.2022.e01293

Nemitz, P. (2018). Constitutional democracy and technology in the age of artificial intelligence. *Philosophical Transactions. Series A, Mathematical, Physical, and Engineering Sciences*, 376(2133), 20180089. Advance online publication. 10.1098/rsta.2018.008930323003

Nenavath, S., & Mishra, S. (2023). Impact of green finance and fintech on sustainable economic growth: Empirical evidence from India. *Heliyon*, 9(5), e16301. 10.1016/j.heliyon.2023.e1630137234625

Nersessian, D., & Mancha, R. (2020). From Automation to Autonomy: Legal and Ethical Responsibility Gaps in Artificial Intelligence Innovation. *SSRN*, 27(55). Advance online publication. 10.2139/ssrn.3789582

Neuman, S. (2021, September 16). *The U.N. Warns That AI Can Pose A Threat To Human Rights*. NPR.org. https://www.npr.org/2021/09/16/1037902314/the-u-n-warns-that-ai-can-pose-a-threat-to-human-rights

Newell, S. (2015). *Recruitment and selection. Managing human resources: personnel management in transition*. Blackwell Publishing.

Nguyen, T., Gosine, R. G., & Warrian, P. (2020). A systematic review of big data analytics for oil and gas industry 4.0. *IEEE Access : Practical Innovations, Open Solutions*, 8, 61183–61201. 10.1109/ACCESS.2020.2979678

Nica, E., Stanciu, C., & Ionu Stan, C. (2021). Internet of Things-based Real-Time Production Logistics, Sustainable Industrial Value Creation, and Artificial Intelligence-driven Big Data Analytics in Cyber-Physical Smart Manufacturing Systems. *Economics, Management, and Financial Markets*, 16(1), 52–62. https://www.ceeol.com/search/article-detail?id=939242

Nica, E., & Stehel, V. (2021). Internet of things sensing networks, artificial intelligence-based decision-making algorithms, and real-time process monitoring in sustainable industry 4.0. *Journal of Self-Governance and Management Economics*, 9(3), 35–47.

Nishant, R., Kennedy, M., & Corbett, J. (2020). Artificial intelligence for sustainability: Challenges, opportunities, and a research agenda. *International Journal of Information Management*, 53(53), 102104. 10.1016/j.ijinfomgt.2020.102104

Njoto, S. (2020) Research paper gendered bots? Bias in the use of artificial intelligence in recruitment.

Noe, R., Hollenbeck, J., Gerhart, B., & Wright, P. (2007). *Fundamentals of Human Resource Management.*

Novak, M. (2021). Pravna argumentacija v teoriji in praksi. Ljubljana: Uradni list RS.

O'Leary, D. E., Bonorris, S., Klosgen, W., Khaw, Y.-T., Lee, H.-Y., & Ziarko, W. (1995). Some privacy issues in knowledge discovery: The OECD personal privacy guidelines. *IEEE Expert*, 10(2), 48–59. 10.1109/64.395352

OECD. (2002). *OECD Guidelines on the Protection of Privacy and Transborder Flows of Personal Data*. OECD Publishing. 10.1787/9789264196391-

Oh, J., Hong, J., Lee, C., Lee, J. J., Woo, S. S., & Lee, K. (2021). Will EU's GDPR act as an effective enforcer to gain consent? *IEEE Access : Practical Innovations, Open Solutions*, 9, 79477–79490. 10.1109/ACCESS.2021.3083897

Olesen-Bagneux, O. (Host). (2024). EU Data Act Special: Decoding the Future Impact of the Act on Technology and Society [Audio podcast]. https://podcast.zeenea.com/eu-data-act-special-w-matthias-niebuhr-decoding-the-future-impact-of-the-act-on-technology-and-society

Oracle & Future Workplace. (2019). From Fear to Enthusiasm. *Artificial Intelligence Is Winning More Hearts and Minds in the Workplace*, 1-18.

Orlitzky, M., Schmidt, F. L., & Rynes, S. L. (2003). Corporate Social and Financial Performance: A Meta-Analysis. *Organization Studies*, 24(3), 403–441. 10.1177/0170840603024003910

Orlitzky, M., & Swanson, D. L. (2006). Socially responsible human resource management. In *J. R. Deckop*. Human Resource Management Ethics.

Oullier, O., & Basso, F. (2010). Embodied economics: How bodily information shapes the social coordination dynamics of decision-making. *Philosophical Transactions of the Royal Society of London. Series B, Biological Sciences*, 365(1538), 291–301. 10.1098/rstb.2009.016820026467

Ozili, P. K. (2023). Assessing global interest in decentralized finance, embedded finance, open finance, ocean finance and sustainable finance. *Asian Journal of Economics and Banking*, 7(2), 197–216. 10.1108/AJEB-03-2022-0029

Compilation of References

Page, M. J., McKenzie, J. E., Bossuyt, P. M., Boutron, I., Hoffmann, T. C., Mulrow, C. D., Shamseer, L., Tetzlaff, J. M., Akl, E. A., Brennan, S. E., Chou, R., Glanville, J., Grimshaw, J. M., Hróbjartsson, A., Lalu, M. M., Li, T., Loder, E. W., Mayo-Wilson, E., McDonald, S., & Moher, D. (2021). The PRISMA 2020 statement: An updated guideline for reporting systematic reviews. *International Journal of Surgery*, 88(March), 105906. Advance online publication. 10.1016/j.ijsu.2021.10590633789826

Panda, G., Dash, M. K., Samadhiya, A., Kumar, A., & Mulat-weldemeskel, E. (2023). Artificial intelligence as an enabler resiliency : past literature, present debate and future research directions. *International Journal of Industrial Engineering and Operations Management*. 10.1108/IJIEOM-05-2023-0047

Pan, K. (2024). Ethics in the Age of AI: Research of the Intersection of Technology, Morality, and Society. *Lecture Notes in Education Psychology and Public Media*, 40(1), 259–262. 10.54254/2753-7048/40/20240816

Pan, Y. (2016). Heading toward artificial intelligence 2.0. *Engineering (Beijing)*, 2(4), 409–413. 10.1016/J.ENG.2016.04.018

Park, E. (2023, February 8). *The AI Bill of Rights: A Step in the Right Direction*. Social Science Research Network. https://papers.ssrn.com/sol3/papers.cfm?abstract_id=4351423

Patel, A., Qassim, Q., & Wills, C. (2010). A survey of intrusion detection and prevention systems. *Information Management & Computer Security*, 18(4), 277–290. 10.1108/09685221011079199

Patterson, M. G., West, M. A., Lawthom, R., & Nickell, S. (1997). *Impact of People Management Practice*. IPD.

Pavčnik, M. (2015). *Teorija prava: prispevek k razumevanju prava*. GV Založba.

Pawar, Y. (2019). *Impact of Artificial Intelligence in Performance Management*.

Pelau, C., Ene, I., & Ionut Pop, M. (2021). The Impact of Artificial Intelligence on Consumers' Identity and Human Skills. *Amfiteatru Economic*, 23(56), 33–45. 10.24818/EA/2021/56/33

Peltola, M., Xue, G., & Yu, Z. (2021). *China-powered ICT Infrastructure: Lessons from Tanzania and Cambodia*. South African Institute of International Affairs.

Penrose, E. (1959). *The Theory of the Growth of the Firm*. Blackwell.

Perenič, A. (1998). Razlaga pravnih aktov. In Uvod v pravoznanstvo (pp. 191-217). Uradni list Republike Slovenije.

Perko, I. (2021). Hybrid reality development-can social responsibility concepts provide guidance? *Kybernetes*, 50(3), 676–693. 10.1108/K-01-2020-0061

Perko, I. (2022). Data sharing concepts: A viable system model diagnosis. *Kybernetes*, 52(9), 2976–2991. 10.1108/K-04-2022-0575

Peyrone, N., & Wichadakul, D. (2023). A formal model for blockchain-based consent management in data sharing. *Journal of Logical and Algebraic Methods in Programming*, 134, 100886. 10.1016/j.jlamp.2023.100886

Pintiliuc, I.-G. (2018). Protection of Personal Data. Logos Universality Mentality Education Novelty. *Law*, 6(1), 37–40. 10.18662/lumenlaw/05

Pislaru, M., Vlad, C. S., Ivascu, L., & Mircea, I. I. (2024). Citizen-Centric Governance: Enhancing Citizen Engagement through Artificial Intelligence Tools. *Sustainability (Basel)*, 16(7), 2686. 10.3390/su16072686

Pivoto, D. G., de Almeida, L. F., da Rosa Righi, R., Rodrigues, J. J., Lugli, A. B., & Alberti, A. M. (2021). Cyber-physical systems architectures for industrial internet of things applications in Industry 4.0: A literature review. *Journal of Manufacturing Systems*, 58, 176–192. 10.1016/j.jmsy.2020.11.017

Pizzi, G., Scarpi, D., & Pantano, E. (2020). Artificial intelligence and the new forms of interaction: Who has the control when interacting with a chatbot? *Journal of Business Research*, 129, 878–890. Advance online publication. 10.1016/j.jbusres.2020.11.006

Pizzi, M., Romanoff, M., & Engelhardt, T. (2020). AI for humanitarian action: Human rights and ethics. *International Review of the Red Cross*, 102(913), 145–180. 10.1017/S1816383121000011

Plathottam, S. J., Rzonca, A., Lakhnori, R., & Iloeje, C. O. (2023). A review of artificial intelligence applications in manufacturing operations. *Journal of Advanced Manufacturing and Processing*, 5(3), e10159. Advance online publication. 10.1002/amp2.10159

Pollach, I. (2011). Online privacy as a corporate social responsibility: An empirical chapter. *Business Ethics (Oxford, England)*, 20(1), 88–102. 10.1111/j.1467-8608.2010.01611.x

Popov, V. V., Kudryavtseva, E. V., Kumar Katiyar, N., Shishkin, A., Stepanov, S. I., & Goel, S. (2022). Industry 4.0 and digitalisation in healthcare. *Materials (Basel)*, 15(6), 2140. 10.3390/ma1506214035329592

Prates, M., Avelar, P., & Lamb, L. C. (2018). Assessing gender bias in machine translation – A case study with google translate. *Neural Computing & Applications*, 32(10), 6363–6381. 10.1007/s00521-019-04144-6

Preethi, A., & Satheesh, N. (2021). Need for a Comprehensive Legislation on Employee's Privacy in India: Comparison with US and EU Models. *Issue 2 Int'l JL Mgmt. &. Human.*, 4, 1449.

Presidência da República. Secretaria-Geral Subchefia para Assuntos Jurídicos. (2018). Lei Geral de Proteção de Dados Pessoais. https://www.planalto.gov.br/ccivil_03/_ato2015-2018/2018/lei/l13709.htm

Puaschunder, J. M. (2023). The Future of Resilient Green Finance. In *The Future of Resilient Finance: Finance Politics in the Age of Sustainable Development* (pp. 185–210). Springer International Publishing. 10.1007/978-3-031-30138-4_6

Compilation of References

Purcell, J., Kinnie, K., Hutchinson, R., Rayton, B., & Swart, J. (2003). *Understanding the People and Performance Link: Unlocking the black box*. CIPD.

Purtova, N. (2018). The law of everything. Broad concept of personal data and future of EU data protection law. *Law, Innovation and Technology*, 10(1), 40–81. 10.1080/17579961.2018.1452176

Radanliev, P. (2024). Cyber diplomacy: defining the opportunities for cybersecurity and risks from Artificial Intelligence, IoT, Blockchains, and Quantum Computing. *Journal of Cyber Security Technology*, 1-51.

Ramsthaler, F., Kettner, M., Gehl, A., & Verhoff, M. A. (2010). Digital forensic osteology: Morphological sexing of skeletal remains using volume-rendered cranial CT scans. *Forensic Science International*, 195(1–3), 148–152. 10.1016/j.forsciint.2009.12.01020074879

Rane, N. (2023). Transformers in Industry 4.0, Industry 5.0, and Society 5.0: Roles and Challenges.

Ransbotham, S., Khodabandeh, S., Kiron, D., Candelon, F., Chu, M., & LaFountain, B. (2020). Expanding AI's impact with organizational learning. *MITSloan Managemen review*.

Rao, P., & Teegen, H. (2009). *Human Resource Management*. Academic Press.

Rath, K. C., Khang, A., & Roy, D. (2024). The Role of Internet of Things (IoT) Technology in Industry 4.0 Economy. In *Advanced IoT Technologies and Applications in the Industry 4.0 Digital Economy* (pp. 1-28). CRC Press.

Raub, M. (2018). Bots, bias and big data: Artificial intelligence, algorithmic bias and disparate impact liability in hiring practices. *Arkansas Law Review*, 71, 529.

Raveendra, P., Satish, Y., & Singh, P. (2020). Changing landscape of recruitment industry: A study on the impact of artificial intelligence on eliminating hiring bias from recruitment and selection process. *Journal of Computational and Theoretical Nanoscience*, 17(9), 4404–4407. 10.1166/jctn.2020.9086

Ravipolu, A. (2017). Role of Artificial Intelligence in Recruitment. *International Journal of Engineering Technology, Management and Applied Sciences*, 115-117.

Ravi, S. K., Chaturvedi, S., Rastogi, N., Akhtar, N., & Perwej, Y. (2022). A Framework for Voting Behavior Prediction Using Spatial Data. *International Journal of Innovative Research in Computer Science & Technology*, 10(2), 19–28. 10.55524/ijircst.2022.10.2.4

Ray, P. P. (2023). ChatGPT: A Comprehensive Review on background, applications, Key challenges, bias, ethics, Limitations and Future Scope. *Internet of Things and Cyber-Physical Systems*, 3(1), 121–154. 10.1016/j.iotcps.2023.04.003

Ray, R. K., Chowdhury, F. R., & Hasan, M. R. (2024). Blockchain Applications in Retail Cybersecurity: Enhancing Supply Chain Integrity, Secure Transactions, and Data Protection. *Journal of Business and Management Studies*, 6(1), 206–214. 10.32996/jbms.2024.6.1.13

Regulation (EU) 2016/679 of the European Parliament and of the Council of 27 April 2016 on the protection of natural persons with regard to the processing of personal data and on the free movement of such data, and repealing Directive 95/46/EC (General Data Protection Regulation), OJ L 119, 4. 5. 2016, 1-88.

Regulation (EU) 2018/1807 of the European Parliament and of the Council of 14 November 2018 on a framework for the free flow of non-personal data in the European Union, OJ L 303, 28. 11. 2018, 59-68.

Regulation (EU) 2023/2854 of the European Parliament and of the Council on harmonised rules on fair access to and use of data and amending Regulation (EU) 2017/2394 and Directive (EU) 2020/1828 (Data Act), OJ L, 2023/2854, 22. 12. 2023.

Reier Forradellas, R. F., & Garay Gallastegui, L. M. (2021). Digital transformation and artificial intelligence applied to business: Legal regulations, economic impact and perspective. *Laws*, 10(3), 70. 10.3390/laws10030070

Richardson, M. (2017). The Right to Privacy. In *The Right to Privacy: Origins and Influence of a Nineteenth-Century Idea*. Cambridge University Press. 10.1017/9781108303972

Rizzello, A., & Kabli, A. (2020). Social finance and sustainable development goals: A literature synthesis, current approaches and research agenda. *ACRN Journal of Finance and Risk Perspectives, 9*.

Robert, L. P., Pierce, C., Marquis, L., Kim, S., & Alahmad, R. (2020). Designing fair AI for managing employees in organizations: A review, critique, and design agenda. *Human-Computer Interaction*, 35(5-6), 545–575. 10.1080/07370024.2020.1735391

Robichaud, F. (2024). *ISO 26000: 7 Core subjects of Corporate Social Responsibility*. Retrieved from Borealis: https://www.boreal-is.com/blog/iso-26000-social-responsibility/

Roser, M. (2022, December 15). *Artificial Intelligence Is Transforming Our World — It Is on All of Us to Make Sure That It Goes Well*. Our World in Data. https://ourworldindata.org/ai-impact

Rothausen, T. J., Henderson, K. E., Arnold, J. K., & Malshe, A. (2015). Should I stay or should I go? Identity and Well-Being in Sensemaking about Retention and Turnover. *Journal of Management*, 43(7), 2357–2385. 10.1177/0149206315569312

Rotimi, F. E., Brauner, M., Burfoot, M., Naismith, N., Silva, C. C., & Mohaghegh, M. (2023). Work environment challenge and the wellbeing of women in construction industry in New Zealand – The mediating role of work morale. *Engineering, Construction, and Architectural Management*. Advance online publication. 10.1108/ECAM-02-2023-0152

Rudin, C. (2019). Stop explaining black box machine learning models for high stakes decisions and use interpretable models instead. *Nature Machine Intelligence*, 1(5), 206–215. 10.1038/s42256-019-0048-x35603010

Russell, S. J., & Norvig, P. (2016). *Artificial intelligence: a modern approach*. Pearson.

Compilation of References

Rygielski, C., Wang, J. C., & Yen, D. C. (2002). Data mining techniques for customer relationship management. *Technology in Society*, 24(4), 483–502. 10.1016/S0160-791X(02)00038-6

Ryngaert, C., & Taylor, M. (2020). The GDPR as Global Data Protection Regulation? *AJIL Unbound*, 114, 5–9. 10.1017/aju.2019.80

Sachs, J. D., Woo, W. T., Yoshino, N., & Taghizadeh-Hesary, F. (2019). Importance of green finance for achieving sustainable development goals and energy security. *Handbook of green finance: Energy security and sustainable development, 10*, 1-10.

Saeed, S., Altamimi, S. A., Alkayyal, N. A., Alshehri, E., & Alabbad, D. A. (2023). Digital transformation and cybersecurity challenges for businesses resilience: Issues and recommendations. *Sensors (Basel)*, 23(15), 6666. 10.3390/s2315666637571451

Sætra, H. S. (2020). A shallow defence of a technocracy of artificial intelligence: Examining the political harms of algorithmic governance in the domain of government. *Technology in Society*, 62, 101283. 10.1016/j.techsoc.2020.10128332536737

Sætra, H. S. (2021). A framework for evaluating and disclosing the ESG related impacts of AI with the SDGs. *Sustainability (Basel)*, 13(15), 8503. 10.3390/su13158503

Salam, R., & Sinurat, M. (2023). Implementation of Artificial Intelligence in Governance: Potentials and Challenges. *Influence: International Journal of Science Review, 5*(1), 243–255. https://influence-journal.com/index.php/influence/article/view/122

Saltman, K. (2020). Artificial intelligence and the technological turn of public education privatization: In defence of democratic education. *London Review of Education*, 18(2), 196–208. 10.14324/LRE.18.2.04

Sánchez-Sotano, A., Cerezo-Narváez, A., Abad-Fraga, F., Pastor-Fernández, A., & Salguero-Gómez, J. (2020). Trends of digital transformation in the shipbuilding sector. In *New Trends in the Use of Artificial Intelligence for the Industry 4.0*. IntechOpen. 10.5772/intechopen.91164

Sandars, J. (2009). The use of reflection in medical education: AMEE Guide No. 44. *Medical Teacher*, 31(8), 685–695. 10.1080/01421590903050374 19811204

Sarkar, S., & Searcy, C. (2016). Zeitgeist or chameleon? A quantitative analysis of CSR definitions. *Journal of Cleaner Production*, 135, 1423–1435. 10.1016/j.jclepro.2016.06.157

Šarotar Žižek, S. (2015). Well-Being as the Basic Aim of Social Responsibility. In *Social responsibility—Range of perspectives per topics and countries (Social responsibility beyond neoliberalism and charity, Vol. 4)* (pp. 49-76). London, Uk: Bentham Books.

Šarotar Žižek, S., & Mulej, M. (2016). Zadostna in potrebna (osebna) celovitost posameznika (ZIPOC). In *Nehajte sovražiti svoje otroke in vnuke. Knj. 1, Družbenoekonomski okvir in osebne lastnosti družbeno odgovornih* (pp. 86-140). Maribor: Kulturni center, zavod za umetniško produkcijo in založništvo.

Šarotar Žižek, S., & Milfelner, B. (2014a). *Vpliv menedžmenta človeških virov na uspešnost*. IRDO.

Šarotar Žižek, S., Mulej, M., & Treven, S. (2014b). *Zagotavljanje zadostne in potrebne osebne celovitosti človeka*. IRDO.

Šarotar Žižek, S., Treven, S., & Mulej, M. (2014c). *Psihično dobro počutje zaposlenih*. IRDO.

Šarotar, Ž. S., Zolak Poljašević, B., Mulej, M., Šket, R., Kovše, K., Preskar, M., ... Hrast, A. (2023a). *Aktualne teme managementa človeških virov. Knj. 1, Izzivi sodobnega časa, poslovanje prihodnosti, trajnostni management človeških virov, blagovna znamka delodajalca*. Ljubljana: IRDO - Inštitut za razvoj družbene odgovornosti.

Šarotar, Ž. S., Zolak Poljašević, B., Mulej, M., Treven, S., Šket, R., & Preskar, M. (2023b). *Aktualne teme managementa človeških virov. Knj. 3, Agilnost in zavzetost zaposlenih, čuječnost vodij in zaposlenih, organizacijska energija, delovna sreča, psihično dobro počutje zaposlenih*. Ljubljana: IRDO - Inštitut za razvoj družbene odgovornosti.

Šarotar, Ž. S., Zolak Poljašević, B., Treven, S., Šket, R., Preskar, M., & Senekovič, P. (2023c). *Aktualne teme managementa človeških virov. Knj. 2, Oblike sodelovanja med zaposlenimi, ključni kazalniki uspešnosti, osebni in osebnostni razvoj človeka*. Ljubljana: IRDO - Inštitut za razvoj družbene odgovornosti.

Schermer, B. W., Custers, B., & van der Hof, S. (2014). The crisis of consent: How stronger legal protection may lead to weaker consent in data protection. *Ethics and Information Technology*. Advance online publication. 10.1007/s10676-014-9343-8

Schermerhorn, J. R., Hunt, J. G., & Osborn, R. N. (2012). *Comportement humain et organisation*. Pearson - Village Mondial.

Schneier, B. (2015). *Secrets and lies: digital security in a networked world*. John Wiley & Sons. 10.1002/9781119183631

Schoenmaker, D., & Volz, U. (2022). *Scaling up sustainable finance and investment in the Global South*. CEPR Press.

Scholz, T. M. (2019). Big data and human resource management. In *Big data: Promise, application and pitfalls* (pp. 69-89). 10.4337/9781788112352.00008

Schulz, K. A., Gstrein, O. J., & Zwitter, A. J. (2020). Exploring the governance and implementation of sustainable development initiatives through blockchain technology. *Futures*, 122, 102611. 10.1016/j.futures.2020.102611

Schulz, K., & Feist, M. (2021). Leveraging blockchain technology for innovative climate finance under the Green Climate Fund. *Earth System Governance*, 7, 100084. 10.1016/j.esg.2020.100084

Schwendicke, F., Samek, W., & Krois, J. (2020). Artificial Intelligence in Dentistry: Chances and Challenges. *Journal of Dental Research*, 99(7), 769–774. 10.1177/0022034520915714 32315260

Sergey & Teteryatnikov. (2021). Artificial Intelligence in Public Governance. *Springer EBooks*, 127–135. 10.1007/978-3-030-63974-7_9

Compilation of References

Shanmugasundaram, M., & Tamilarasu, A. (2023). The impact of digital technology, social media, and artificial intelligence on cognitive functions: A review. *Frontiers in Cognition*, 2, 1203077. Advance online publication. 10.3389/fcogn.2023.1203077

Sharari, S., & Faqir, R. (2014). Protection of Individual Privacy under the Continental and Anglo-Saxon Systems: Legal and Criminal Aspects. *Beijing Law Review*, 5(3), 184–195. 10.4236/blr.2014.53018

Sharma, A., & Singh, B. (2022). Measuring Impact of E-commerce on Small Scale Business: A Systematic Review. *Journal of Corporate Governance and International Business Law*, 5(1).

Sharma, S., & Dwivedi, R. (2024). A survey on blockchain deployment for biometric systems. *IET Blockchain*, 4(2), 124–151. 10.1049/blc2.12063

Sharma, V. (2011). *Information technology law and practice*. Universal Law Publishing.

Shaw, J. (2019). Artificial intelligence and ethics. *Perspect: Policy Pract High Educ*, 30, 1–11.

Shih, C., Gwizdalski, A., & Deng, X. (2023). *Building a Sustainable Future: Exploring Green Finance*. Regenerative Finance, and Green Financial Technology.

Shi, Z., Xie, Y., Xue, W., Chen, Y., Fu, L., & Xu, X. (2020). Smart factory in Industry 4.0. *Systems Research and Behavioral Science*, 37(4), 607–617. 10.1002/sres.2704

Shrestha, Y. R., Ben-Menahem, S. M., & von Krogh, G. (2019). Organizational Decision-Making Structures in the Age of Artificial Intelligence. *California Management Review*, 61(4), 66–83. 10.1177/0008125619862257

Shuford, J. (2024). Interdisciplinary Perspectives: Fusing Artificial Intelligence with Environmental Science for Sustainable Solutions. *Journal of Artificial Intelligence General Science*, 1(1), 1–12. 10.60087/jaigs.v1i1.p12

Siau, K., & Wang, W. (2018). Building Trust in Artificial Intelligence, Machine Learning, and Robotics. *Cutter Business Technology Journal*, 31(2), 47–53.

Siman, E., Abiodun, J., Timothy, G. & Nandom, S. S. (2023). IoT-Driven Smart Cities: Enhancing Urban Sustainability and Quality of Life. 7th International Interdisciplinary Research & Development Conference.

Sima, V., Gheorghe, I. G., Subić, J., & Nancu, D. (2020). Influences of the industry 4.0 revolution on the human capital development and consumer behavior: A systematic review. *Sustainability (Basel)*, 12(10), 4035. 10.3390/su12104035

Sinclair, B. (2017). *IoT Inc.: how your company can use the Internet of things to win in the outcome economy*. McGraw-Hill Education.

Singh Rajawat, A., Bedi, P., Goyal, S. B., Shukla, P. K., Zaguia, A., Jain, A., & Monirujjaman Khan, M. (2021). Reformist framework for improving human security for mobile robots in industry 4.0. *Mobile Information Systems*, 2021, 1–10. 10.1155/2021/4744220

Singh, B. (2023). Blockchain Technology in Renovating Healthcare: Legal and Future Perspectives. In *Revolutionizing Healthcare Through Artificial Intelligence and Internet of Things Applications* (pp. 177-186). IGI Global.

Singh, B. (2019). Profiling Public Healthcare: A Comparative Analysis Based on the Multidimensional Healthcare Management and Legal Approach. *Indian Journal of Health and Medical Law*, 2(2), 1–5.

Singh, B. (2020). Global science and jurisprudential approach concerning healthcare and illness. *Indian Journal of Health and Medical Law*, 3(1), 7–13.

Singh, B. (2022). COVID-19 Pandemic and Public Healthcare: Endless Downward Spiral or Solution via Rapid Legal and Health Services Implementation with Patient Monitoring Program. *Justice and Law Bulletin*, 1(1), 1–7.

Singh, B. (2022). Relevance of Agriculture-Nutrition Linkage for Human Healthcare: A Conceptual Legal Framework of Implication and Pathways. *Justice and Law Bulletin*, 1(1), 44–49.

Singh, B. (2022). Understanding Legal Frameworks Concerning Transgender Healthcare in the Age of Dynamism. *Electronic Journal of Social and Strategic Studies*, 3(1), 56–65. 10.47362/EJSSS.2022.3104

Singh, B. (2023). Federated Learning for Envision Future Trajectory Smart Transport System for Climate Preservation and Smart Green Planet: Insights into Global Governance and SDG-9 (Industry, Innovation and Infrastructure). *National Journal of Environmental Law*, 6(2), 6–17.

Singh, B. (2024). Legal Dynamics Lensing Metaverse Crafted for Videogame Industry and E-Sports: Phenomenological Exploration Catalyst Complexity and Future. *Journal of Intellectual Property Rights Law*, 7(1), 8–14.

Singh, H. (2024). Navigating the Legal Landscape of Information Governance. In Helge, K., & Rookey, C. (Eds.), *Creating and Sustaining an Information Governance Program* (pp. 217–235). IGI Global. 10.4018/979-8-3693-0472-3.ch012

Singh, R. P., Hom, G. L., Abramoff, M. D., Campbell, J. P., & Chiang, M. F. (2020). Current Challenges and Barriers to Real-World Artificial Intelligence Adoption for the Healthcare System, Provider, and the Patient. *Translational Vision Science & Technology*, 9(2), 45–45. 10.1167/tvst.9.2.4532879755

Singh, R., Vatsa, M., & Ratha, N. (2021). Trustworthy ai. *Proceedings of the 3rd ACM India Joint International Conference on Data Science & Management of Data (8th ACM IKDD CODS & 26th COMAD)*, 449-453.

Sirohi, M. N. (2015). *Transformational Dimensions of Cyber Crime*. Vij Books India Pvt Ltd.

SITNFlash. (2024, March 4). *A Sky Full of Data: Weather forecasting in the age of AI*. Science in the News. https://sitn.hms.harvard.edu/flash/2024/ai_weather_forecasting/

Compilation of References

Slapnik, T., Hrast, A., & Mulej, M. (2018). Nacionalna strategija družbene odgovornosti v Sloveniji - idejni osnutek. In *Uvod v politično ekonomijo družbeno odgovorne družbe* (pp. 295-328). Maribor: Kulturni center, zavod za umetniško produkcijo in založništvo.

Smouter-Umans, K. L. (2017). Research, GDPR, and the DPO how GDPR changes the game for those conducting research and the data supervisors. International Journal for the Data Protection Officer *Privacy Officer and Privacy Counsel*, 1(1), 26.

Song, M., Xing, X., Duan, Y., Cohen, J., & Mou, J. (2022). Will artificial intelligence replace human customer service? The impact of communication quality and privacy risks on adoption intention. *Journal of Retailing and Consumer Services*, 66, 102900. 10.1016/j.jretconser.2021.102900

Speer, A. B., Dutta, S., Chen, M., & Trussel, G. (2019). Here to stay or go? Connecting turnover research to applied attrition modeling. *Industrial and Organizational Psychology: Perspectives on Science and Practice*, 12(3), 277–301. 10.1017/iop.2019.22

Sreenivasan, A., & Suresh, M. (2024). Start-up sustainability: Does blockchain adoption drives sustainability in start-ups? A systematic literature reviews. *Management Research Review*, 47(3), 390–405. 10.1108/MRR-07-2022-0519

Stahl, B. C., Schroeder, D., & Rodrigues, R. (2023). AI for good and the SDGs. In *Ethics of artificial intelligence: Case studies and Options for addressing ethical challenges* (pp. 95-106). Springer International Publishing. 10.1007/978-3-031-17040-9_8

State of California, Department of Justice. (2018). California Consumer Privacy Act of 2018. https://oag.ca.gov/privacy/ccpa

Strban, G. (2022). Metodologija primerjalne razlage prava socialne varnosti. Pravne panoge in metodologija razlage prava. GV Založba.

Strine, L. E.Jr. (2010). One fundamental corporate governance question we face: Can corporations be managed for the long term unless their powerful electorates also act and think long term? *Business Lawyer*, 1–26.

Studiawan, H., & Sohel, F. (2021). Anomaly detection in a forensic timeline with deep autoencoders. *Journal of Information Security and Applications*, 63, 103002. 10.1016/j.jisa.2021.103002

Stupp, C. (2019). Fraudsters Used AI to Mimic CEO's Voice in Unusual Cybercrime Case. The Wall Street Journal. Retrieved from https://www.wsj.com/articles/fraudsters-use-ai-to-mimic-ceos-voice-in-unusualcybercrime-case-11567157402

Suleman, D., Rusiyati, S., Sabil, S., Hakim, L., Ariawan, J., Wianti, W., & Karlina, E. (2022). The impact of changes in the marketing era through digital marketing on purchase decisions. *International Journal of Data and Network Science*, 6(3), 805–812. 10.5267/j.ijdns.2022.3.001

Szabo, N. (1997). The Idea of Smart Contracts.

Tambe, P., Cappelli, P., & Yakubovich, V. (2019). Artificial intelligence in human resources management: Challenges and a path forward. *California Management Review*, 61(4), 15–42. 10.1177/0008125619867910

Tambe, P., Cappelli, P., & Yakubovich, V. (2019). Artificial intelligence in human resources management: Challenges and a path forward. *California Management Review, 61*(4), 15–42. Upadhyay AK, Khandelwal K (2018) Applying artificial intelligence: implications for recruitment. *Strategic HR Review*, 17(5), 255–258.

Tammpuu, P., & Masso, A. (2019). Transnational digital identity as an instrument for global digital citizenship: The case of Estonia's E-residency. *Information Systems Frontiers*, 21(3), 621–634. 10.1007/s10796-019-09908-y

Tanaka, H. (2019). Impact of the GDPR on Japanese companies. *Business Law International*, 20(2), 137–146.

Tao, F., Akhtar, M. S., & Jiayuan, Z. (2021). The future of artificial intelligence in cybersecurity: A comprehensive survey. *EAI Endorsed Transactions on Creative Technologies*, 8(28), e3–e3. 10.4108/eai.7-7-2021.170285

Taylor, L., van der Sloot, B., & Floridi, L. (2017). Conclusion: what do we know about group privacy? In *Group Privacy: New Challenges of Data Technologies* (pp. 225–237). 10.1007/978-3-319-46608-8_12

Taylor, L. (2017). What is data justice? The case for connecting digital rights and freedoms globally. *Big Data & Society*, 4(2), 1–14. 10.1177/2053951717736335

Taylor, P. J. (2005). Leading world cities: Empirical evaluations of urban nodes in multiple networks. *Urban Studies (Edinburgh, Scotland)*, 42(9), 1593–1608. 10.1080/00420980500185504

Taylor, P. J., Dargahi, T., Dehghantanha, A., Parizi, R. M., & Choo, K. K. R. (2020). A systematic literature review of blockchain cyber security. *Digital Communications and Networks*, 6(2), 147–156. 10.1016/j.dcan.2019.01.005

Thangavel, J. (2014). Digital Signature: Comparative study of its usage in developed and developing countries.

The Economic Times. (2024, April 30). US newspapers sue OpenAI for copyright infringement over AI training. *The Economic Times*. https://economictimes.indiatimes.com/tech/technology/us-newspapers-sue-openai-for-copyright-infringement-over-ai-training/articleshow/109737033.cms?from=mdr

The Grainger College of Engineering Coordinated Science Laboratory. (2016, September 6). Klara Nahrstedt. *Archives*.library.illinois.edu. https://csl.illinois.edu/directory/faculty/klara

The Toronto Declaration. (2018). https://www.torontodeclaration.org/

Compilation of References

The United Nations Educational, Scientific and Cultural Organization. (2017). *AI and the Rule of Law: Capacity Building for Judicial Systems | UNESCO*. Www.unesco.org. https://www.unesco.org/en/artificial-intelligence/rule-law/mooc-judges

The White House. (2022, October). *Blueprint for an AI Bill of Rights*. The White House; The White House. https://www.whitehouse.gov/ostp/ai-bill-of-rights/

Thiebes, S., Lins, S., & Sunyaev, A. (2021). Trustworthy artificial intelligence. *Electronic Markets*, 31(2), 447–464. 10.1007/s12525-020-00441-4

Thompson, M. (2002). *High Performance Work Organization in UK Aerospace*. The Society of British Companies.

Thouvenin, F., & Tamò-Larrieux, A. (2021). Data Ownership and Data Access Rights: Meaningful Tools for Promoting the European Digital Single Market? In Burri, M. (Ed.), *Big Data and Global Trade Law* (pp. 316–339). Cambridge University Press. 10.1017/9781108919234.020

Truby, J. (2020). Governing Artificial Intelligence to benefit the UN Sustainable Development Goals. *Sustainable Development (Bradford)*, 28(4), 946–959. Advance online publication. 10.1002/sd.2048

Tseng, M. L., Tran, T. P. T., Ha, H. M., Bui, T. D., & Lim, M. K. (2021). Sustainable industrial and operation engineering trends and challenges Toward Industry 4.0: A data driven analysis. *Journal of Industrial and Production Engineering*, 38(8), 581–598. 10.1080/21681015.2021.1950227

Tucci, V., Saary, J., & Doyle, T. E. (2021). Factors influencing trust in medical artificial intelligence for healthcare professionals: A narrative review. *Journal of Medical Artificial Intelligence*, 0. Advance online publication. 10.21037/jmai-21-25

Türk, V. (2023, July 12). *Artificial intelligence must be grounded in human rights, says High Commissioner*. OHCHR; Uniter Nations. https://www.ohchr.org/en/statements/2023/07/artificial-intelligence-must-be-grounded-human-rights-says-high-commissioner

Tursunbayeva, A., Di Lauro, S., & Pagliari, C. (2018). People analytics—A scoping review of conceptual boundaries and value propositions. *International Journal of Information Management*, 43(1), 224–247. 10.1016/j.ijinfomgt.2018.08.002

Udeagha, M. C., & Ngepah, N. (2023). The drivers of environmental sustainability in BRICS economies: Do green finance and fintech matter? *World Development Sustainability*, 3, 100096. 10.1016/j.wds.2023.100096

Udvaros, J., & Forman, N. (2023). Artificial intelligence and Education 4.0. *INTED2023 Proceedings*, 6309–6317. 10.21125/inted.2023.1670

Ulrich, D. (1997). *Human resource champions: The next agenda for adding value and delivering results*. Harvard Business Press.

United Nation. (2024, April 11). *UN: AI resolution must prioritise human rights and sustainable development*. https://www.article19.org/resources/un-ai-resolution-must-prioritise-human-rights-and-sustainable-development/

United Nations. (2015). *Transforming Our World: The 2030 Agenda for Sustainable Development*. Division for Sustainable Development Goals.

Upadhyay, A. K., & Khandelwal, K. (2018). Applying artificial intelligence: Implications for recruitment. *Strategic HR Review*, 17(5), 255–258. 10.1108/SHR-07-2018-0051

Ursić, H. (2018). The Failure of Control Rights in the Big Data Era: Does a Holistic Approach Offer a Solution? In Personal Data in Competition, Consumer Protection and Intellectual Property Law. MPI Studies on Intellectual Property and Competition Law, vol 28. Springer. 10.1007/978-3-662-57646-5_4

Valle-Cruz, D., Criado, J. I., Sandoval-Almazán, R., & Ruvalcaba-Gomez, E. A. (2020). Assessing the public policy-cycle framework in the age of artificial intelligence: From agenda-setting to policy evaluation. *Government Information Quarterly*, 37(4), 101509. 10.1016/j.giq.2020.101509

van den Broek, E., Sergeeva, A., & Huysman, M. (2021). When the machine meets the expert: An ethnography of developing AI for hiring. *Management Information Systems Quarterly*, 45(3), 1557–1580. 10.25300/MISQ/2021/16559

van der Broek, E., Sergeeva, A., & Huysman, M. (2019). Hiring Algorithms: An Ethnography of Fairness in Practice. *ICIS 2019 Proceedings, 6*.

van der Merwe, J., & Achkar, Z. A. (2022). Data responsibility, corporate social responsibility, and corporate digital responsibility. *Data & Policy*, 4, e12. 10.1017/dap.2022.2

Van der Meulen, N. (2013). Diginotar: Dissecting the first dutch digital disaster. *Journal of Strategic Security*, 6(2), 46–58. 10.5038/1944-0472.6.2.4

Van Ooijen, I., & Vrabec, H. U. (2019). Does the GDPR Enhance Consumers' Control over Personal Data? An Analysis from a Behavioural Perspective. *Journal of Consumer Policy*, 42(1), 91–107. 10.1007/s10603-018-9399-7

Veeramani, S., Rong, R. P. N., & Singh, S. (2019). Digital Transformation and Corporate Governance in India: A Conceptual Analysis.

Verbin, I. (2020). *Corporate Responsibility in the Digital Age: A Practitioner's Roadmap for Corporate Responsibility in the Digital Age*. Routledge. 10.4324/9781003054795

Verhulst, S. G. (2023). Operationalising digital self-determination. *Data & Policy*, 5, e14. 10.1017/dap.2023.11

Vinuesa, R., Azizpour, H., Leite, I., Balaam, M., Dignum, V., Domisch, S., . . . Fuso Nerini, F. (2020). The role of artificial intelligence in achieving the Sustainable Development Goals. Academic Press.

Compilation of References

Vinuesa, R., Luna, M., & Cachafeiro, H. (2016). Simulations and experiments of heat loss from a parabolic trough absorber tube over a range of pressures and gas compositions in the vacuum chamber. *Journal of Renewable and Sustainable Energy*, 2(8). Advance online publication. 10.1063/1.4944975

Vogt, J. (2021). Where is the human got to go? Artificial intelligence, machine learning, big data, digitalisation, and human–robot interaction in Industry 4.0 and 5.0: Review Comment on: Bauer, M.(2020). Preise kalkulieren mit KI-gestützter Onlineplattform BAM GmbH, Weiden, Bavaria, Germany. *AI & Society*, 36(3), 1083–1087. 10.1007/s00146-020-01123-7

von Gravrock, E. (2022, March 31). *Artificial intelligence design must prioritize data privacy.* World Economic Forum; World Economic Forum. https://www.weforum.org/agenda/2022/03/designing-artificial-intelligence-for-privacy/

W. (2019). Mitigating gender bias in Natural Language Processing: A literature review. *Proceeding of 57thAnnual Meeting of Association for Computational Linguistics.* 10.18653/v1/P19-1159

Walker, J., Pekmezovic, A., & Walker, G. (2019). *Sustainable Development Goals: Harnessing Business to Achieve the SDGs through Finance, Technology and Law Reform.* John Wiley & Sons. 10.1002/9781119541851

Wamba-Taguimdje, S. L., Fosso Wamba, S., Kala Kamdjoug, J. R., & Tchatchouang Wanko, C. E. (2020). Influence of artificial intelligence (AI) on firm performance: The business value of AI-based transformation projects. *Business Process Management Journal*, 26(7), 1893–1924. 10.1108/BPMJ-10-2019-0411

Wang, F. Y. (2020). Cooperative Data Privacy: The Japanese model of data privacy and the EU-Japan GDPR adequacy agreement. *Harvard Journal of Law & Technology*, 33(2), 661–692.

Wang, W. S., Li, X., Qiu, X. Q., Zhang, X., Brusic, V., & Zhao, J. D. (2023). A privacy preserving framework for federated learning in smart healthcare systems. *Information Processing & Management*, 60(1), 23. 10.1016/j.ipm.2022.103167

Wan, J., Yang, J., Wang, Z., & Hua, Q. (2018). Artificial intelligence for cloud-assisted smart factory. *IEEE Access : Practical Innovations, Open Solutions*, 6, 55419–55430. 10.1109/ACCESS.2018.2871724

Warren, S. D., & Brandeis, L. D. (1890). The Right to Privacy. *Harvard Law Review*, 4(5), 193–220. 10.2307/1321160

WBCSD. (1999). *Corporate social responsibility: Meeting Changing Expectations, World Business Council for Sustainable Development.* World Business Council for Sustainable Development.

Wegren, S. K. (2018). The "left behind": Smallholders in contemporary Russian agriculture. *Journal of Agrarian Change*, 18(4), 913–925. 10.1111/joac.12279

West, D., & Allen, J. (2018, April 24). *How Artificial Intelligence Is Transforming the World.* Brookings; The Brookings Institution. https://www.brookings.edu/articles/how-artificial-intelligence-is-transforming-the-world/

West, M. A., Borrill, C. S., Dawson, C., Scully, J., Carter, M., Anclay, S., & Waring, J. (2002). The link between the management of employees and patient mortality in acute hospitals. *International Journal of Human Resource Management*, 13(8), 1299–1310. 10.1080/09585190210156521

Wike, R., Fetterolf, J., Schumacher, S., & Moncus, J. J. (2021, October 21). Appendix A: Classifying democracies. *Pew Research Center's Global Attitudes Project*. https://www.pewresearch.org/global/2021/10/21/spring-2021-democracy-appendix-a-classifying-democracies/

Wilkes, J. (2014). The creation of HIPAA culture: Prioritizing privacy paranoia over patient care. *Brigham Young University Law Review*, 2014(5), 1213–1250.

William, W., & William, L. (2019, November). Improving corporate secretary productivity using robotic process automation. In *2019 International Conference on Technologies and Applications of Artificial Intelligence (TAAI)* (pp. 1-5). IEEE. 10.1109/TAAI48200.2019.8959872

Wilson, H. J., & Daugherty, P. R. (2018). Collaborative intelligence: Humans and AI are joining forces. *Harvard Business Review*, 96(4), 114–123.

Wingard, D. (2019). Data-driven automated decision-making in assessing employee performance and productivity: Designing and implementing workforce metrics and analytics social sciences, sociology, management and complex organizations. *Psychosociological Issues in Human Resource Management*, 7(2), 13–18. 10.22381/PIHRM7220192

Winn, P. A. (2010). Older than the Bill of Rights: The Ancient Origins of the Right to Privacy. https://ssrn.com/abstract=1534309

Wirtz, B. W., Weyerer, J. C., & Kehl, I. (2022). Governance of artificial intelligence: A risk and guideline-based integrative framework. *Government Information Quarterly*, 39(4), 101685. 10.1016/j.giq.2022.101685

Wong, J., Henderson, T., & Ball, K. (2022). Data protection for the common good: Developing a framework for a data protection-focused data commons. *Data & Policy*, 4, e3. 10.1017/dap.2021.40

Wooldridge, M. (1999). Intelligent agents. Multiagent systems: A modern approach to distributed artificial intelligence, 1, 27-73.

World Economic Forum (WEF). (2018). Harnessing Artificial Intelligence for the Earth. *Fourth Industrial Revolution for the Earth Series Harnessing Artificial Intelligence for the Earth*.

Wu, W., Huang, T., & Gong, K. (2019). Ethical principles and governance technology development of AI in China. *Engineering (Beijing)*, 6(3), 302–309. 10.1016/j.eng.2019.12.015

Wu, Y. (2023). Data Governance and Human Rights: An Algorithm Discrimination Literature Review and Bibliometric Analysis. *Journal of Humanities. Arts and Social Science*, 7(1), 128–154. 10.26855/jhass.2023.01.018

Xing, K., Cropley, D. H., Oppert, M. L., & Singh, C. (2021). Readiness for digital innovation and industry 4.0 transformation: studies on manufacturing industries in the city of salisbury. *Business Innovation with New ICT in the Asia-Pacific: Case Studies*, 155-176.

Compilation of References

Xu, L. D., Xu, E. L., & Li, L. (2018). Industry 4.0: State of the art and future trends. *International Journal of Production Research*, 56(8), 2941–2962. 10.1080/00207543.2018.1444806

Yan, C., Siddik, A. B., Yong, L., Dong, Q., Zheng, G. W., & Rahman, M. N. (2022). A two-staged SEM-artificial neural network approach to analyze the impact of FinTech adoption on the sustainability performance of banking firms: The mediating effect of green finance and innovation. *Systems*, 10(5), 148. 10.3390/systems10050148

Yang, Y., Su, X., & Yao, S. (2021). Nexus between green finance, fintech, and high-quality economic development: Empirical evidence from China. *Resources Policy*, 74, 102445. 10.1016/j.resourpol.2021.102445

Yin, R. K. (2009). Case Chapter Research: Design and Methods. *Sage (Atlanta, Ga.)*.

Your Europe. (2024). Prosti pretok neosebnih podatkov. https://europa.eu/youreurope/business/running-business/developing-business/free-flow-non-personal-data/index_sl.htm

Yue, Y., & Shyu, J. Z. (2024). A paradigm shift in crisis management: The nexus of AGI-driven intelligence fusion networks and blockchain trustworthiness. *Journal of Contingencies and Crisis Management*, 32(1), e12541. 10.1111/1468-5973.12541

Zhang, J. (2021). *AI: can it be regulated?* FIFDH. https://fifdh.org/en/festival/program/2024/forum/ai-can-it-be-regulated/

Zhang, Y., Zhang, M., Li, J., Liu, G., Yang, M. M., & Liu, S. (2021). A bibliometric review of a decade of research: Big data in business research – Setting a research agenda. *Journal of Business Research, 131*(December), 374–390. 10.1016/j.jbusres.2020.11.004

Zhang, C., & Lu, Y. (2021). Study on Artificial Intelligence: The State of the Art and Future Prospects. *Journal of Industrial Information Integration*, 23(23), 100224. 10.1016/j.jii.2021.100224

Zhang, J., & Chen, Z. (2023). Exploring human resource management digital transformation in the Digital Age. *Journal of the Knowledge Economy*. Advance online publication. 10.1007/s13132-023-01214-y

Zhang, W., Sun, L. H., Wang, X. P., & Wu, A. B. (2022). The influence of AI word-of-mouth system on consumers' purchase behaviour: The mediating effect of risk perception. *Systems Research and Behavioral Science*, 39(3), 516–530. 10.1002/sres.2871

Zhang, Y., Xu, S., Zhang, L., & Yang, M. (2021). Big data and human resource management research: An integrative review and new directions for future research. *Journal of Business Research*, 133(April), 34–50. 10.1016/j.jbusres.2021.04.019

Zhao, W. (2021). Artificial Intelligence and ISO 26000 (Guidance on Social Responsibility). In A. a.-I. Directions. IntechOpen.

Zhao, W. W. (2018). Improving social responsibility of artificial intelligence by using ISO 26000. *Iop conference series: Materials science and engineering, 428*(1).

Zhao, W. W. (2019). Research on social responsibility of artificial intelligence based on ISO 26000. *Recent Developments in Mechatronics and Intelligent Robotics:Proceedings of International Conference on Mechatronics and Intelligent Robotics (ICMIR2018)*, 130-137.

Zhong, R. Y., Xu, X., Klotz, E., & Newman, S. T. (2017). Intelligent manufacturing in the context of industry 4.0: A review. *Engineering (Beijing)*, 3(5), 616–630. 10.1016/J.ENG.2017.05.015

Zhou, J., Zhang, S., Lu, Q., Dai, W., Chen, M., Liu, X., . . . Herrera-Viedma, E. (2021). A survey on federated learning and its applications for accelerating industrial internet of things. *arXiv preprint arXiv:2104.10501*.

Zlatanović, D., Mulej, M., & Ženko, Z. (2022). Corporate Social Responsibility Considered With Two Systems Theories: A Case from Serbia. *Naše gospodarstvo, 68*(3), 10-17.

Zore, M. (2018). Družbena odgovornost podjetij: soavtorska monografija. In *Uvod v politično ekonomijo družbeno odgovorne družbe* (pp. 153-192). Maribor: Kulturni center, zavod za umetniško produkcijo in založništvo.

Zou, R. P., Lv, X. X., & Zhao, J. S. (2021). SPChain: Blockchain-based medical data sharing and privacy-preserving eHealth system. *Information Processing & Management*, 58(4), 18. 10.1016/j.ipm.2021.102604

About the Contributors

Maja Pucelj completed her first PhD at Alma Mater Europaea - ISH in the field of Humanities and completed her second doctorate at the Faculty of Government and European Studies in the field of International Studies with a focus on Human Rights. Prior to joining Faculty of organisational studies, she worked as an advisor to the Minister of Education, Science and Sports in the areas of pre-school education, primary education, secondary and higher education, adult education and quality of education, and as an undersecretary in the Service for the Implementation of Cohesion Policy at the Ministry of Education, Science and Sports. She is the author of numerous works on various current social challenges, connected with the topic of human rights.

Rado Bohinc is a full professor of Economic, Labor and European law at the EMUNI University in Piran and its president and scientific councillor at FDV and the Scientific Research Centre Koper; he researches corporate and institutional law as well as public and corporate governance and social responsibility. His earlier scientific works are the books Property and Management, 1988 (GV) and Delniška družba 1990, (GV). At the turn of the millennium, his books Corporate Governance between Europe and the USA, 2001 (FDV) and Legal Persons, 2003 (GV) were published. His most important scientific monographs published abroad are: Corporations and Partnerships, 4 editions, 2006, 2009, 2012, 2018, 2020 (Kluwer), Comparative Corporate law, 2010 (Verlag DM, Saarbrücken), Insurance Law (co-authored)), 2018 (Kluwer), Media Law, 2014, 2019 (Kluwer). The fundamental scientific works published in Slovenia are: Corporations, 2009 (Nebra), Comparative Corporate Management, 2011 (UP, FM) Social Responsibility 2017 (FDV), University and State, 2021. He is also the editor and co-author of several collections and joint monographs, e.g.: For social responsibility, 2018 (FDV), Corporate social responsibility, 2016 (FDV), Corporate governance as a tool for economic growth, 2016 (FDV). He lectured at many universities abroad (USA - Fullbright, Spain, Italy - Erasmus, Austria, Sweden - Tempus, Norway, Russia, India, Nepal, South Korea, New Zealand, Italy, Germany, Czech Republic) and appeared as a speaker at several than 30 domestic and international scientific conferences and symposiums.

* * *

Neema Gupta is an Associate Professor in the Management Department of University School of Business, Chandigarh University, Mohali, Punjab. She has a teaching experience of over 17 years and her area of specialization is Human Resource Management. Her papers have been published in a wide array of reputed journals.

About the Contributors

Vishal Jain is presently working as an Associate Professor at Department of Computer Science and Engineering, School of Engineering and Technology, Sharda University, Greater Noida, U. P. India. Before that, he has worked for several years as an Associate Professor at Bharati Vidyapeeth's Institute of Computer Applications and Management (BVICAM), New Delhi. He has more than 15 years of experience in the academics. He obtained Ph.D (CSE), M.Tech (CSE), MBA (HR), MCA, MCP and CCNA. He has authored more than 100 research papers in reputed conferences and journals, including Web of Science and Scopus. He has authored and edited more than 45 books with various reputed publishers, including Elsevier, Springer, IET, Apple Academic Press, CRC, Taylor and Francis Group, Scrivener, Wiley, Emerald, NOVA Science, IGI-Global and River Publishers. His research areas include information retrieval, semantic web, ontology engineering, data mining, ad hoc networks, and sensor networks. He received a Young Active Member Award for the year 2012–13 from the Computer Society of India, Best Faculty Award for the year 2017 and Best Researcher Award for the year 2019 from BVICAM, New Delhi.

Siddharth Kanojia is an Asst. Professor of Law at O.P Jindal Global University. He is also a visiting professor at Indian Institute of Management Amritsar and Sirmaur. He holds a Ph.D in Corporate Law & Governance, LL.M in International Business and Commercial Law, LL.B and BBS in Finance. Before commencing the academic stints, he was associated with U.K Intellectual Property Management Company as Legal Consultant and Manchester based Start-up Company as Legal Advisor. Till date, he has various publication in the SCOPUS and WoS indexed journals. He is also an active member in Human Development and Capability Association, USA and Development Studies Association, U.K.

Abhishek Kumar possesses a master's degree in political science with grade 'O' and a first class graduation degree in B.A with political science, history and sociology as the core subject, apart from these he has completed his Bachelors degree in Law (3 year course) from RTM Nagpur University and Masters degree in Constitutional and Administrative Law from RTM Nagpur University. He is currently pursuing PhD in international relations from Symbiosis Centre for Research and Innovation (SCRI) Symbiosis International (Deemed) University, Pune. Apart from Academics Abhishek Kumar Also has a teaching experience of 9 years.

Matjaž Mulej is a double Ph.D., Prof. Emeritus, member of 3 International Academies of Science, about 2.500 (mostly co-authored) publications in close to 50 countries, 2.800 citations, 110.000 reads. Author of the Dialectical Systems Theory, the concept Innovative Sustainable Socially Responsible Society, etc.; visiting professor abroad for 15 terms.

Gal Pastirk is an LLM student at the Faculty of Law, University of Maribor, with a focus on EU law, data governance, and sustainable development. He has gained practical insights into legal practice through internships at both courts and law firms. His active participation in university projects and international conferences underlines his involvement in advancing knowledge in these areas. He takes a sincere interest in exploring the complex legal aspects of EU matters, reflecting his commitment to a thorough understanding of European affairs.

Igor Perko presently occupies the role of Assistant Professor at the Faculty of Economics and Business, University of Maribor, Slovenia. He also serves as the Director General of the World Organisation of Systems and Cybernetics (WOSC) and acts as a Co-Editor for the journal Kybernetes. His professional background is rich in the development of information systems for the financial sector, specialising in the creation and implementation of business intelligence systems, predictive analytics, and risk management systems. These systems are enhanced through cooperation, multimedia and virtual reality. In his research, Igor delves into the synergy of systems thinking, cybernetics, multimedia, artificial intelligence, intelligent agents, and Hybrid reality. He aims to craft sustainable strategies for organisations by applying multidisciplinary methods. His current research focus is on uncovering universal principles of interactions. Igor Perko is a prolific author and editor, having contributed to a wide array of professional and scientific research reports, special issues, monographs, and multimedia materials, underscoring his dedication to contributing to the advancement of his field. For further information about Igor, please visit his LinkedIn Page.

About the Contributors

Andreja Primec, Ph.D., began her judicial career at the High Court in Maribor. After passing the national judicial examination, she joined the Chair of Commercial Law at the Faculty of Economics and Business, University of Maribor. She holds the associate professor of law position at the same institution and the Faculty of Management, University of Primorska. Her academic and professional pursuits include commercial, business, European Union, and intellectual property law. Dr. Primec has authored and co-authored numerous scientific and professional publications she disseminates through national and international journals. Additionally, she presents her research at various scientific and international conferences in her home country and abroad. Her contributions demonstrate her expertise in interdisciplinary integration and her adeptness at addressing complex legal and societal issues.

Shikha Saloni is a Research scholar in the Management department of University school of Buisness, Chandigarh University,Mohali, Punjab. Her area of specialization is Workplace Incivility, Turnover Intention, Knowledge Management. She has published various research papers in reputed journals.

Simona Šarotar Žižek holds PhD in Economic and Business Sciences. She has completed her theoretical knowledge permanently through practical work. She is head of the Institute of Management and Organization and Head of the University education program (3+) "Business Management in Organization". She is the author and/or co-author of 90 scientific articles published in international and Slovenian journals and of +100 papers presented at scientific and expert conferences. She is also the author and/or co-author of 4 scientific monographs and +104 chapters in scientific and expert monographs.

Bhupinder Singh is working as Professor at Sharda University, India. Also, Honorary Professor in University of South Wales UK and Santo Tomas University Tunja, Colombia. His areas of publications as Smart Healthcare, Medicines, fuzzy logics, artificial intelligence, robotics, machine learning, deep learning, federated learning, IoT, PV Glasses, metaverse and many more. He has 3 books, 139 paper publications, 163 paper presentations in international/national conferences and seminars, participated in more than 40 workshops/FDP's/QIP's, 25 courses from international universities of repute, organized more than 59 events with international and national academicians and industry people's, editor-in-chief and co-editor in journals, developed new courses. He has given talks at international universities, resource person in international conferences such as in Nanyang Technological University Singapore, Tashkent State University of Law Uzbekistan; KIMEP University Kazakhstan, All'ah meh Tabatabi University Iran, the Iranian Association of International Criminal law, Iran and Hague Center for International Law and Investment, The Netherlands, Northumbria University Newcastle UK,

Hemendra Singh is a Lecturer of Law at Jindal Global Law School, situated within O.P. Jindal Global University, Sonipat. He pursued his LL.M. in Intellectual Property Rights from National Law University, Jodhpur, following his graduation with a B.A. LL.B. (Hons.) from Rajiv Gandhi National University of Law, Punjab. With a focus on corporate law, intellectual property law, criminal law, and constitutional law, he has authored numerous research articles and book chapters published in both national and international journals and books.

Index

A

Artificial Intelligence 2, 4, 5, 21, 23, 28, 29, 30, 31, 32, 33, 35, 36, 37, 38, 42, 43, 44, 48, 50, 51, 52, 53, 54, 55, 56, 57, 58, 59, 60, 61, 62, 63, 64, 65, 66, 67, 68, 76, 77, 78, 79, 80, 82, 83, 84, 90, 92, 94, 95, 97, 102, 111, 130, 131, 132, 133, 134, 135, 136, 137, 138, 139, 140, 141, 142, 143, 144, 145, 146, 148, 149, 151, 157, 159, 163, 165, 166, 168, 169, 170, 176, 179, 185, 190, 192, 194, 195, 196, 199, 200, 201, 205, 207, 211, 212, 214, 215, 216, 217, 218, 219, 220, 221, 222, 223, 224, 225, 230, 233, 237, 245, 263

B

Bibliometric Review 27, 32

C

ChatGPT 37, 46, 51, 59, 191
Corporate Governance 216, 223, 233, 234, 235, 236, 242, 243, 244, 245
Corporate Social Responsibility (CSR) 120

D

Data 1, 2, 4, 9, 11, 13, 15, 17, 21, 22, 23, 24, 25, 26, 27, 28, 30, 32, 33, 34, 35, 36, 39, 40, 41, 42, 43, 44, 45, 46, 47, 48, 49, 50, 52, 53, 56, 59, 60, 62, 64, 66, 67, 68, 70, 71, 72, 73, 74, 80, 82, 90, 91, 92, 93, 94, 95, 96, 97, 98, 100, 101, 106, 107, 108, 109, 110, 111, 114, 118, 119, 123, 129, 130, 133, 134, 136, 138, 139, 141, 143, 144, 145, 147, 148, 154, 155, 156, 157, 158, 159, 160, 161, 162, 163, 164, 165, 166, 167, 168, 169, 170, 171, 172, 173, 174, 175, 176, 177, 178, 179, 180, 181, 182, 183, 184, 185, 186, 187, 188, 189, 190, 191, 192, 193, 194, 195, 196, 197, 198, 199, 200, 201, 202, 203, 204, 205, 206, 207, 208, 209, 210, 211, 212, 213, 214, 221, 222, 225, 227, 228, 229, 230, 231, 232, 233, 234, 235, 236, 237, 238, 239, 240, 241, 243, 244, 245, 246, 247, 248, 249, 250, 251, 252, 253, 254, 255, 256, 257, 258, 259, 260, 261, 262, 263, 264, 265, 266, 267, 268, 269, 270, 271, 272, 273, 274, 278, 279, 280
Data Accessibility 22, 108, 238
Data Breach 207, 234, 240, 243
Data Collecting 183, 196, 204, 207
Data Ownership 175, 181, 195, 197, 256
Data Protection 34, 47, 56, 68, 109, 110, 144, 145, 147, 148, 154, 155, 156, 157, 159, 160, 162, 163, 164, 165, 166, 169, 170, 171, 172, 173, 174, 175, 177, 179, 180, 181, 186, 187, 189, 190, 191, 193, 194, 195, 197, 200, 208, 209, 210, 211, 222, 231, 239, 240, 241, 243, 244, 246, 252, 254, 255, 260, 261, 264, 267, 271, 280
Data Sharing 170, 175, 177, 178, 179, 181, 186, 187, 191, 193, 196, 197, 243
Democracy 31, 36, 37, 38, 40, 41, 42, 43, 44, 58, 118, 134, 143, 148, 187, 235, 254, 278
Digital Age 6, 23, 24, 29, 81, 114, 140, 146, 147, 148, 161, 162, 168, 169, 170, 177, 199, 203, 204, 207, 212, 213, 247, 248, 253, 254, 279
Digital Connectivity 246, 280
Digital Ecosystem 171, 228, 250, 275
Digital Ethics 1, 2, 3, 4, 5, 6, 7, 11, 12, 14, 15, 17, 19, 20, 21, 22, 23, 24, 25, 26, 27, 28, 33, 144, 148, 154, 155, 156, 163, 274
Digital Human Rights 28
Digitalized Economic System 228
Digital Realm 4, 34, 147, 248
Digital Technology 2, 3, 4, 21, 28, 33, 60, 64, 147, 184, 200, 202, 218, 227, 228, 252, 253, 279

E

ECHR 36, 173
Employees 29, 45, 46, 66, 68, 69, 70, 84, 85, 86, 87, 88, 89, 90, 92, 93, 94, 95, 96, 98, 99, 100, 101, 102, 107, 108, 109, 110, 111, 113, 121, 123, 125, 126, 127, 128, 129, 130, 138, 141, 143, 240, 256, 262, 274
Ethical Considerations 24, 28, 79, 83, 146, 151, 153, 156, 157, 159, 162, 163, 165, 166, 183, 247, 251, 252, 265, 266, 272, 279, 280
Ethical Use 131, 144, 150, 157, 159, 160, 163, 279
EU Artificial Intelligence Act 145, 166

F

Future of Employment 131

G

Gender Bias 63, 69, 70, 71, 72, 74, 79, 80
Gender Prejudice 70
General Data Protection Regulation 68, 145, 147, 154, 157, 159, 164, 165, 169, 173, 174, 189, 194, 208, 209, 211, 231, 244, 252, 271
Government 29, 41, 43, 44, 45, 48, 55, 56, 57, 60, 61, 62, 121, 122, 127, 132, 147, 150, 162, 163, 169, 173, 178, 191, 204, 209, 211, 227, 233, 234, 238, 242, 261, 263, 265

H

HRM 64, 73, 74, 78, 82, 83, 84, 85, 86, 87, 88, 89, 90, 92, 93, 94, 95, 96, 97, 100, 101, 102, 106, 107, 111, 117, 121, 126, 127, 128, 129, 130, 131, 133, 143
Human Resource Management 30, 31, 32, 64, 78, 81, 82, 84, 85, 87, 92, 94, 95, 97, 122, 126, 131, 132, 133, 134, 135, 137, 138, 139, 141, 143
Human Rights 1, 2, 3, 4, 5, 6, 7, 11, 12, 14, 15, 17, 19, 20, 21, 22, 23, 24, 25, 26, 27, 28, 30, 32, 33, 34, 35, 36, 37, 41, 43, 53, 58, 59, 61, 73, 99, 106, 107, 108, 109, 110, 111, 112, 117, 118, 119, 123, 127, 144, 145, 148, 151, 152, 153, 161, 163, 173, 177, 187, 190, 192, 199, 200, 201, 202, 203, 205, 206, 207, 209, 210, 211, 212, 246, 247, 248, 251, 255, 257, 259, 260, 262, 265, 269, 273, 279, 280

I

Implementation Gap 246, 280
Informative Society 228
Internet of Things 59, 170, 181, 189, 195, 197, 199, 200, 216, 221, 222, 224, 226, 242

L

Legal Framework 36, 174, 175, 187, 197, 224, 232

M

Machine Learning 23, 35, 36, 41, 45, 46, 49, 50, 51, 79, 90, 91, 100, 134, 136, 138, 145, 155, 157, 166, 176, 187, 193, 196, 199, 207, 211, 214, 216, 225

O

OECD 55, 144, 148, 149, 151, 152, 153, 154, 161, 164, 173, 193, 277

P

Privacy 2, 4, 5, 23, 27, 32, 33, 34, 36, 43, 45, 47, 48, 52, 56, 58, 60, 62, 72, 78, 93, 94, 97, 102, 106, 108, 109, 110, 114, 130, 144, 145, 146, 147, 148, 150, 154, 155, 156, 159, 160, 162, 163, 164, 165, 166, 167, 168, 169, 170, 171, 172, 173, 174, 178, 187, 189, 190, 192, 193, 194, 195, 196, 197, 199, 200, 201, 202, 203, 204, 205, 206, 207, 208, 209, 210, 211,

212, 213, 227, 228, 229, 230, 231, 232, 233, 234, 235, 236, 237, 238, 239, 240, 241, 242, 243, 244, 245, 246, 247, 249, 252, 254, 255, 257, 258, 259, 260, 261, 262, 263, 264, 265, 266, 267, 269, 270, 271, 272, 273, 274, 276, 278, 279, 280

R

Recruitment 63, 64, 65, 66, 67, 68, 70, 71, 73, 74, 76, 77, 78, 79, 80, 86, 94, 95, 96, 97, 98, 99, 100, 113, 125, 128, 129, 130, 134, 138, 140, 143

Right to Data Self-Determination 171, 197, 198

Right to Privacy 168, 170, 171, 172, 173, 174, 190, 192, 194, 195, 197, 200, 204, 212, 240, 276

Rule of Law 36, 37, 41, 43, 58, 61, 99, 114, 123, 148

S

Sabah 1

T

Technological Advancements 21, 38, 146, 200, 227, 228, 237, 253, 273

Technology 2, 3, 4, 5, 7, 15, 17, 18, 19, 21, 24, 25, 27, 28, 29, 30, 31, 33, 35, 36, 37, 38, 40, 43, 46, 51, 53, 54, 56, 58, 59, 60, 61, 64, 65, 66, 68, 72, 73, 74, 76, 77, 79, 80, 83, 84, 90, 91, 92, 94, 96, 98, 100, 111, 113, 117, 118, 119, 127, 129, 131, 135, 138, 140, 142, 145, 147, 156, 159, 163, 166, 167, 168, 170, 172, 175, 177, 178, 184, 186, 192, 193, 194, 196, 197, 200, 201, 202, 204, 209, 211, 212, 214, 215, 218, 219, 221, 222, 223, 224, 227, 228, 230, 231, 233, 234, 235, 237, 239, 241, 243, 244, 245, 246, 252, 253, 256, 260, 262, 266, 269, 271, 274, 278, 279, 280

Transparency 32, 36, 41, 45, 93, 95, 99, 102, 106, 107, 108, 111, 112, 114, 117, 118, 123, 128, 130, 143, 145, 146, 148, 149, 150, 151, 152, 153, 154, 155, 156, 157, 158, 159, 160, 161, 163, 191, 212, 231, 234, 235, 237, 248, 257, 275

U

UDHR 36

W

Work Performance 68, 82

Workplace 30, 69, 70, 85, 90, 93, 94, 128, 130, 132, 133, 134, 137

Publishing Tomorrow's Research Today

Uncover Current Insights and Future Trends in Business & Management
with IGI Global's Cutting-Edge Recommended Books

Print Only, E-Book Only, or Print + E-Book.
Order direct through IGI Global's Online Bookstore at **www.igi-global.com** or through your preferred provide

ISBN: 9798369306444
© 2023; 436 pp.
List Price: US$ **230**

ISBN: 9798369300084
© 2023; 358 pp.
List Price: US$ **250**

ISBN: 9781668458594
© 2023; 366 pp.
List Price: US$ **240**

ISBN: 9781668486344
© 2023; 256 pp.
List Price: US$ **280**

ISBN: 9781668493243
© 2024; 318 pp.
List Price: US$ **250**

ISBN: 9798369304181
© 2023; 415 pp.
List Price: US$ **250**

Do you want to stay current on the latest research trends, product announcements, news, and special offers?
Join IGI Global's mailing list to receive customized recommendations, exclusive discounts, and more.
Sign up at: **www.igi-global.com/newsletters.**

Scan the QR Code here to view more related titles in Business & Management.

Ensure Quality Research is Introduced to the Academic Community

Become a Reviewer for IGI Global Authored Book Projects

  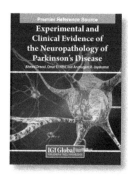 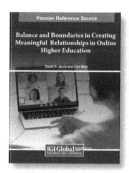

The overall success of an authored book project is dependent on quality and timely manuscript evaluations.

Applications and Inquiries may be sent to:
development@igi-global.com

Applicants must have a doctorate (or equivalent degree) as well as publishing, research, and reviewing experience. Authored Book Evaluators are appointed for one-year terms and are expected to complete at least three evaluations per term. Upon successful completion of this term, evaluators can be considered for an additional term.

If you have a colleague that may be interested in this opportunity, we encourage you to share this information with them.

www.igi-global.com

IGI Global Open Access Journal Program

Publishing Tomorrow's Research Today
IGI Global's Open Access Journal Program

Including Nearly 200 Peer-Reviewed, Gold (Full) Open Access Journals across IGI Global's Three Academic Subject Areas: Business & Management; Scientific, Technical, and Medical (STM); and Education

Consider Submitting Your Manuscript to One of These Nearly 200 Open Access Journals for to Increase Their Discoverability & Citation Impact

Web of Science Impact Factor **6.5**
JOURNAL OF
Organizational and End User Computing

Web of Science Impact Factor **4.7**
JOURNAL OF
Global Information Management

Web of Science Impact Factor **3.2**
INTERNATIONAL JOURNAL ON
Semantic Web and Information Systems

Web of Science Impact Factor **2.6**
JOURNAL OF
Database Management

Choosing IGI Global's Open Access Journal Program Can Greatly Increase the Reach of Your Research

Higher Usage
Open access papers are 2-3 times more likely to be read than non-open access papers.

Higher Download Rates
Open access papers benefit from 89% higher download rates than non-open access papers.

Higher Citation Rates
Open access papers are 47% more likely to be cited than non-open access papers.

Submitting an article to a journal offers an invaluable opportunity for you to share your work with the broader academic community, fostering knowledge dissemination and constructive feedback.

Submit an Article and Browse the IGI Global Call for Papers Pages

We can work with you to find the journal most well-suited for your next research manuscript. For open access publishing support, contact: journaleditor@igi-global.com

Are You Ready to Publish Your Research?

IGI Global — Publishing Tomorrow's Research Today

IGI Global offers book authorship and editorship opportunities across three major subject areas, including Business, STM, and Education.

Benefits of Publishing with IGI Global:

- Free one-on-one editorial and promotional support.
- Expedited publishing timelines that can take your book from start to finish in less than one (1) year.
- Choose from a variety of formats, including Edited and Authored References, Handbooks of Research, Encyclopedias, and Research Insights.
- Utilize IGI Global's eEditorial Discovery® submission system in support of conducting the submission and double-blind peer review process.
- IGI Global maintains a strict adherence to ethical practices due in part to our full membership with the Committee on Publication Ethics (COPE).
- Indexing potential in prestigious indices such as Scopus®, Web of Science™, PsycINFO®, and ERIC – Education Resources Information Center.
- Ability to connect your ORCID iD to your IGI Global publications.
- Earn honorariums and royalties on your full book publications as well as complimentary content and exclusive discounts.

Join Your Colleagues from Prestigious Institutions, Including:

Australian National University
Massachusetts Institute of Technology
Johns Hopkins University
Harvard University
Tsinghua University
Columbia University in the City of New York

Learn More at: www.igi-global.com/publish
or by Contacting the Acquisitions Department at: acquisition@igi-global.com

Individual Article & Chapter Downloads
US$ 37.50/each

Easily Identify, Acquire, and Utilize Published Peer-Reviewed Findings in Support of Your Current Research

- Browse Over **170,000+ Articles & Chapters**
- **Accurate & Advanced** Search
- Affordably Acquire **International Research**
- **Instantly Access** Your Content
- Benefit from the **InfoSci® Platform Features**

THE UNIVERSITY of NORTH CAROLINA at CHAPEL HILL

" *It really provides an excellent entry into the research literature of the field. It presents a manageable number of highly relevant sources on topics of interest to a wide range of researchers. The sources are scholarly, but also accessible to 'practitioners'.* "

- Ms. Lisa Stimatz, MLS, University of North Carolina at Chapel Hill, USA

Milton Keynes UK
Ingram Content Group UK Ltd.
UKHW052235120824
446789UK00009B/122